Molecular Strategies of
Pathogens and Host Plants

Suresh S. Patil   Seiji Ouchi
Dallice Mills   Carroll Vance
Editors

Molecular Strategies of Pathogens and Host Plants

With 65 Illustrations

Springer–Verlag
New York Berlin Heidelberg London
Paris Tokyo Hong Kong Barcelona

Suresh S. Patil
Department of Plant Pathology
  and the Biotechnology Program
University of Hawaii
Honolulu, HA
USA

Seiji Ouchi
Laboratory of Plant Pathology
Kinki University
Nara
Japan

Dallice Mills
Department of Botany and
  Plant Pathology
Oregon State University
Corvallis, OR
USA

Carroll Vance
Department of Agronomy and
  Plant Genetics, USDA-ARS
University of Minnesota
St. Paul, MN
USA

Cover illustration: Scanning electron micrograph of a freeze-fractured maize seedling whorl showing infection by *Ustilago maydis*. See page 113.

Library of Congress Cataloging-in-Publication Data
Molecular strategies of pathogens and host plants / edited by Suresh
  S. Patil . . . [et al.].
      p.   cm.
  Includes bibliographical references and index.
  ISBN-13: 978-1-4612-7791-0   e-ISBN-13: 978-1-4612-3084-7
  DOI: 10.1007/978-1-4612-3084-7

    1. Plant-pathogen relationships — Molecular aspects — Congresses.
  I. Patil, Suresh S.
  SB732.7.M66   1991
  632'.3 – dc20   90-22921

Printed on acid-free paper.

© 1991 Springer-Verlag New York Inc.

Camera-ready copy provided by the editors.

9 8 7 6 5 4 3 2 1

This book is dedicated to the memory of Professors Tsune Kosuge and Syoyo Nishimura for their remarkable contributions to the science of physiological plant pathology and their unswerving commitment to the US-Japan seminar series.

In Professor Kosuge's 30 year career in plant pathology he was a moving force in leading the science into the molecular era. Professor Kosuge's studies of phenolics in relation to plant disease and his research on the molecular regulation of hyperplasia diseases will be instrumental in shaping theory for years to come.

Professor Nishimura's career spanned some 30 years. He was recognized as a world authority on toxins in plant disease, particularly those associated with *Alternaria*. In addition, his studies of *Rhiztonia* diseases are classics in phytopathology literature. Professor Nishimura's laboratory brought toxin research from the descriptive stage to the modern analytical structure-function arena.

The legacies of both of these eminent scientists will live not only through their research contributions, but also through the numerous graduate students and postdoctoral fellows who trained in their laboratories. We will all miss their keen intellect and gentle spirits.

Preface

Fifteen years ago at an international gathering of host-parasite physiologists, one of the scientists plaintively asked "Where art thou O specificity?" Although in the intervening years enormous progress has been made in strengthening the base of knowledge, particularly in the areas of ultrastructure, biochemistry and chemistry of host-pathogen interactions, the answer to the question of what constitutes the basis of specificity still eludes us. However, the recent development of direct, precise and powerful tools of molecular biology offers substantial hope that our quest to decipher the secrets of specificity will be realized in a not too distant future. Indeed, if the progress made during the past several years in understanding signal transduction and in the isolation of genes involved in plant pathogenesis (and their regulation) is any guide, our hope is justified.

Under the auspices of the "U.S.-Japan Cooperative Science Seminar" program, research scientists interested in plant-pathogen specificity have met six times during the last 25 years. These gatherings have allowed workers from the two countries to review the latest developments in the field, to share research results, to discuss future direction of research, to strengthen old collaborative links and to forge new ones.

The most recent meeting in the U.S.-Japan series was entitled "Molecular Strategies of Pathogens and Host Plants." It was held April 15-20, 1990, at the East-West Center in Honolulu, Hawaii and this volume is based on its proceedings. This meeting, which marked the 25th anniversary of the U.S.-Japan seminar series, constituted a turning point in that it was devoted primarily to the molecular biology of host-pathogen interactions. In the five years that have elapsed since the last meeting was held (in Japan), increasing numbers of scientists from both countries have availed themselves of the very powerful techniques of molecular biology. The chapters in this volume bear witness to the rapid pace of progress they have achieved in this short period of time in dissecting the molecular strategies of pathogens and host plants. The summaries of discussion that followed each lecture and the synopsis by Richard Durbin, a veteran plant pathologist and strong supporter of the U.S.-Japan Seminar series from its inception, allow us to take stock of what has been accomplished and remind us that we still have a long way to go before we understand the basis of plant-pathogen specificity.

We thank the National Science Foundation and the Japan Society for the Promotion of Science for their financial support and the East-West Center and Mr. J. McMahon for providing excellent facilities for the meeting. We are also grateful to the President of the University of Hawaii and to the Dean of the College of Tropical Agriculture and Human Resources, University of Hawaii, the University of Minnesota, the U.S. Department of Agriculture, the Monsanto Company, Heinz U.S.A., Calgene Inc., and Hawaii Biotechnology Group, Inc., for their generous support. We thank the American Phytopathological Society and the Phytopathological Society of Japan for the recognition extended to the seminar. We also gratefully acknowledge those who helped in the organization of the seminar as well as those who helped edit some of the chapters in this volume. We extend a special aloha to Dr. Stephen Saul of the Department of Entomology, without whose patience and support it would not have been possible to submit the camera-ready book manuscript to the publishers in a timely manner. Special thanks are also extended to Dayle Sasaki and Helen Cho for their expert help in word processing and to Nancy Preston and Bonnie Read for secretarial and administrative support.

November 1990

Suresh S. Patil
Seiji Ouchi
Dallice Mills
Carroll Vance

Contents

Preface vii
Participants xiii
Contributors xv

Overviews

1. Molecular Strategies in the Interaction Between *Agrobacterium*
   and Its Hosts 3
   E.W. Nester and M.P. Gordon

2. Molecular Biology of Fungal Host-Parasite Interactions 15
   S. Ouchi

Bacterial Strategies

3. Distinct Induction of Pectinases as a Factor Determining Host
   Specificity of Soft-Rotting *Erwinia* 31
   S. Tsuyumu, M. Miura, and S. Nishio

4. Regulation of the Production of Pectinases and Other
   Extracellular Enzymes in the Soft-Rotting *Erwinia* spp. 45
   A.K. Chatterjee, J.L. McEvoy, H. Murata, and A. Collmer

5. Characterization and Function of Bacterial Avirulence Genes 59
   N.T. Keen

6. Organization and Function of Pathogenicity Genes of *Pseudomonas*
   syringae Pathovars *phaseolicola* and *syringae* 69
   D. Mills, P. Mukhopadhyay, Y. Zhao, and M. Romantschuk

7. The Role of Indoleacetic Acid Biosynthetic Genes in
    Tumorigenicity 83
    T. Yamada, T. Nishino, T. Shiraishi, T. Gaffney, F. Roberto,
    C.J. Palm, H. Oku, and T. Kosuge

8. Molecular Analysis of Phaseolotoxin Production in
    Pseudomonas Syringae pv. *phaseolicola* 95
    S.S. Patil, K.B. Rowley, H.V. Kamdar, D. Clements, M. Mandel,
    and T. Humphreys

Fungal Strategies

9. Molecular Analysis of Pathogenesis in *Ustilago maydis* 107
    S.A. Leong, J. Wang, J. Kronstad, D. Holden, A. Budde,
    E. Froeliger, T. Kinscherf, P. Xu, W.A. Russin, D. Samac,
    T. Smith, S. Covert, B. Mei, and C. Voisard

10. Molecular Analysis of Genes for Pathogenicity of *Alternaria*
    alternata Japanese Pear Pathotype, a Host-Specific Toxin
    Producer 119
    T. Tsuge and H. Kobayashi

11. Strategies for Characterizing and Cloning Host Specificity
    Genes in *Magnaporthe grisea*, the Rice Blast Fungus 131
    F. G. Chumley and B. Valent

12. Role of Host-Specific Toxins in the Pathogenesis of
    Alternaria alternata 139
    H. Otani, K. Kohmoto, M. Kodama, and S. Nishimura

13. Suppressor Production as a Key Factor for Fungal Pathogenesis 151
    T. Shiraishi, T. Yamada, H. Oku, and H. Yoshioka

Plant's Response

14. Molecular Aspects of Elicitation of Host Defense Reactions 165
    M. Yoshikawa and Y. Takeuchi

15. Genetic Fine Structure Analysis of a Maize Disease Resistance
    Gene 177
    J.L. Bennetzen, S.H. Hulbert, and P.C. Lyons

16. Recognition of Fungal Nonpathogens by Plant Cells at the
    Prepenetration Stage 189
    H. Kunoh, I. Kobayashi, and N. Yamaoka

17. Role of Phytoalexins in Host Defense Reactions 203
    S. Mayama, A.P.A. Bordin, T. Morikawa, and T. Tani

18. Regulation of Nodule Gene Expression in Plant-Controlled
    Ineffective Alfalfa 215
    M.A. Egli, C.P. Vance, and R.J. Larson

Breeding of Disease–Resistant Plants

19. The Use of Somaclonal Variation for the Breeding of Disease-
    Resistant Plants 229
    H. Toyoda and S. Ouchi

Poster Abstracts

Abstracts presented at the U.S.-Japan Cooperative Science Seminar
(Molecular Strategies of Pathogens and Host Plants)
East-West Center, Honolulu, Hawaii, April 15-20, 1990 241

Synopsis 263
R.D. Durbin

Index 265

Participants

US-Japan Cooperative Science Seminar, East-West Center, Honolulu, HI, USA, April 15-20, 1990.

Alexander, D.	Calgene Inc. 1920 Fifth Street, Davis, CA 95616, USA
Bennetzen, J.L.	Department of Biological Sciences, Purdue University, West Lafayette, IN 47907, USA
Bushnell, W.R.	Department of Plant Pathology, University of Minnesota, St. Paul, MN 55108, USA
Chatterjee, A.K.	Department of Plant Pathology, University of Missouri, Columbia, MO 65211, USA
Chumley, F.G.	Central Research and Development Department, E.I. DuPont DeNemours and Company, Wilmington, DE 19880-0402, USA
Durbin, R.D.	Plant Disease Research Unit USDA-ARS and Department of Plant Pathology, University of Wisconsin, Madison, WI 53706, USA
Essenberg, M.K.	Department of Biochemistry, Oklahoma State University, Stillwater, OK 74074, USA
Hammerschmidt, R.	Department of Botany and Plant Pathology and in Cell and Molecular Biology Program, Michigan State University, East Lansing, MI 48824, USA
Hashiba, T.	Department of Plant Pathology, Faculty of Agriculture, Tohoku University, Sendai, Miyagi 981, Japan
Keen, N.T.	Department of Plant Pathology, University of California, Riverside, CA 92521-0122, USA
Kohmoto, K.	Department of Plant Pathology, Tottori University, Tottori 680, Japan
Kunoh, H.	Faculty of Bioresources, Mie University, Tsu City 514, Japan
Lamb, C.J.	Plant Biology Laboratory, Salk Institute for Biological Studies, 10010 North Torrey Pines Road, LaJolla, CA 92307, USA
Leong, S.A.	USDA-ARS Plant Disease Resistance Research Unit and Department of Plant Pathology, University of Wisconsin, Madison, WI 53706, USA
Macko, V.	Boyce Thompson Institute for Plant Research, Cornell University, Ithaca, NY 14853, USA
Matsuda, Y.	Laboratory of Plant Pathology, Kinki University, 3327-204 Nakamachi, Nara 631, Japan

Mayama, S.   Faculty of Agriculture, Kagawa University, Miki-cho, Kagawa 761-07, Japan

Mills, D.   Department of Botany and Plant Pathology, Oregon State University, Corvallis, OR 97331 USA

Morris, R.O.   Department of Biochemistry, University of Missouri, Columbia, MO 65211, USA

Nester, E.W.   Department of Microbiology, University of Washington, Seattle, WA 98195, USA

Oku, H.   Laboratory of Plant Pathology, Faculty of Agriculture, Okayama University, Okayama 700, Japan

Ouchi, S.   Laboratory of Plant Pathology, Faculty of Agriculture, Kinki University, 3327-204 Nakamachi, Nara 631, Japan

Patil, S.S.   Department of Plant Pathology and the Biotechnology Program, University of Hawaii, 3050 Maile Way, Gilmore 410, Honolulu, HI 96822, USA

Shiraishi, T.   Laboratory of Plant Pathology, Faculty of Agriculture, Okayama University, Tsushima, Okayama 700, Japan

Toyoda, H.   Laboratory of Plant Pathology, Faculty of Agriculture, Kinki University, 3327-204 Nakamachi, Nara 631, Japan

Tsuge, T.   Laboratory of Plant Pathology, Faculty of Agriculture, Nagoya University, Chikusa, Nagoya 464, Japan

Tsuyumu, S.   Laboratory of Plant Pathology, Faculty of Agriculture, Shizuoka University, Shizuoka 422, Japan

Van Alfen, N.K.   Department of Plant Pathology and Microbiology, Texas A&M University, College Station, TX 77843-2132, USA

Vance, C.P.   Department of Agronomy and Plant Genetics, USDA-ARS, University of Minnesota, St. Paul, MN 55108, USA

Yamada, T.   Laboratory of Plant Pathology & Genetic Engineering, College of Agriculture, Okayama University, Tsushima, Okayama 700, Japan

Yamamoto, H.   Laboratory of Plant Pathology, Faculty of Agriculture, Kagawa University, Miki-cho, Kagawa 761-07, Japan

Yokoyama, K.   Obihiro University of Agriculture and Veterinary Medicine, Inada-Cho, Obihiro, Hokkaido 080, Japan

Yoshikawa, M.   Laboratory of Plant Pathology, Faculty of Agriculture, Kyoto Prefectural University, Shimogamo, Kyoto 606, Japan

Contributors

Bennetzen, J.L. Department of Biological Sciences, Purdue University, West Lafayette, IN 47907, USA

Bordin, A.P.A. Laboratory of Genetics and Breeding, Faculty of Agriculture, Osaka Prefecture University, Sakai 591, Japan

Budde, A. USDA-ARS Plant Disease Resistance Research Unit and Department of Plant Pathology, University of Wisconsin, Madison, WI 53706, USA

Bushnell, W.R. Department of Plant Pathology, University of Minnesota, St. Paul, MN 55108, USA

Chatterjee, A.K. Department of Plant Pathology, University of Missouri, Columbia, MO 65211, USA

Chumley, F.G. Central Research and Development Department, E.I. DuPont DeNemours and Company, Wilmington, DE 19880-0402, USA

Clements, D. Department of Biochemistry, University of Hawaii, Honolulu, HI 96822, USA (present address: Hawaii Biotechnology Group Inc., P. O. Box 1057, Aiea, HI, 96701, USA

Collmer, A. Department of Plant Pathology, Cornell University, Ithaca, NY 14853-5908, USA

Covert, S. USDA-ARS Plant Disease Resistance Research Unit and Department of Plant Pathology, University of Wisconsin, Madison, WI 53706, USA (present address: USDA Forest Products Laboratory, Madison, WI, USA)

Durbin, R.D. Plant Disease Research Unit USDA-ARS and Department of Plant Pathology, University of Wisconsin, Madison, WI 53706, USA

Egli, M.A. USDA-ARS, Plant Science Research Unit and the Department of Agronomy and Plant Genetics, University of Minnesota 55108, USA

Essenberg, M.K. Department of Biochemistry, Oklahoma State University, Stillwater, OK 74074, USA

Froeliger, E. USDA-ARS Plant Disease Resistance Research Unit and Department of Plant Pathology, University of Wisconsin, Madison, WI 53706, USA

Gaffney, T.   CIBA-GEIGY Biotechnology, P.O. Box 12257, Research
                     Triangle Park, NC 27709, USA

Gordon, M.P.   Department of Microbiology, University of Washington,
                     Seattle, WA 98195, USA

Hashiba, T   Department of Plant Pathology, Faculty of Agriculture,
                     Tohoku University, Sendai, Miyagi 981, Japan

Holden, D.   USDA-ARS Plant Disease Resistance Research Unit and
                     Department of Plant Pathology, University of Wisconsin,
                     Madison, WI 53706, USA

Hulbert, S.H.   Department of Biological Sciences, Purdue University,
                     West Lafayette, IN 47907, USA

Humphreys, T.   Pacific Biomedical Research Center, Kewalo Marine
                     Laboratory, 41 Auhi Street, University of Hawaii,
                     Honolulu, HI 96813, USA

Kamdar, H.V.   Department of Plant Pathology, University of Hawaii,
                     Honolulu, HI 96822, USA (present address: Department
                     of Plant Pathology, University of California, Davis, CA
                     95616, USA

Keen, N.T.   Department of Plant Pathology, University of California,
                     Riverside, CA 92521-0122, USA

Kinscherf, T.   USDA-ARS Plant Disease Resistance Research Unit and
                     Department of Plant Pathology, University of Wisconsin,
                     Madison, WI 53706, USA

Kobayashi, H.   Laboratory of Plant Pathology, Faculty of Agriculture,
                     Nagoya University, Chikusa, Nagoya 464, Japan

Kobayashi, I.   Laboratory of Plant Pathology, Faculty of Bioresources,
                     Mie University, Tsu-city, 514, Japan

Kodama, M.   Laboratory of Plant Pathology, Faculty of Agriculture,
                     Tottori University, Tottori 680, Japan

Kohmoto, K.   Department of Plant Pathology, Tottori University,
                     Tottori 680, Japan

Kosuge, T.   Department of Plant Pathology, College of Agriculture,
                     University of California, Davis, CA 95616, USA
                     (deceased March 13, 1988)

Kronstad, J.   USDA-ARS Plant Disease Resistance Research Unit and
                     Department of Plant Pathology, University of Wisconsin,
                     Madison, WI 53706, USA (present address:
                     Biotechnology Laboratory, University of British
                     Columbia, Vancouver, B.C., Canada)

Kunoh, H.   Faculty of Bioresources, Mie University, Tsu City 514,
                     Japan

Larson, R.J.   USDA-ARS, Plant Science Research Unit and
                     Department of Agronomy and Plant Genetics, University
                     of Minnesota, St. Paul, MN 55108, USA

Leong, S.A.	USDA-ARS Plant Disease Resistance Research Unit and Department of Plant Pathology, University of Wisconsin, Madison, WI 53706, USA
Lyons, P.C.	Department of Biological Sciences, Purdue University, West Lafayette, IN 47907, USA
Mandel, M	Department of Biochemistry, University of Hawaii, Honolulu, HI 96822, USA
Mayama, S.	Faculty of Agriculture, Kagawa University, Miki-cho, Kagawa 761-07, Japan
McEvoy, J.L.	Department of Plant Pathology, 108 Waters Hall, University of Missouri-Columbia, Columbia, MO 65211, USA
Mei, B.	USDA-ARS Plant Disease Resistance Research Unit and Department of Plant Pathology, University of Wisconsin, Madison, WI 53706, USA
Mills, D.	Department of Botany and Plant Pathology, Oregon State University, Corvallis, OR 97331, USA
Miura, M.	Laboratory of Plant Pathology, Faculty of Agriculture, Shizuoka University 836 Ohya, Shizuoka city, 422, Japan
Morikawa, T.	Laboratory of Plant Pathology, Faculty of Agriculture, Kagawa University, Kagawa 761-07, Japan
Morris, R.O.	Department of Biochemistry, University of Missouri, Columbia, MO 65211, USA
Mukhopadhyay, P.	Crop Protection Department, Monsanto Agricultural Co., St. Louis, MO 63198, USA
Murata, H.	Department of Plant Pathology, 108 Waters Hall, University of Missouri-Columbia, Columbia, MO 65211, USA
Nester, E.W.	Department of Microbiology, University of Washington, Seattle, WA 98195, USA
Nishimura, S.	Laboratory of Plant Pathology, Faculty of Agriculture, Nagoya University, Nagoya 464, Japan (deceased 27 May, 1989)
Nishino, T.	Laboratory of Plant Pathology & Genetic Engineering, College of Agriculture, Okayama University, Tsushima, Okayama 700, Japan
Nishio, S.	Laboratory of Plant Pathology, Faculty of Agriculture, Shizuoka University 836 Ohya, Shizuoka city, 422, Japan
Oku, H.	Laboratory of Plant Pathology, Faculty of Agriculture, Okayama University, Okayama 700, Japan
Otani, H.	Laboratory of Plant Pathology, Faculty of Agriculture, Tottori University, Tottori 680, Japan
Ouchi, S.	Laboratory of Plant Pathology, Faculty of Agriculture, Kinki University, 3327-204 Nakamachi, Nara 631, Japan

Palm, C.J.	Institute of Biological Chemistry, Washington State University, Pullman, WA 99164-6340, USA
Patil, S.S.	Department of Plant Pathology and the Biotechnology Program, University of Hawaii, 3050 Maile Way, Gilmore 410, Honolulu, HI 96822, USA
Roberto, F.	Idaho National Engineering Laboratory, P.O. Box 1625, Idaho Falls, ID 83415-2203, USA
Romantschuk, M.	Department of General Microbiology, University of Helsinki, Mannerheimintie 172, SF-00300, Helsinki, Finland.
Rowley, K.B.	Department of Plant Pathology and the Biotechnology Program, University of Hawaii, 3050 Maile Way, Honolulu, HI 96822, USA
Russin, W.A.	USDA-ARS Plant Disease Resistance Research Unit and Department of Plant Pathology, University of Wisconsin, Madison, WI 53706, USA (present address: Department of Botany, University of Wisconsin, Madison, WI 53706, USA
Samac, D.	USDA-ARS Plant Disease Resistance Research Unit and Department of Plant Pathology, University of Wisconsin, Madison, WI 53706, USA (present address: Monsanto Co., St.Louis, MO, USA
Shiraishi, T.	Laboratory of Plant Pathology, Faculty of Agriculture, Okayama University, Tsushima, Okayama 700, Japan
Smith, T.	USDA-ARS Plant Disease Resistance Research Unit and Department of Plant Pathology, University of Wisconsin, Madison, WI 53706, USA (present address: Genencor Inc., San Francisco, CA, USA)
Takeuchi, Y.	Laboratory of Plant Pathology, Faculty of Agriculture, Kyoto Prefectural University, Shimogamo, Kyoto 606, Japan
Tani, T.	Laboratory of Plant Pathology, Faculty of Agriculture, Kagawa University, Kagawa 761-07, Japan
Toyoda, H.	Laboratory of Plant Pathology, Faculty of Agriculture, Kinki University, 3327-204 Nakamachi, Nara 631, Japan
Tsuge, T.	Laboratory of Plant Pathology, Faculty of Agriculture, Nagoya University, Chikusa, Nagoya 464, Japan
Tsuyumu, S.	Laboratory of Plant Pathology, Faculty of Agriculture, Shizuoka University, Shizuoka 422, Japan
Valent, B	E.I. Du Pont de Nemours and Co., Inc. Central Research and Development Department Experimental Station, P.O. Box 80402 Wilmington, DE 19880-0402, USA
Vance, C.P.	USDA-ARS, Department of Agronomy and Plant Genetics, University of Minnesota, St. Paul, MN 55108, USA

Voisard, C.	USDA-ARS Plant Disease Resistance Research Unit and Department of Plant Pathology, University of Wisconsin, Madison, WI 53706, USA
Wang, J.	USDA-ARS Plant Disease Resistance Research Unit and Department of Plant Pathology, Unviersity of Wisconsin, Madison, WI 53706, USA (present address: La Trobe University, Bundoora, Victoria, Austrailia)
Xu, P.	USDA-ARS Plant Disease Resistance Research Unit and Department of Plant Pathology, University of Wisconsin, Madison, WI 53706, USA
Yamada, T.	Laboratory of Plant Pathology & Genetic Engineering, College of Agriculture, Okayama University, Tsushima, Okayama 700, Japan
Yamaoka, N.	Laboratory of Plant Pathology, Faculty of Bioresources, Mie University, Tsu city, 514, Japan
Yoshikawa, M.	Laboratory of Plant Pathology, Faculty of Agriculture, Kyoto Prefectural University, Shimogamo, Kyoto 606, Japan
Yoshioka, H.	Laboratory of Plant Pathology and Genetic Engineering, College of Agriculture, Okayama University, Okayama 700, Japan
Zhao, Y.	Genetics Program, Oregon State University, Corvallis, OR 97331-2902, USA

Overviews

Chapter 1
Molecular Strategies in the Interaction Between *Agrobacterium* and Its Hosts

Eugene W. Nester and Milton P. Gordon

It is a great pleasure and honor for me to present the Tsune Kosuge Memorial Lecture. Not only was Professor Kosuge a good friend and colleague, but his scientific contributions in the biochemical-genetic analysis of *Pseudomonas syringae* pathovar *savastanoi* (*P. savastanoi*) pathogenesis of olive and oleander plants served as a model for our own studies in crown gall disease caused by *Agrobacterium*. Indeed, in a broader sense, Tsune's studies serve as a model for all other plant pathogen interactions that are currently under investigation. His demonstration of the association of indoleacetic acid (IAA) with the tumorous growths caused by *Pseudomonas*, in 1963, was the first identification of a virulence factor in plants (Magie et al., 1963). In the next 25 years, his laboratory elucidated the pathway for IAA production in *P. savastanoi* and then showed that the genes which encoded this pathway mapped on a plasmid. Over ten years ago, when the use of genetics in studies in plant pathology was in its infancy, Tsune was employing mutants that either did not accumulate or overproduced IAA (Smidt et al., 1978). He and his colleagues further demonstrated that bacterially synthesized cytokinin which maps to the same plasmid also plays an important role in this disease. His most recent studies, which were continued after his death by his colleagues, involved the metabolism of IAA by *Pseudomonas* (Roberto et al., 1988). In these studies, they demonstrated that the plasmid that encodes enzymes of auxin and cytokinin biosynthesis also encodes an enzyme which converts IAA to IAA-lysine. This represents the first characterization of genes involved in the conjugation of IAA. This IAA conjugate has less biological activity than IAA, which has significance for the virulence of the organism. Thus, when *P. savastanoi* strains that infect oleander were mutagenized with transposon insertions in the locus that was responsible for IAA-lysine production, they became less virulent and demonstrated a reduced ability to grow within tissues of the host plant.

In all of his studies in plant-microbe interactions, Tsune used his training as a biochemist who appreciated and skillfully utilized the power of genetics. He always aimed to understand the biology of the system in molecular terms.

Other writers in this volume cover the strategies that a variety of bacterial and fungal pathogens employ in interacting with their hosts. My own presentation will focus on the interaction of *Agrobacterium* with a wide range of dicotyledonous plants. As far as we are aware, the *Agrobacterium*-plant interaction is unique in that the end result is the transfer and integration of a piece of bacterial DNA into the plant chromosome. However, the interaction of this organism with the plant can serve as a model system for many other bacterial-plant interactions, since it involves such features as the attachment of the bacteria to their host cells, the activation of genes required for pathogenicity by plant signal molecules, and the production of phytohormones, auxin and cytokinin, as the virulence factors which result in the gall-like symptoms of the disease. All of these aspects can be found in other bacterial-plant interactions.

Overall Features of Crown Gall Tumor Formation

Agrobacterium tumefaciens induces a disease, crown gall, in a wide variety of dicotyledonous plants by transferring a piece of its tumor-inducing (Ti-) plasmid into the plant cell where it becomes integrated and functions in the plant. The overall features of this disease are illustrated in Figure 1.

The transferred and integrated DNA (T-DNA) codes for the synthesis of the two growth regulators, auxin and cytokinin, as well as for a group of amino acid derivatives termed "opines." The expression of the genes for phytohormone synthesis, which are not subject to regulation by the plant, gives rise to the symptoms of crown gall tumor formation. The transfer of the T-DNA

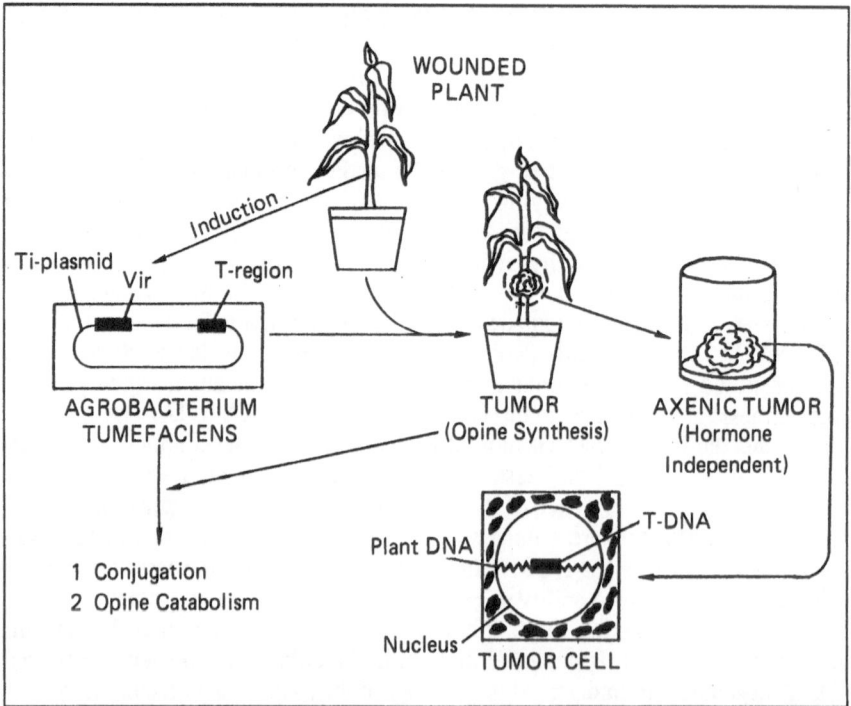

Figure 1. Overall features of crown gall tumor formation by *Agrobacterium*. It is possible to delete the T-DNA and insert useful genes under the control of plant promoters. The expression of these genes integrated into the plant DNA confers desired properties on the plant. This technology forms the basis of genetic engineering of plants using *Agrobacterium*.

requires the expression of a variety of other genes on the Ti-plasmid, the virulence (*vir*) genes. These genes, which are involved in the processing and transfer of the T-DNA, are not expressed when *Agrobacterium* grows in the absence of plant cells, but are activated by plant cell metabolites synthesized by the wounded plant.

Early Events in the Transfer of T-DNA into Plant Cells

1. Attachment of Bacteria to Plant Cells

Considerable evidence exists that *Agrobacterium* must bind to plant cells in order to cause crown gall tumors. However, the molecular basis of this attachment process remains elusive. A number of mutants of *Agrobacterium* have been identified which map to three loci, all of which map to the chromosome and are termed *Chv*. None of these mutants are capable of attaching, and all are avirulent. Furthermore, all of these mutants are involved with the synthesis or transport of a low molecular weight polysaccharide, ß-1,2-glucan (Fig. 2).

The *Chv*A codes for a protein that is necessary for the transport of the ß-1,2-glucan into the periplasm; the *Chv*B gene codes for a 235-kd membrane-associated protein which converts glucose into the cyclic ß-1,2-glucan. The *exo*C locus codes for an enzyme which converts glucose 6-phosphate to glucose 1phosphate, a step required for the synthesis of cellulose as well as ß-1,2-

BIOCHEMICAL ANALYSIS OF β-1,2-GLUCAN MUTANTS

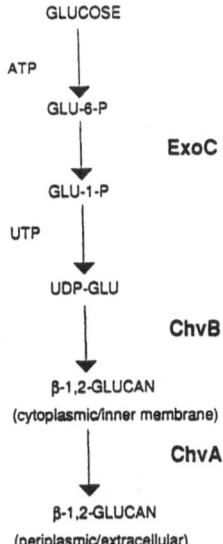

Figure 2. Identification of the site of mutations involving the synthesis and transport of ß-1,2-glucan.

glucan. How ß1,2-glucan functions in attachment is not clear. It has not been possible to demonstrate that the addition of concentrated cell supernatants from ß-1,2-glucan synthesizing cells to ß-1,2-glucan negative cells can complement the negative cells in attachment (J. Cangelosi, unpublished observation). Significantly, the ß-1,2-glucan negative cells are pleiotrophic and some property other than ß-1,2-glucan synthesis may be responsible for the cells' inability to attach.

2. Ti-plasmid

Although a number of genes necessary for tumor formation have been identified that map to the chromosome, it appears that most of the genes required for tumor formation are located on the large 180-kb Ti-plasmid (Fig. 3). In addition to the two sets of genes required for tumor formation, the T-DNA and the *vir* genes, other genes which play no significant role in tumor formation have also been identified. These include genes for the metabolism of the opines, as well as housekeeping genes concerned with the replication of the plasmid.

3. Designation of vir Genes

The virulence genes are comprised of approximately 35-kb of DNA and are essential for tumor formation, although they are not transferred into the plant. They can be divided into six highly studied transcription units, oroperons. These include *vir*A, *vir*G, *vir*B, *vir*C, *vir*D, and *vir*E. Two additional *vir*

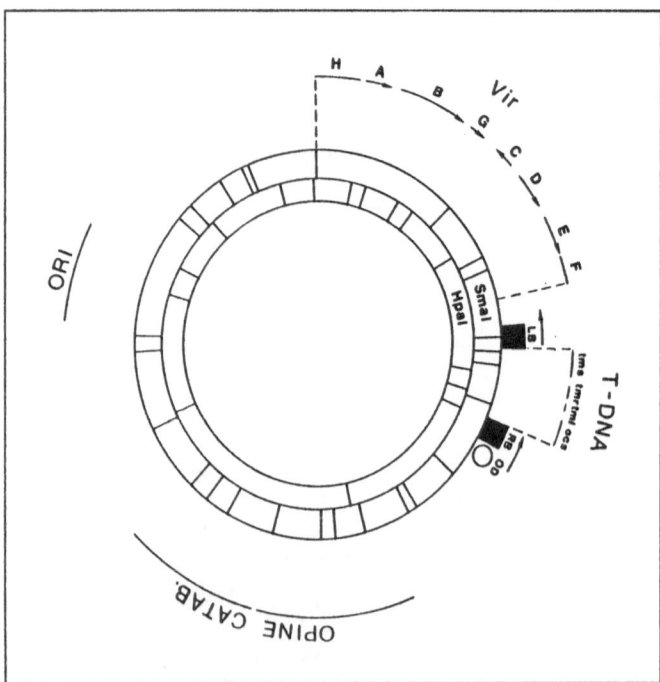

Figure 3. Restriction enxyme map of Ti-plasmid indicating restriction sites produced by SmaI and HpaI.

operons have been identified in the octopine Ti *vir* regions (*vir*H and *vir*F), but they have not been studied extensively. The most relevant data on the *vir* genes are summarized in Figure 4.

It seems likely that additional *vir* genes may be identified in the future. These probably will be recognized as genes which are required for tumor formation on some, but not other, plants, and/or as genes which are required for optimal tumor formation. Different strains of *Agrobacterium* have different *vir* genes. For example, strains of *Agrobacterium* that induce tumors that synthesize the opine octopine have the *vir*H region, whereas strains which induce nopaline synthesizing tumors do not. However, the nopaline-inducing strains have a *vir* gene coding for the synthesis of isopentenyl adenosine monophosphate at the same relative map position. There are other less well-identifiable differences between these two types of Ti-plasmids.

4. Chemical Inducer of vir Genes

The induction of *vir* genes depends upon the synthesis of plant signal molecules synthesized by the wounded plant. One such molecule which has been identified is acetosyringone (3,5dimethoxy4hydroxy acetophenone) (Stachel et al., 1985). This compound, which seems to be a derivative of a precursor of lignin biosynthesis, has been found in a wide variety of plants following wounding. A number of other compounds chemically related to acetosyringone also have inducing activity (Spencer et al., 1988).

5. Activation of VirA Protein

Two of the *vir* genes, *vir*A and *vir*G are required for the expression of all *vir* genes, since mutations in either locus eliminate the expression of all other *vir* genes (Winans et al., 1986). Protein homology searches have shown that the VirA and VirG proteins are homologous with a large number of two-component regulatory systems including EnvZ/OmpR, NtrB/NtrC and CheA/CheY (Winans et al., 1986). The first protein of each pair detects a particular environmental signal, and then transfers this information to a second component which in turn, in most cases, activates the expression of a series of genes whose gene products respond to the environment. The VirA protein is a transmembrane protein (Winans, et al., 1989). Two hydrophobic regions anchor the N-terminal portion of the protein to the cytoplasmic membrane with a region protruding into the periplasmic space and the C-terminal domain remaining in the cytoplasm. Presumably in some undefined way, the VirA protein interacts with the plant signal molecule, thereby becoming activated, and in turn activates the VirG protein. The mechanism of VirA protein activation is now being clarified. It has been shown that the VirA protein has an autophosphorylating activity which is the most likely mechanism for VirA protein activation (Jin et al., 1990). The role that acetosyringone plays in autophosphorylation is unclear, since the addition of γ-labeled ATP with purified VirA protein but without acetosyringone still results in the labeling of the VirA protein.

A histidine residue located in the highly conserved block of amino acids which is found in all of the VirA homologues is the amino acid phosphorylated. If this histidine is mutated to a glutamine residue which cannot be

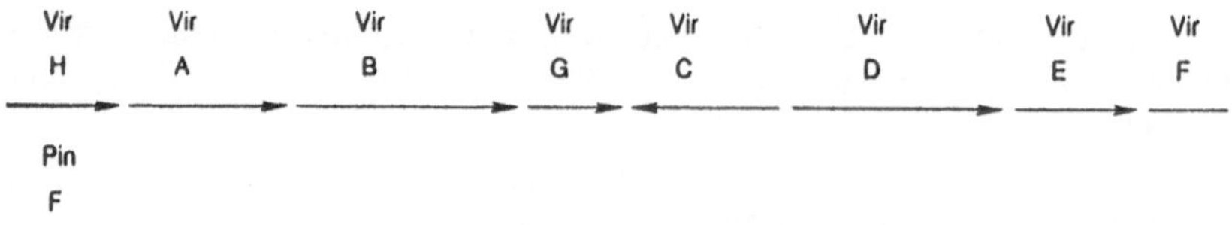

Vir	Vir	Vir	Vir	Vir	Vir	Vir	Vir
H	A	B	G	C	D	E	F

Pin
F

Vir	Inducibility	Size (Kb)	ORF's	Function
A	+	2.8	1	Plant Signal Sensor
G	+	1.0	1	Transcriptional Activator
B	+	9.5	11	Pore?
D	+	4.5	4	Processing of T-DNA-Endonuclease
C	+	1.5	2	Processing of T-DNA
E	+	2.2	2	Single-Strand DNA Binding Protein
H	+	3.4	2	Cytochrome P450 Enzyme
F	?	?	?	?

Figure 4. *vir* gene order and relevant information.

phosphorylated, then the *vir* genes are not inducible nor is the strain capable of inducing crown gall tumors. This demonstrates that phosphorylation of the VirA protein is essential for its proper functioning.

6. *Interaction of* VirA *and* VirG

The phosphate residue of the phosphorylated histidine in the VirA protein is transferred directly to the VirG protein (S. Jin, unpublished observation). Thus, when the phosphorylated VirA protein was mixed with unphosphorylated VirG protein and samples withdrawn at various times, with increasing time of incubation, the phospho-VirA signal got weaker while the phospho-VirG signal got stronger. The phosphate transfer occured very rapidly since 5 s after the VirA and VirG proteins were mixed together, transfer could be observed.

The phosphate bond in the VirG protein is highly unstable to both acid and base, which suggests that either an aspartic acid or glutamic acid is being phosphorylated. By appropriate techniques, the unstable phosphate-amino acid bond was stabilized and the protein was then cleaved and the cleavage products sequenced. From these studies, we conclude that a specific aspartic acid residue is phosphorylated in the VirG protein (S. Jin et al., unpublished observation). When this particular aspartic acid was mutated in vitro to asparagine by site-directed mutagenesis, the mutant was no longer able to induce *vir* genes nor was it able to induce tumors on kalanchoe leaves. This indicates that phosphorylation of the VirG protein is essential for its biological function.

7. *Interaction of the* VirG *Protein with the "vir Box"*

The VirG protein binds specifically to a 12 bp conserved sequence called the "*vir* box" which is located upstream of each of the *vir* genes (Winans et al., 1987). Footprinting analysis of each of the promoter regions indicates that the

VirG protein covers the "*vir* box" on both strands of the DNA (S. Jin et al., 1990). However, phosphorylation does not seem to be required for the binding of the VirG protein to the "*vir* box." The VirG protein was isolated from *E. coli* cells into which the gene had been cloned, and alkaline phosphatase was added to remove any phosphate from the protein. This preparation of the VirG protein still bound specifically to the "*vir* box." However, these data do not rule out the possibility that the phosphorylated VirG protein may have a higher affinity for the "*vir* box" than the nonphosphorylated form. Alternately, the action of the phosphorylated VirG protein may be beyond the binding step, such as the activation of transcription by RNA polymerase.

The overall features of the role of the VirA and VirG interaction with one another and the interaction of the VirG protein with the "*vir* box" are shown in Figure 5.

8. Role of Sugars in <u>vir</u> Gene Induction

It is now clear that in addition to acetosyringone, sugars may also be involved in *vir* gene induction in some way. This conclusion is based on the observation that a mutation in a particular locus on the chromosome reduced induction

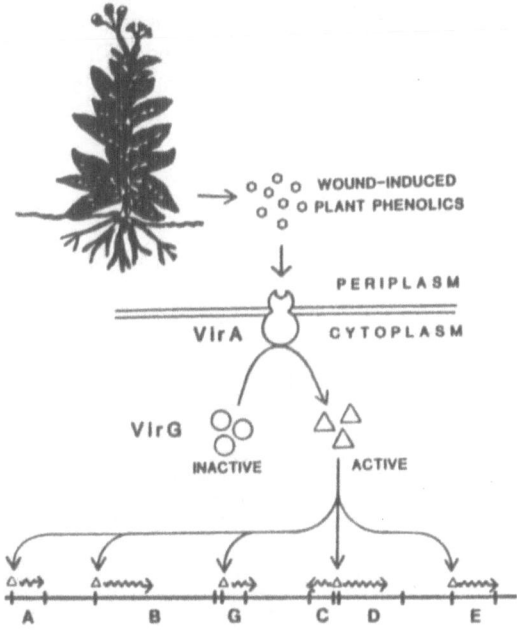

Figure 5. Model for the interaction of the VirA and VirG proteins with each other and with the "*vir* box."

by acetosyringone significantly and thereby attenuated virulence (Huang et al., 1990). When this locus was cloned and sequenced, it was shown to code for a protein which is homologous with a glucose/galactose binding protein in *E. coli*. This suggests that glucose, and perhaps other sugars, are required for maximum *vir* gene induction.

Vir Gene Products

Following activation of the *vir* genes, the T-DNA is processed for transfer into plant cells. A number of steps have been identified. One of the early stages in T-DNA processing involves the conversion of the supercoiled Ti-plasmid into a relaxed form by a topoisomerase, the product of the VirD1 protein (Ghai et al., 1989). The next step is a site-specific cleavage in the two 25 bp direct repeats which bound the T-DNA, the right and left borders. This is accomplished by an endonuclease, the product of the VirD2 protein (Yanofsky et al., 1986). Considerable evidence exists that the VirD2 protein cleaves at virtually identical sites in the bottom strand of the right and left border sequences, thereby resulting in the formation of a single-stranded intermediate. Following cleavage, the VirD2 protein remains covalently bound to the 5′ end of the T-DNA (Young et al., 1988). The *virE2* gene codes for a single-stranded DNA binding protein which associates with the T-DNA and probably protects this DNA in both the bacteria and the plant cell (Christie et al., 1988). The transfer intermediate can be looked upon as a single-stranded DNA molecule attached to the VirD2 gene product at the 5′ end, and coated by a single-stranded DNA

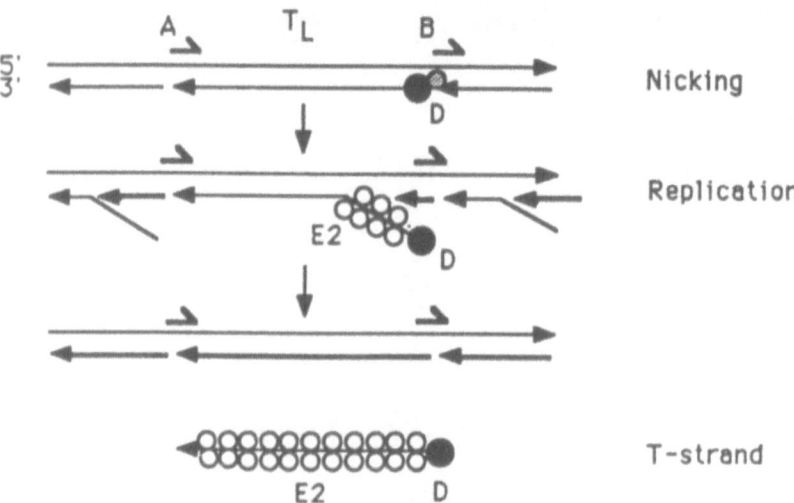

Figure 6. Proposed model for formation of T strand intermediate in transfer process. A and B refer to the left and right borders, respectively. TL refers to the leftward T-DNA. The E2 is a single strand DNA binding protein, and is a product of the *virE2* gene. D refers to the product of the *vir* D2 locus which binds tightly to the 5′ end of the T-strand.

binding protein. There is some speculation that the VirD2 protein may also serve to target the DNA to the plant cell nucleus. The early steps in T-DNA processing are illustrated in Figure 6.

The exit of the T-DNA from the bacterial cell probably occurs through pores in the bacterial cell envelope formed by the gene products coded by the *vir*B operon (11 open reading frames) (Ward et al., 1988). Most of these gene products have signal sequences and are associated with the bacterial cell envelope. Of considerable interest is the fact that the last open reading frame in this operon (no. 11) is phosphorylated and has ATPase activity (Christie et al., 1989). Furthermore, this protein is very similar to one required for the development of competence in the *Bacillus subtilis* DNA transformation system (Albano et al., 1989).

From these studies, we envision that *Agrobacterium* conjugates with plant cells and transfers single-stranded DNA from the bacterium to the plant cell. Once inside the plant cell, the DNA presumably recombines by a nonhomologous mechanism into a variety of sites of the plant chromosome. Once integrated, the T-DNA genes behave and function as plant genes.

Expression of the T-DNA

The T-DNA codes for a number of different transcripts, some of which have been functionally identified. Two of the transcripts best-characterized code for two enzymes of auxin synthesis and another codes for an enzyme of cytokinin synthesis (Binns et al., 1988). There is considerable interest that these three genes are highly homologous to their counterparts in *P. savastonoi* (Yamada et al., 1985). These latter studies, carried out in Tsune Kosuge's laboratory, strongly suggest that the structural genes of the T-DNA are probably of bacterial origin and have not been "captured" from plants by bacteria. Other well-characterized genes of the T-DNA code for enzymes of opine synthesis.

Summary

The formation of hyperplasias termed "galls" produced by *P. savastanoi* and the formation of crown gall tumors by *Agrobacterium tumefaciens* have a number of features in common. In both cases, the unregulated production of IAA and cytokinin at the site of infection is required for gall formation. Thus, both compounds, the product of bacterial genes, are virulence factors in both diseases, and, therefore, it is not surprising that the disease symptoms are similar. Furthermore, both of the virulence factors are coded by genes on large plasmids which have a remarkable degree of homology with each other. However, in *Pseudomonas*, the genes are constitutively expressed in the bacterium, whereas in *Agrobacterium* expression occurs only after the genes are integrated into the plant chromosome. Apparently, the structural component of the gene is prokaryotic in nature, whereas the upstream regulatory sequences are eukaryotic and respond to plant transcription and translation machinery.

Disease production by *P. savastanoi* appears to be a much simpler process than that by *Agrobacterium*. In the former case, the two phytohormones are synthesized by the bacteria at the site of infection, and living bacteria are necessary for continued symptomatalogy. On the other hand, *Agrobacterium* has evolved a very complex mechanism for introducing a piece of its plasmid into plant cells where it functions. Consequently, the continued growth of the bacteria at the wound site is not necessary for tumor formation. The *vir* genes of *Agrobacterium* apparently do not have a counterpart in *Pseudomonas*, although its virulence plasmid has not been subjected to extensive mutagenesis.

It is also instructive to compare the reasons why these two organisms induce galls. The most reasonable explanation as to why *Agrobacterium* induces tumors is that T-DNA codes for the synthesis of opines which can serve as a source of carbon, nitrogen, and energy for the inducing strain of *Agrobacterium* but very few other organisms in the soil environment (Tempé et al., 1982). These opines further promote the transfer of the T-DNA from plasmid-containing to non-plasmid-containing *Agrobacterium* strains in the environment, thereby promoting the spread of the Ti-plasmid. It has been postulated that *Pseudomonas* forms galls in order to provide a protected environment for the bacteria during the hot, dry summers characteristic of regions where its hosts, olive and oleander, flourish. Only in this protected environment do the bacteria grow to high concentrations.

Acknowledgments

We would like to thank the members of our research groups both past and present for providing much of the data presented in this paper. This work was supported by NIH grant 5RO1GM32618, NSF no. DMB8704292, and USDA no. 88-37234-3618.

References

Albano, M., Breitling, R., and Dubnau, D., 1989, Nucleotide sequence and genetic organization of the *Bacillus subtilis comG* operon. *J. Bacteriol.*, 171:5286-5404.

Binns, A., and Thomashow, M., 1988, Cell biology of *Agrobacterium* infection and transformation of plants, *Annu. Rev. Microbiol.* 42:575-606.

Christie, P., et al., 1988, The *Agrobacterium tumefaciens virE$_2$* gene product is a single-stranded-DNA-binding protein that associates with T-DNA, *J. Bacteriol.* 170:2659-2667.

Christie, P., et al., 1989, A gene required for transfer of T-DNA to plants encodes an ATPase with autophosphorylating activity, *Proc. Natl. Acad. Sci. USA*, 86:9677-9681.

Ghai, J., and Das, A., 1989, The VirD operon of *Agrobacterium tumefaciens* Ti-plasmid encodes a DNA-relaxing enzyme, *Proc. Nat. Acad. Sci. USA*, 89:3109-3113.

Huang, M-LW., et al., 1990, A chromosomal *Agrobacterium tumefaciens* gene required for effective plant signal transduction, *J. Bacteriol.* 172:1814-1822.

Jin, S., et al., 1990, The VirA protein of *Agrobacterium tumefaciens* is autophosphorylated and is essential for *vir* gene regulation, *J. Bacteriol.* 172:525-530.

Jin, S., et al., 1990, The regulatory VirG protein specifically binds to a cis-acting regulatory sequence involved in transcriptional activation of *Agrobacterium tumefaciens* virulence genes, *J. Bacteriol.* **172**:531-537.

Magie, A.R., Wilson, E.E., and Kosuge, T., 1963, Indoleacetamide as an intermediate in the synthesis of indoleacetic aicd in *Pseudomonas savastanoi*, *Science* **141**:1281-1282.

Roberto, I., and Kosuge, T., 1988, Aspects of phytohormone metabolism in *Pseudomonas savastanoi* physiology and biochemistry of plant-microbial interaction. *In* Physiology and Biochemistry of Plant-Microbiol Interactions (Keen, N.T., and Walling, L., eds.) Amer. Soc. of Plant Phys., Riverside, CA pp. 31-39.

Smidt, M., and Kosuge, T., 1978, The role of indole-3-acetic acid accumulation by alpha methyl tryptophan-resistant mutants of *Pseudomonas savastanoi* in gall formation on oleanders, *Physiol. Plant Pathol.* **13**:203-214.

Spencer, P. and Towers, G.H.N., 1988, Specificity of signal compounds detected by *Agrobacterium tumefaciens*, *Phytochemistry* **27**:2781-2785.

Stachel, S., et al., 1985, Identification of the signal molecules produced by wounded plant cells that activate T-DNA transfer in *Agrobacterium tumefaciens*, *Nature* (London) **318**:624-629.

Tempé, J. and Goldman, A., 1982, Occurrence and biosynthesis of opines, *In* Molecular Biology of Plant Tumors (Kahl, G. and Schell, J., eds.), Academic Press, New York, p. 615.

Ward, J., et al., 1988, Characterization of the *vir*B operon from an *Agrobacterium tumefaciens* Ti-plasmid, *J. Biol. Chem.* **263**:5804-5814.

Winans, S., et al., 1986, A gene essential for *Agrobacterium* virulence is homologous to a family of positive regulatory loci, *Proc. Natl. Acad. Sci. USA* **83**:8278-8282.

Winans, S., et al., 1987, The role of virulence regulatory loci in determining *Agrobacterium* host range, *In* Plant Molecular Biology (Von Wettstein, D. and Chua, N., eds.), Plenum Publishing Corp., New York, pp. 573-582 .

Winans, S., et al., 1989, A protein required for transcriptional regulation of *Agrobacterium* virulence genes spans the cytoplasmic membrane, *J. Bacteriol.* **171**:1616-1622.

Yamada, T., et al., 1985, Nucleotide sequences of the *Pseudomonas savastanoi* indole acetic acid pathway genes show homology with *Agrobacterium tumefaciens* T-DNA, *Proc. Natl. Acad. Sci. USA* **82**:6522-6526.

Yanofsky, M., et al., 1986, The *vir*D operon of *Agrobacterium tumefaciens* encodes a site-specific endonuclease, *Cell* **417**:471-477.

Young, C., and Nester E., 1988, Association of the VirD$_2$ protein with the 5' end of T strands in *Agrobacterium tumefaciens*, *J. Bacteriol.* **170**:3367-3374.

Summary of Discussion of Nester's Paper

Bennetzen initiated the questioning by asking whether there was anything known about the binding of the *vir*G product in vivo. *Nester* replied that they had no information but were setting up an in vitro transcription/translation system to see if there is specific transcription of *vir* genes only in the presence of VirG. They have mutants of *vir*G that are nonphosphorylatable and do not function in induction or virulence. *Chumley* then asked about the status of VirA phosphorlyation in vivo and whether or not it shows any dependency on acetosyringone or other phenolics. *Nester* replied that they have no information on that and that their experiments are all done in vitro using a truncated *vir*A. *Durbin* asked what the system might have to do with the fact that monocots are generally not hosts and whether this relates to his system. *Nester* responded that if they could better understand the mechanism of transfer of DNA and how these genes function they should be able to construct better vectors. Work in other labs leaves little question that *Agrobacterium* can transfer some DNA to monocots because when maize streak virus was inserted between repeats of the T-DNA and *Agrobacterium* was used to transmit it, viral symptoms appeared on the corn plant. If you mutate any of the *vir* genes or use only isolated DNA from the bacterium, *Agrobacterium* infection is abolished and no symptoms develop. This has been done with other monocots and it is a sensitive way to measure transfer, but it says nothing about integration. The octopine strains do not cause infection, but nopaline or *rhizogenes* strains cause 95 percent infection indicating that there is some difference between strains. *Margaret Bolton* has shown that the effect is the result of the Ti-plasmid, and there are differences in the *vir* gene region. *Roy Morris* has shown that a locus called tzs for transribosyl zeatin synthase, which is a *vir* gene in that it is activated by plant signal molecules in nopaline and *rhizogenes* strains; it is not present in octopine strains. One might be able to modify the *vir* genes and see if this results in gene transfer into monocots. This suggests that *vir* genes are relevant in gene transfer into monocots, and perhaps this system will offer some insight into why some strains transfer DNA while others do not. *Chatterjee* then asked whether, for example, what is called a glucose/galactose binding protein could have overlapping substrate specificity so that it takes in the phenolics. *Nester* responded that it is possible, but not much is known about the uptake of phenolics although radiolabeled phenolics are poorly taken up.

Toyoda asked about the mechanism for the activation of VirG by VirA. *Nester* responded that there is transfer of the phosphate from a histidine of VirA to an aspartic acid on VirG. *Bushnell* asked what actually moves into the host and could this part of the process be a problem with monocots. *Nester* replied that we know what moves into the host. He said the dogma is that it is only the T-DNA (ca. 15 kb) bound up with a single-stranded DNA binding protein. The bacteria and plasmid remain outside the plant cell. The model many people entertain is that there is conjugation between the bacteria and the plant. *Bennetzen* ended the questioning by asking how far one can go with the analysis of the agro-system with respect to the host genetics? *Nester* responded that that was a good point and that they are looking at mutants in *Arabodopsis* that can no longer form tumors.

Chapter 2
Molecular Biology of Fungal Host-Parasite Interactions

Seiji Ouchi

During evolution, along with their adaptation to the environment, plants acquired versatile defense mechanisms against pathogens. Physical barriers such as the cuticular layer fortified with wax, cutin or suberin, lignified cell walls, and modified pectins play a significant role in the containment of pathogens at the sites of attempted penetration, and chemical barriers such as preformed compounds and post-recognitionally synthesized phytoalexins play fungistatic and fungicidal roles in the defense reactions of the host or non-host plants. The boundaries of these arbitrarily classified barriers are, however, ambiguous, but these can be discussed within a unitary concept of molecular biology. In fact, considerable information is available on the molecular mechanisms of pathogenicity and host defenses which can be viewed in terms of interactions between host and pathogen genes.

On the basis of the concept that infection is the consequence of a series of gene interactions of pathogens and host plants, it is reasonable to assume that disease specificities are determined at different stages from the pathogenicity phase including pathogen contact, germination, penetration, and establishment of a pseudosymbiotic relationship, to the virulence phase including multiplication or extensive growth of the pathogen in the invaded tissues leading to characteristic symptoms.

In this treatise, some of the recent progress in molecular strategies used by fungal pathogens in their establishment and host plant strategies involved in their defense against fungal pathogens will be discussed with a special emphasis on gene interactions.

Gene Interactions

Although the observation of disease specificity dates back to Theophrastus (370-286 B.C.), and breeding of disease resistant cultivars had been practiced before gene theory was established, Mendelian segregation of a disease resistance gene was elucidated by Biffen (1905) only at the beginning of this century, and the subsequent breeding of varieties of crop plants eventually led Flor (1955) to propose the gene-for-gene theory which was the beginning of studies on genetic aspects of host-parasite interactions. As exemplified by the so-called quadratic

---

This overview is dedicated to the late Professor Syoyo Nishimura.

check, the essence of the gene-for-gene hypothesis is that resistance is expressed only when the resistance genes of host plants interact with the avirulence genes of pathogens. Molecular aspects of gene interactions in plant pathogenesis have been summarized in several excellent books (Day, 1974; Day et al., 1987; Palacios et al., 1988; Kosuge et al., 1984, 1987, 1989), and molecular models have been proposed to explain these interactions. The principles on which these models are based are simple and basically assume, like the hormone-receptor systems in higher animals, that the signal molecules produced by pathogens (elicitors) are recognized, in an all-or-none fashion by the corresponding receptor molecules of the host plant.

In fact, recent progress in the isolation of avirulence genes from some pathovars of the phytobacterium *Pseudomonas syringae* (Keen, this volume) supports the basic observation in most disease interactions, that the avirulence genes of pathogens and the resistance genes of host plants are dominant and epistatic over virulence and non-resistance genes, respectively. In fungal diseases, however, most of the elicitors so far characterized are non-specific, even though they come from genetically defined races of pathogens.

In fungal diseases in which host-specific toxins determine disease specificity, gene interactions are entirely reversed i.e., in most cases, susceptibility to the toxin producer is dominant over resistance. In these diseases, the toxins and their specific receptors play a determinative role in the expression of specificity.

Why the molecular basis of disease specificity of pathogens which elaborate host-specific toxins evolved in a diametrically opposite fashion from those involved in gene-for-gene systems is not known. Once the detailed architecture of gene expression in host-pathogen systems is understood, the reasons for these different specificities will be elucidated. In this respect, the strategies the pathogens utilize for the establishment of basic compatibility will offer a clue to the comprehensive understanding of pathogenesis. Synergistic, cooperative, or competitive interactions of genes involved in the determination of disease specificity should also be taken into consideration for establishing the unitary concept, because the phenotypes are determined by a threshold derived from complicated interactions of many genes.

Strategies of Pathogens

In order to be successful pathogens have to overcome the versatile defense mechanisms which plants have acquired during their evolution. In fact, they produce an array of compounds that elicit or suppress the molecular processes involved in the expression of resistance by the host plants. Because of space limitations, the discussion here will be confined to a few examples that have significant genetic implications.

1. Enzymes

Fungal pathogens produce a variety of enzymes to establish themselves in plant tissues. Some of them have been isolated in almost pure form and characterized as pathogenesis related enzymes.

a) Cellulose-degrading enzymes

The cell wall is a unique plant structure. Although its structure and function are very complex (McNeil et al., 1984), pathogens capable of establishing intracellular parasitism produce several types of cellulose-degrading enzymes. Many fungal pathogens are known to produce endo-1,4-β-D-glucanase, 1,4-β-D-glucancellobiohydrolase, and other hydrolytic enzymes that degrade the cellulose backbone. In some cases, endoglucanases of host cells seem to play a significant role in the release of a specific elicitor from fungal mycelium (Yoshikawa et al., this volume). However, in contrast to the progress in the analysis of cellulases produced by bacterial pathogens, such as β-1,4-endoglucanase of *Pseudomonas solanacearum* (Roberts et al., 1988), no significant progress has been made in clarifying the role of cellulases in fungal diseases, except for those of the wood decaying fungi.

b) Cutin-degrading enzymes

The cuticular layer is the first barrier of aerial parts of plants. The pathogens need to traverse this barrier before they can enter the plant cell. The degradation of cutins by plant pathogens had been chemically determined in the 1960s (Heinen, 1962; Linskens et al., 1965; Shishiyama et al., 1970), but that some fungal pathogens possess cutin degrading enzymes was only recently elucidated. A series of intensive studies by Kolattukudy et al., 1987 and his colleagues, and others revealed that cutinases are indeed essential for the primary penetration by some fungal pathogens.

Several lines of evidence have been presented in support of this contention; e.g., low virulence of cutinase deficient mutants on intact tissues (Dickman et al., 1986), establishment of infection by these mutants in the presense of added cutinase, inhibition of infection by the addition of cutinase inhibitors such as diisopropyl-fluorophosphate or anti-cutinase antibodies to the spore suspension, and apparent secretion of the enzyme by the conidia at the penetration stage as demonstrated by immunosorbent electron microscopy (Kolattukudy et al., 1987).

A gene for cutinase was cloned from mRNA isolated from the mycelia of *Fusarium solani pisi* which had been cultured in the presence of cutin. Clones with the largest fragment corresponded with the length of mRNA and thus were subjected to sequence analysis. An open reading frame of this fragment was found to code a protein of M.W. 23,951 and assigned amino acid sequence partially coincided with that of cutinase indicating that the cloned fragment is indeed the cutinase gene. It is interesting to note that high cutinase producers contained two copies of this gene while hypovirulent isolates had only one copy (Kolattukady et al., 1987). Insertion of this cutinase gene into *Mycosphaerella spp.*, a wound pathogen of papaya fruit, made the latter capable of invading the intact fruit. Transformants with higher cutinase activity produced larger lesions on intact fruits, similar to those incited by *Colletotrichum gloeosporioides*, an authentic pathogen of papaya fruits (Dickman et al., 1989). The cutinase gene of *C. gloeosporioides* has considerable homology with that of *F. solani pisi* (Ettinger et al., 1987).

c) Pectin-degrading enzymes

Degradation of pectic polymers which are predominant in the middle lamella, is one of the prerequisites for pathogens for penetrating the primary barrier, and pectic enzymes are the first wall-degrading enzymes secreted by fungal plant pathogens (Cooper, 1983; Jones et al., 1972). Induction of these enzymes was found to be repressed by sugars (Patil et al., 1968; Holz et al., 1986). Endo- and exo-types of lyases and hydrolases have been identified from various fungal pathogens and their role in cell wall degradation has been well documented (Bateman et al., 1976; Keon et al., 1987).

Genetic manipulation of genes for these enzymes, however, was not attempted until recently when Dean et al., (1989) cloned the pectate lyase gene, *pel*A, of *Aspergillus nidulans*. The *pel*A gene, cloned from a cDNA library, contained at least two short introns and encoded a 1,300 nucleotide mRNA that was detected in the mycelium cultured in a polygalacturonic acid containing medium, but not in mycelium grown in media containing glucose or acetate. Thus, the production of PelA in this fungus was regulated at the transcriptional level (Dean et al., 1989).

The availability of molecular techniques used in studies of these types together with the recent advances in the transformation of plant pathogenic fungi (Rodriguez et al., 1987; Parsons et al., 1987) will help us in cloning and characterizing the pathogenicity genes of fungal parasites. Successful cloning of the *pda* gene in *N. haematococca* and the transformation of *C. heterostrophus* by this gene has proven the feasibility of this strategy (Ciuffetti et al., 1988; Schaffer et al., 1988).

d) Phytoalexin-degrading enzymes

Phytoalexins are, by definition, low molecular weight antimicrobial compounds produced by plants after invasion by microbes. It has been suggested that they play an important role in plant defense against pathogens, even though there are some contradictory opinions as to their specific roles.

. It follows that successful pathogens must be able to degrade (or escape) these antibiotic compounds during their colonization of plant tissues. One way to negate the antimicrobiological effect of these compounds is to metabolize them to inactive forms (this could be the major mechanism behind the so-called tolerance) (VanEtten et al., 1989). Mutation toward simple tolerance, e.g., alteration of the target protein is another way for the pathogens to circumvent this chemical barrier. The third, and still more comprehensive and sophisticated way is to suppress phytoalexin synthesis by either interferring with the recognition phase or blocking the synthetic pathway. Uehara (1964) first demonstrated that *Ascochyta pisi* metabolized pisatin, a pea phytoalexin, to unidentified compounds. He suggested that one of the pathogen's qualifications is its ability to detoxify the host's phytoalexins. Subsequent studies chemically determined the degradation products of many phytoalexins, and some of the enzymes involved in these metabolic detoxifications have been thoroughly characterized (VanEtten et al., 1989). The gene for pisatin demethylase of *Nectria haematococca* (*Fusarium solani*) was characterized by conventional crossing, and a 3.2 kb fragment encoding a 1.8 kb transcript was used for transformation of a Pda⁻ mutant of the same fungus and also the maize pathogen, *Cochliobolus heterostrophus*. Both transformants were more tolerant to pisatin than the parent organism. Some transformants of the pea pathogen showed increased virulence and those of the maize pathogen produced larger lesions on pea leaves, suggesting that the ability to degrade or tolerate phytoalexins determines the virulence or pathogenicity of pathogens (Ciuffetti et al., 1988; Schaffer et al., 1988). Recent advances in studies on the metabolism of other phytoalexins by fungal pathogens have been reviewed by VanEtten et al., (1989). The isolation and characterization of the genes involved in these detoxifications remain for the future.

2. Toxins

Fungal pathogens produce a variety of toxic metabolites; some of them are host cultivar specific (host-specific or host-selective toxins, HST) while others are not (non-specific toxins). Some of the latter play a significant role in the virulence of pathogens. As far as host specificity is concerned, studies on HSTs are the most advanced in terms of structure and function.

Tanaka (1933) first demonstrated that *Alternaria kikuchiana* (*A. alternata*, Japanese pear pathotype) produces a toxic metabolite specific to cv. Nijisseiki. The significance of such products in plant pathogenesis, however, was not elucidated until the early 1960s (Scheffer, 1989). In depth studies on the host-specific toxins conducted during the last three decades have provided us with much information on the chemical structures and biological activities associated with specificity of these toxins. Professor Nishimura and his associates have made major contributions to this field without which our present understanding of specificity of host-specific toxins would not have been possible. The isolation and structure determination of HSTs from some pathotypes of *A. alternata*, analysis of structure-activity relationships, and practical application of these toxins to agriculture are but a few examples of their achievements on which our current understanding of the molecular basis of specificity (Nishimura, 1987) is based. The details of their contributions will be summarized by Otani et al. and Tsuge et al., in this volume. R.P. Scheffer, H.E. Wheeler, J.M. Daly, V. Macko, and their colleagues as well as others in the United States have also added much, to our knowledge of host specificity in other disease systems. Recently, a specific receptor for HV toxin was isolated from oat plants (Wolpert et al., 1989), and the presence of a similar receptor for AK toxin in pear leaves has recently been suggested.

Genes that encode HSTs will have to be elucidated before we can develop an understanding of the molecular architecture of specificity. Research to this end is now in progress (Yoder, 1989).

3. Suppressors

Some fungal pathogens produce still another class of compounds that helps them establish in plant tissues; they are called suppressors. The active production of suppressors by fungal pathogens was expected from the concept of accessibility induction, which derived from observations of mutual interactions between races of powdery mildew fungi and their hosts, and the assumption that pathogens possess molecular strategies for establishling a pseudosymbiotic relationship with host plant cells (Ouchi, 1983).

This class of compounds do not, by definition, cause any detectable damage to host cells, but rather induce in the invaded cells accessibility to the producer. The production of suppressors by some fungal pathogens of the genus *Mycosphaerella* have experimentally verified this concept. Oku et al. (1987) demonstrated that they are the specific determinants of pathogenicity. The suppressor of *M. pinodes*, F5, causes a 3-h delay in the transcription of the PAL (phenylalanine ammonia-lyase) and CHS (chalcone synthase) genes, and a 3- to 6-h delay in the accumulation of pisatin in treated pea leaves. These delays are long enough for the fungus to establish infection, because by the time pisatin accumulates, the fungus is no longer sensitive to pisatin.

The mode of action of F5 as discussed by Shiraishi et al., (this volume) apparently involves damage to membrane function. This brings into contention the accepted definition of toxins.

4. Elicitors

As stated in the previous section, pathogens produce both specific as well as non-specific elicitors. One of the hepta-β-glucoside alditols from the mycelium of *Phytophthora megasperma* f.sp. *glycinea* has extremely high activity in eliciting glyceollins in soybean cotyledons (Darvill et al., 1984). Elicitors with much higher activity were recently isolated from the mycelium of the same fungus by the action of the host β-glucanase. The chemical structure and the mode of action of these elicitors will be discussed in detail by Yoshikawa et al., in this volume.

Since elicitors, according to the original definition, trigger phytoalexin synthesis in plant tissues, they are, at first glance, products that have an adverse effect on the producer. However, they may play an additional role that involves keeping the balance between the host and the pathogen in their coexistence, as is often recognized as a consequence of coevolution in nature. Thus, the real function of elicitors in host-pathogen interactions needs further analysis.

Strategies of Host Plants

Although plants have no antibody-mediated defense mechanisms, they have acquired many physical and chemical strategies to defend themselves from pathogens. Some of these mechanisms will be briefly summarized here.

1. Physical Barriers

a) Papillae
Papillae which form in some Gramineaceous plants are one of the most extensively studied structural barriers against fungal penetration. Along with the analysis of chemical components, the rapidity of their formation and their density have been correlated with the success or failure of infection, especially in the powdery mildew and stem rust of barley and wheat (Aist et al., 1988; Kogel et al., 1988; Moerschback et al., 1988). However, little is known about the function of host genes in the expression of this resistance, although the *ml-o*

gene in barley was recently implicated in the formation of the papilla (Aist et al., 1988). Hopefully, using recombinant DNA techniques, this gene will be characterized in the near future.

b) lignification of cell wall

Lignification of the cell wall has been implicated to play, in both monocotyledonous and dicotyledonous plants, a significant role in defense reactions (Sherwood et al., 1982; Asada et al., 1987). Time course studies of infection-specific peroxidase isozymes in downy mildew infected radish root suggested that the initiation of lignification was accompanied by the transcription of specific genes. Inhibition of transcription and translation by actinomycin D and blasticidin S, respectively, blocked de novo synthesis of PAL and basic isoperoxidases as well as lignification of the cell wall of radish root tissues. A similar induction of specific peroxidase isozymes was reported in suspension cultures of castor bean treated with pectic fragments (Bruce et al., 1989). Chitin oligosaccharides are also known to induce lignification in wheat leaves (Barber et al., 1989). A non-specific lignification inducing factor (LIF) was isolated from infected as well as stressed tissues and was found to induce extensive lignification of walls in leaves of crucifers, sweet potato tubers, and leaves and fruits of cucumber. LIF not only induced the lignin, a guaiacyl lignin, in these tissues, but also protected locally and systemically cucumber leaves from challenge inoculated *Colletotrichum lagenarium* (Asada et al., 1987). In view of the close association between lignification and induced resistance, cloning of genes involved in this or other physical or chemical defenses would be very worthwhile.

2. Chemical Barriers

a) Enzymes and proteins

Plants produce versatile, proteinaceous compounds in response to injury, stresses, or microbial infections, and the expression of genes involved in these responses has been rigorously investigated (Collinge et al., 1987). Chitinases and β-1,3-glucanases are common constitutive or inducible enzymes in many higher plants, and they have been implicated in defense reactions against fungal pathogens because of their inhibitory activity on fungal growth (Schlumbaum et al., 1986).

Cytological studies revealed the subcellular localization of ethylene-induced chitinase as being in the vacuole and β-1,3-glucanase as being in the vacuoles and cell walls, and suggested that the β-1,3-glucanase localized at the cell wall functions in the recognition phase, releasing signal molecules from the invading fungus, whereas both enzymes in the vacuole are involved in the last line of defense, fungal growth inhibition (Mauch et al., 1989). In French bean, chitinase is encoded by a multigene family consististing of at least three members, and at least two of these are induced by ethylene. Induction of these genes is regulated at the mRNA level (Broglie et al., 1986; Vogeli et al., 1988). One of the ethylene inducible chitinase genes, *ch5B*, was cloned and shown to encode a protein of M.W. 35,400. The 984 bp open reading frame specifies 301 amino acids of the mature enzyme and a 27-residue amino terminal sequence that presumably determines the vacuolar localization (Broglie et al., 1986). The nucleotide sequence of the 4.7 kb fragment containing the *ch5*B gene was recently determined (Broglie et al., 1989).

Pathogenesis-related proteins (PR-proteins) are another group of proteins produced in plants in response to viral, bacterial, and fungal infections. These proteins have been well characterized in terms of their molecular sizes, isoelectric points, modes of induction, enzymatic nature, and their significance in defense reactions. Genes for some of these proteins have been cloned and their amino acid sequences have been determined (Somssich et al., 1984; Lucas et al., 1985; van Huijsduijnen et al., 1986). The role of these proteins in pathogenesis, however, has yet to be elucidated.

Proteinase inhibitors (PI) and the polygalacturonase-inhibiting proteins (PGIPs) should be briefly mentioned here. The former are detected in various plants and display multiple functions; as storage proteins, as regulators of endogenous proteases (serine proteases such as trypsin and chymotrypsin), and as inhibitiors of endopeptidases of insects or microbial pathogens, which protect plants against these parasites. The PIs are induced by the proteinase inhibitor-inducing factor (PIIF) which was recently identified, in tomato plants, to be a pectic polysaccharide of M.W. 5,000-10,000 (Bishop et al., 1989). The PIs are also induced by elicitors of *Phytophthora parasitica* var. *nicotianae* in suspension cultured tobacco cells (Rickauer et al., 1989). The PGIP isolated from *Phaseolus vulgaris* inhibits endopolygalacturonases of *Aspergillus niger* and *Fusarium moniliforme* to the extent that elicitor-active oligogalacturonides are not further hydrolyzed by these enzymes (Cervone et al., 1989).

b) Phytoalexin synthesis

Phytoalexins (PAs) accumulate in many plant species in response to infection or elicitor treatment. They have been implicated in plant defense against various pathogens, because of a high correlation between their accumulation and expression of resistance. The chemical structures of PAs, their biological activities, and their biosynthetic pathways have been well delineated.

In many cases, the induced production of PA is characterized by the de novo synthesis of terminal enzymes of their biosynthetic pathways. Recent progress in cDNA technology has made it possible to easily analyze the accumulation of mRNA encoding these enzymes, and, in fact, genes for some key enzymes have been cloned and their DNA sequences have been described (Ryder et al., 1984; Ebel, 1986; Collinge et al., 1987; Bowels et al., 1989; Dhawale et al., 1989; Tepper et al., 1989). Progress in in situ hybridization techniques has provided another means to analyze the expression of these genes during the early host cell-pathogen interactions (Cuypers et al., 1988).

Concluding Remarks

Physiological and biochemical analysis of host plant-pathogen interactions in the last three decades has elucidated many versatile molecular strategies displayed by pathogens and host plants during their interactions. Remarkable advances in analytical instrumentation have made it possible to determine chemical structures of products of pathogens and host plants, as well as the biochemical processes involved in their synthesis. Thus, the metabolic pathways and their significance in host-parasite interactions can now be discussed at several levels including purified enzymes, proteins, and other constituents, and morphological and biochemical events observed at the interface can be critically examined on the basis of the new scientific knowledge.

It is now possible to examine the biochemical processes involved in host-pathogen interactions involved in gene-for-gene systems at the level of gene expression. Progress in the technology of gene manipulation in the last decade, has helped us to develop research in this direction. Although gene cloning in eukaryotic plant pathogens has progressed more slowly than in prokaryotic plant pathogens, pathogenicity genes of fungal pathogens and the gene expression in their host plants will soon be elucidated at the DNA base sequence level as has been done in the prokaryote-plant interactions.

Acknowledgements

This work was supported in part by the grant in aid for scientific research from the Ministry of Education, Science and Culture of Japan (Nos. 01440009, 01304014) and Project Research of Kinki University (P-0300).

References

Aist, J.R., et al., 1988, Evidence that molecular compounds of papillae may be involved in *ml-o* resistance to barley powdery mildews, *Physiol. Mol. Pl. Pathol.* **33**:17-32.

Asada, Y., and Matsumoto, I., 1987, Induction of disease resistance in plants by a lignification-inducing factor, *In* Molecular Determinants of Plant Disease (Nishimura, S., et al., eds.), Japan Sci. Soc. Press, Tokyo/Springer-Verlag, Berlin, pp. 223-233.

Barber, M.S., Bertram, R.E., and Ride, J.P., 1989, Chitin oligosaccharides elicit lignification in wounded wheat leaves, *Physiol. Mol. Pl. Pathol.* **34**:3-12.

Bateman, D.F., and Basham, H.G., 1976, Degradation of plant cell walls and membranes by microbial enzymes, *In* Physiological Plant Pathology (Heitefuss, R., and Williams, P.H., eds.), Springer-Verlag, New York, pp. 316-355.

Biffen, R.H., 1905, Mendel's laws of inheritance and wheat breeding, *J. Agr. Sci.* **1**:4-48.

Bishop, P.D., et al., 1981, Proteinase inhibitor-inducing factor activity in tomato leaves resides in oligosaccharides enzymically released from cell walls, *Proc. Natl. Acad. Sci. USA* **78**:3536-3540.

Bowles, D., Hogg, J., and Small, H., 1989, The effects of elicitor treatment on translatable mRNAs of carrot cells in suspension culture, *Physiol. Mol. Pl. Pathol.* **34**: 463-470.

Broglie, K.E., Gaynor, J.J., and Brogrie, R.M., 1986, Ethylene-regulated gene expression: Molecular cloning of the genes encoding an endochitinase from *Phaseolus vulgaris, Proc. Natl. Acad. Sci. USA* **83**:6820-6824.

Broglie, K.E., et al., 1989, Functional analysis of DNA sequences responsible for ethylene regulation of a bean chitinase gene in transgenic tobacco, *Plant Cell* **1**:599-607. ·

Bruce, R.J., and West, C.A., 1989, Elicitation of lignin biosynthesis and isoperoxidase activity by pectic fragments in suspension cultures of castor bean, *Plant Physiol.* **91**:889-897.

Cervone, F., et al., 1989, Host-pathogen interactions, XXXIII. A plant protein converts a fungal pathogenesis factor into an elicitor of plant defense responses, *Plant Physiol.* **90**:542-548.

Ciuffetti, L.M.W., et al., 1988, Transformation of *Nectria haematococca* with a gene for pisatin demethylating activity and the role of pisatin detoxification in virulence, *J. Cell. Biol.* **12C**:287(Abstr.).

Collinge, D.B., and Slusarenko, A.J., 1987, Plant gene expression in response to pathogens, *Pl. Mol. Biol.* **9**:389-410.

Cooper, R.M., 1983, The mechanisms and significance of enzymic degradation of host cell walls by parasite, *In* Biochemical Plant Pathology (Collow, J.A., ed.), J. Wiley, Chichester, pp. 101-135.

Cuypers, B., Schmelzer, E., and Hahlbrock, K., 1988, *In situ* localization of rapidly accumulated phenylalanine ammonia-lyase mRNA around penetration sites of *Phytophthora infestans* in potato leaves, *Mol. Pl.-Microbe. Intr.* **1**:157-160.

Darvill, A.G., and Albersheim, P., 1984, Phytoalexins and their elicitors-A defense against microbial infection in plants, *Annu. Rev. Pl. Physiol.* **35**:143-175.

Day, P.R., 1974, Genetics of Host-Parasite Interaction, Freeman, San Francisco p. 238

Day, P.R. and Jellis, G.J., 1987, Genetics and Plant Pathogenesis, Blackwell Sci. Pub., Oxford p. 352.

Dean, R.A., and Timberlake, W.E., 1989, Production of cell wall-degrading enzymes by *Aspergillus nidulans*: A model system for fungal pathogenesis of plants, *Plant Cell*, **1**:265-273.

Dean, R.A., and Timberlake, W.E., 1989, Regulation of the *Aspergillus nidulans* pectate lyase gene (pelA), *Plant Cell* **1**:275-284.

Dhawale, S., Souciet, G., and Kuhn, D.N., 1989, Increase of chalcone synthase mRNA in pathogen-inoculated soybeans with race-specific resistance is different in leaves and roots, *Plant Physiol.* **91**:911-916.

Dickman, M.B., and Patil, S.S., 1986, Cutinase defecient mutants of *Colletotrichum gloesporioides* are non-pathogenic to papaya fruit, *Physiol. Pl. Pathol.* **28**:235-242.

Dickman, M.B., Podila, G.K., and Kolattukudy, P.E., 1989, Insertion of cutinase gene into a wound pathogen enable it to infect intact host, *Nature* **342**:446-448.

Dixon, R.A., et al., 1989, Elicitor-active components from French bean hypocotyls, *Physiol. Mol. Pl. Pathol.* **34**:99-115.

Ebel, J., 1986, Phytoalexin synthesis: The biochemical analysis of the induction process, *Annu. Rev. Phytopathol.* **24**:235-264.

Ettinger, W.F., Thukral, S.K. and Kolattukudy, P.E., 1987, Structure of cutinase gene, cDNA, and the derived amino acid sequence from phytopathogenic fungi. *Biochemistry* **26**:7883-7902.

Flor, H.H., 1955, Host-parasite interaction in flax rust: its genetics and other implications, *Phytopathology*, **45**:680-685.

Heinen, W., 1962, Uber den enzymatischen Cutin-Abbau. III. Die enzymatischen Ausrustung von *Penicillium spinulosum* zum Abbau der Cuticular-Bestandteile, *Arch. Microbiol.* **41**:268-281.

Holz, G., and Knox-Davies, P.S., 1986, Possible involvement of apoplast sugar in endo-pectin-transeliminase synthesis and onion bulb rot caused by *Fusarium oxysporum* f.sp. *cepae*, *Physiol. Mol. Pl. Pathol.* **28**:403-410.

Hooft van Huijsduijnen, R.A.M., Van Loon, L.C., and Bol, J.F., 1986, cDNA cloning of six mRNAs induced by TMV infection of tobacco and a characterization of their translation products, *EMBO J.* **5**:2057-2061.

Jones, T.M., Anderson, A.J., and Albersheim, P., 1972, Host-pathogen interactions. IV. Studies on the polysaccharide degrading enzymes secreted by *Fusarium oxysporium* f.sp. *lycopersici*, *Physiol. Mol. Pl. Pathol.* **2**:153-166.

Keon, J.P.R., Byrd, R.J.W., and Cooper, R.M., 1987, Some aspects of fungal enzymes that degrade plant cell walls, *In* Fungal Infection of Plants (Pegg, G.F., and Ayers, P.G., eds.), Cambridge Univ. Press, Cambridge, pp. 133-157.

Kogel, G., Beissman, B., and Reisener, H.J., 1988, A single glycoprotein from *Puccinia graminis* f.sp. *trittici* cell walls elicits the hypersensitive lignification response in wheat, *Physiol. Mol. Pl. Pathol.* **33**:173-185.

Kolattukudy, P.E., and Crawford, M.S., 1987, The role of polymer degrading enzymes in fungal pathogenesis, *In* Molecular Determinants of Plant Diseases (Nishimura, S. , et al. , eds.), Japan Sci. Soc. Press. Tokyo/Springer-Verlag, Berlin, pp.75-95.

Kosuge, T. and Nester, E.W., 1984, Molecular and genetic Perspectives, *In* Plant-Microbe Interactions (Kosuge, T., and Nester, E.W., eds.), Macmillan, New York, vol.1, p. 444.

Kosuge, T. and Nester, E.W., 1987, Molecular and genetic perspectives, *In* Plant-Microbe Interactions (Kosuge, T., and Nester, E.W., eds.), Macmillan, New York, vol.2, p. 448.

Kosuge, T, and Nester, E.W., 1989, Molecular and genetic perspectives, *In* Plant-Microbe Interactions (Kosuge, T., and Nester, E.W., eds.), McGraw-Hill, New York, vol.3, p. 480.

Linskens, H.F., Heinen, W., and Stoffers, A.L., 1965, Cuticula of leaves and the residue problem, *Residue Rev.* **8**:136-178.

Lucas, J., et al., 1985, Amino acid sequence of the "pathogenesis-related" leaf protein p14 from viroid-infected tomato reveals a new type of structurally unfamiliar proteins, *EMBO J.* **4**:2745-2749.

Mauch, F., and Staehelin, A., 1989, Functional implications of the sub-cellular localization of ethylene-induced chitinase and β-1,3-glucanase in bean leaves, *Plant Cell* **1**:447-457.

McNeil, M., et al., 1984, Structure and function of the primary cell walls of plants, *Annu. Rev. Biochem.* **53**:625-663.

Moerschback, B., et al., 1988, Lignin biosynthetic enzymes in stem rust infected, resistant and susceptible near-isogenic wheat lines, *Physiol. Mol. Pl. Pathol.* **33**:33-46.

Nishimura, S., 1987, Recent development of host-specific toxin research in Japan and its agricultural use, *In* Molecular Determinants of Plant Diseases (Nishimura, S., et al., eds.), Japan Sci. Soc. Press, Tokyo/Springer-Verlag, Berlin, pp.11-26.

Oku, H., Shiraishi, T., and Ouchi, S., 1987, Role of specific suppressors in pathogenicity of *Mycosphaerella* species, *In* Molecular Determinants of Plant Diseases (Nishimura, S., et al., eds.), Japan Sci. Press, Tokyo/ Springer-Verlag, Berlin, pp. 145-156.

Ouchi, S., 1983, Induction of resistance or susceptibility, *Annu. Rev. Phytopathol.* **21**:289-315.

Palacios, R. and Verma, D.P.S., eds 1988, Molecular Genetics of Plant-Microbe Interactions, p. 401, APS Press, St. Paul.

Parsons, K.A., et al., 1987. Genetic transformation of the fungal pathogen responsible for rice blast disease, *Proc. Natl. Acad. Sci. USA*, **84**:4161- 4165.

Patil, S.S., and Dimond, A.E., 1968, Repression of polygalacturonase synthesis in *Fusarium oxysporum* f.sp. *lycopersici* by sugars and its effect on symptoms development in infected tomato plants, *Phytopathology* **58**:676-682.

Rice, D., et al., 1988, Molecular characterization of the cutinase gene from *Colletotrichum gloeosporioides*, *In* Molecular Genetics of Plant-Microbe Interactions (Palacios, R., and Verma, D.P.S.), APS Press, St. Paul, pp. 299-300.

Rickauer, M., Fournier, J., and Esquirre-Tugaye, M.T., 1989, Induction of proteinase inhibitors in tobacco cell suspension culture by elicitors of *Phytophthora parasitica* var. *nicotianae*, *Plant Physiol.* **90**:1065-1070.

Ride, J.P., 1983, Cell walls and other structural barriers in defense, *In* Biochemical Plant Pathology (Callow, J.A. , ed.), John Wiley, Chichester, pp.215-236.

Roberts, D.P., Denney, T.P., and Schell, M.A., 1988, Cloning of the *egl* gene of *Pseudomonas solanacearum* and analysis of its role in phytopathogenicity, *J. Bacteriol.* **170**:1445-1451.

Rodriguez, R.J., and Yoder, O.C., 1987, Selectable genes for transformation of the fungal pathogen *Glomerella cingulata* f.sp. *phaseoli* (*Colletotrichum lindemuthianum*). *Gene* **54**:73-81.

Ryder, T. B., et al., 1984, Elicitor rapidly induces chalcone synthose mRNA in *Phaseolus vulgaris* cells, *Proc. Natl. Acad. Sci. USA*, **81**:5724-5728.

Schafer, W., VanEtten, H.D., and Yoder, O.C., 1988, Transformation of the maize pathogen *Cochliobolus heterostrophus* to a pea pathogen by insertion of a PDA gene of *Nectria haematococca*, *J. Cell Biol.* **12c**:291 (abstr.).

Scheffer, R.P., 1989, Host-specific toxins in phytopathology: origin and evolution of the concept, *In Host-Specific Toxins-Recognition* and *Specificity Factors in Plant Diseases* (Kohmoto, N., and Durbin, R.D., eds.), Tottori Univ., Tottori, pp.1-17.

Schlumbaum, A., et al., 1986, Plant chitinases are potent inhibitors of fungal growth, *Nature* **324**:365-367.

Sherwood, R.T., and Vance, C.P., 1982, Initial events in the epidermal layer during penetration, *In* Plant Infection-The Physiological and Biochemical Basis (Asada, Y. et al. , eds.), Japan Sci. Soc. Press, Tokyo/Springer-Verlag, Berlin, pp. 27-44.

Shishiyama, J., Araki, F., and Akai, S., 1970, Studies on cutin esterase. I. Preparation of cutin and its fatty acid component from tomato fruit peel, *Plant & Cell Physiol.* **11**:323-334.

Somssich, I.E., et al., 1984, Rapid activation by fungal elicitor of genes encoding "pathogenesis-related" proteins in cultured parsley cells, *Proc. Natl. Acad. Sci. USA*, **83**:2427-2430.

Tanaka, S., 1933, Studies on black spot disease of Japanese pears (*Pirus serotina* Rehd.), *Mem. Cell. Agr. Kyoto Imp. Univ.* **28**:1-31.

Tepper, C.S., Albert, F.G., and Anderson, A.J., 1989, Differential mRNA accumulation in three cultivars of bean in response to elicitors from *Colletotrichum lindemuthinum*, *Physiol. Mol. Pl. Pathol.* **34**:85-98.

Uehara, K., 1964, Relationship between host specificity of pathogen and phytoalexin, *Ann. Phytopathol. Soc. Jpn.* **29**:103-110.

VanEtten, H.D., Matthews, D.E., and Matthews, P.S., 1989, Phytoalexin detoxification: Importance for pathogenicity and practical implications, *Annu. Rev. Phytopathol.* **27**:143-164.

Vogeli, U., Meins, F., and Boller, T., 1988, Co-ordinated regulation of chitinase and β-1,3-glucanase in bean leaves, *Planta* **174**:364-372.

Weltring, K.M., et al., 1988, Isolation of a phytoalexin-detoxifying gene from the plant pathogenic fungus *Nectria haematococca* by detecting its expression in *Aspergillus nidulans*, *Gene* **68**:335-344.

Wolpert, T.J., and Macko, V., 1989, Victorin binding to proteins in susceptible and resistant oat genotypes, *In* Phytotoxins and Plant Pathogenesis (Graniti, A., et al.,
eds.), Springer-Verlag, Berlin, p.439.

Yoder, O.C., et al., 1989, Genetic analysis of toxin production in fungi, *In* Phytotoxins
and Plant Pathogenesis (Graniti, A. et al., eds.), Springer-Verlag, Berlin, pp. 43-60.

Summary of Discussion of Ouchi's Paper

Durbin initiated the questioning by asking whether there is interest in Japan to develop *Arabidopsis* as a potential model system. *Ouchi* responded that he was unaware of any interest at this time. *Vance* asked if *Ouchi* could comment on whether any of the suppressors described by the Japanese researchers have been purified and their identity fully established. *Ouchi* responded that they have not been crystallized or purified yet but they do know the amino acid composition and something about the sugar component. *Bushnell* asked whether that was with regard to *Mycospharella*. *Ouchi* answered affirmatively and said they thought that by definition suppressors should not produce any visible damage to the host cell. They actually inhibited membrane-bound ATPases, which probably means damage to the membrane system, and the function of the membrane might have been changed by the suppressor molecule they have isolated. *Bushnell* asked whether by definition suppressors are not harmful. *Ouchi* said that toxins and suppressors induce diseases and the toxins by definition are toxic to the cells. But suppressors, as far as they have tested, produce no visible changes in the cell structure. He continued that *Oku*, *Shiraishi*, and *Ouchi* coined the term "supressors" for compounds produced by the pathogen that do not damage the plant cell. He said that we probably have to go a little further to define the real function of suppressor molecules. *Nester* then asked whether *Ouchi* could tell us a bit about the 27-amino acid necrosis-inducing peptide. *Ouchi* responded that in those combinations of tomato cultivars and races that produce incompatible reactions, the pathogens produce this necrosis-inducing factor which has been determined to ba a 27-amino acid peptide, but he was unaware as to whether the gene encoding this peptide has been cloned. *Chumley* commented that a 63-amino acid protein is processed; it is unknown where the processing occurs, and whether it is associated with plant extracts or fungal extracts. The most interesting component is that lines that do not produce this race-specific necrosis factor do not have DNA homologous to it. In this regard it seems to be similar to the case with some of the avirulence genes of bacteria. It is not the allelic difference that accounts for this but rather the presence or absence of a gene that determines whether the compound is synthesized.

Bacterial Strategies

Chapter 3
Distinct Induction of Pectinases as a Factor Determining Host Specificity of Soft-Rotting *Erwinia*

Shinji Tsuyumu, Masahiko Miura, and Shouichi Nishio

Soft-rotting *Erwinia* produce several different pectinases. Among these endo-pectate lyase (PL), which randomly cleaves the preferentially deesterified pectic substance (i.e., pectate) in the transeliminative manner, has been suggested as the major disease-inducing factor in these pathogens. This suggestion is primarily based on the finding that a purified endo-PL can cause maceration and cellular death of plant tissues (Basham et al., 1975). However, endopectin lyase (PNL) and polygalacturonase (PG) have also been shown to cause maceration (Tsuyumu et al., 1984; Lei et al., 1985). Besides these pectinases, exo-type pectinases (Collmer et al., 1981) and pectin methylesterase (PME) (Goto et al., 1962) have also been reported to be produced by soft-rotting *Erwinia*. However, the significance of the ability of soft-rotting *Erwinia* to produce such a variety of pectin-degrading enzymes in their pathogenicity is not fully understood.

Most of the pathogenicity tests for plant diseases, especially soft-rotting *Erwinia*, have been performed by putting the inoculum of the pathogen onto wounded plant tissues. Such tests do not reveal the initial process which leads

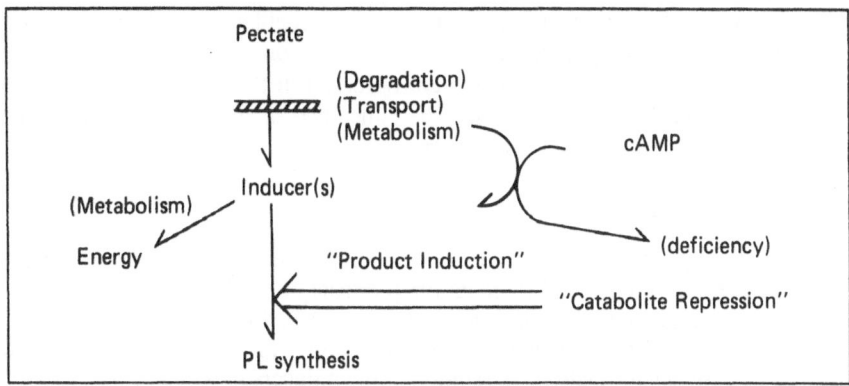

Figure 1. A model for basic regulation of endopectate lyase synthesis.

to the propagation of the pathogen and to the induced production of the pectinases (this process may be termed as "elicitation of soft-rotting disease").

In this contribution, we summarize the regulatory mechanisms of synthesis and the biochemical nature of the pectinases in soft-rotting *Erwinia*, followed by the possible significance of distinct induction of each pectinase in the development of soft-rotting disease with special emphasis on the elicitation step.

Regulation of the Synthesis of Endopectate Lyase

1. Basic Regulatory Mechanism

Previous physiological studies suggested that the production of PL in *E. carotovora* subsp. *carotovora* (Eca) was controlled by "product induction" (switch-on mechanism by the degradative product of pectate) (Tsuyumu, 1977) and "self-catabolite repression" (turn-off mechanism by catabolite repression due to the accumulation of the metabolic intermediates of pectate) (Tsuyumu, 1979) as summarized in Figure 1. This basic regulatory scheme has since been shown to be common in many strains of soft-rotting *Erwinia* (see reviews by Collmer et al., 1987 and by Kotoujansky, 1988).

2. Variation of Basic Regulatory Mechanism

It is important to note that PL in soft-rotting *Erwinia* consists of several isozymes. The genes for each isozyme have been cloned and sequenced (Keen et al., 1986; Lei et al., 1987; Itoh et al., 1988). Considerable homology was found among the isozymes with near isoelectric points, even from different species of soft-rotting *Erwinia*. However, even in these cases, the sequences of the operator regions were distinguishable in Eca and Ech. By mobility shift DNA-binding assays, we found that the protein which binds specifically to the 56-mer oligonucleotide containing the suspected operator(s) for *pel*E of Ech strain EC16 was present only in the extracts of Ech strains but not in Eca strains (Fig. 2). Thus, the repressor for the regulation of PL synthesis may be distinguishable in these two representatives of soft-rotting *Erwinia* species. Furthermore, the isolation of different types of regulatory mutants for each isozyme in Ech suggested that finite regulation of each isozyme is distinguishable. Two shifted spots were observed in the extracts of Ech EC16 and D12 strains (Fig. 2). There may be either two operators in this region or two types of binding to a single operator.

3. Additional Regulatory Mechanism

Based on their genetic studies of *E. chrysanthemi*, Collmer et al., (1981) suggested that of the several degradative products of pectate, 4-deoxy-L-threo-5-hexosulose uronate (DTH) or 3-deoxy-D-glycero-2, 5-hexodiulosonate (DHG) which are the first and the second metabolic products produced from unsaturated digalacturonate (UDG), are the true inducers of endopectate lyase.

(+)

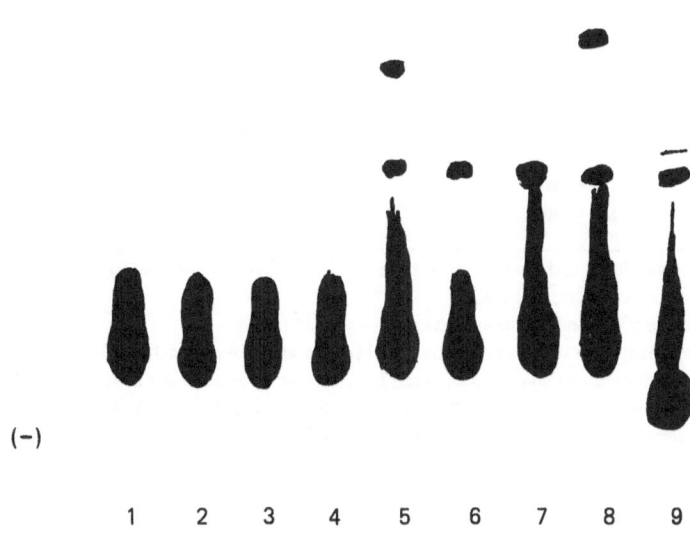

(−)

| 1 | 2 | 3 | 4 | 5 | 6 | 7 | 8 | 9 |

Figure 2.   Mobility shift assay by the binding of operator-like palindrome(s) and the sequence specific protein(s).   Oligonucleotide (56 mer) which contains the suspected operator(s) of *pelE* of Ech strain EC16 was electrophoresed in low-ionic strength polyacrylamide gel after mixing it with the sonic extracts of soft-rotting *Erwinia* strains.   1, the only oligomer by itself; 2, Eca EC1; 3, Eca DB71; 4, Eca DB192; 5, Ech EC16; 6, Ech ALE8292p; 7, Ech D7; 8, Ech D12; 9, *E. coli* HB101.

However, 2-keto-3-deoxy-D gluconate (KDG), which is not only the third product from UDG but also the product from galacturonate and glucuronate *via* different pathways, was also shown to be the true inducer of the enzyme in other strains of Ech (Condemine et al., 1986).

In the case of *E. coli* clone (HPL16) which contains both the structural gene and its own regulatory regions of *pelE* from Ech strain EC16, the specific activity of PLe was elevated by the addition of either galacturonate or glucuronate in the later growth phase (Table 1). This result seems to support the hypothesis of Condemine et al. (1986) that KDG is the true inducer. In the donor, strain EC16, however, the induction of PL by DGH and/or DTH has been well established (Chatterjee et al., 1985). Also, it is important to note that this increase in PLe specific activity in the later growth phase was hard to explain in terms of the lack of catabolite repression. A clone (YEJ-16 in Table 1) in which PLe synthesis is regulated by the foreign promoter and operator did not show such growth-phase dependent increase. Thus, the effects of addition of uronates at the later growth stage seem to be specific for the increase of PLe specific activity. These data suggest that there may be two independent induction

Table 1. Effect of additions of galacturonate and of glucuronate into the cultures of *E. coli*.

Clones[a]	Media[b]	OD_{600}	In-PL	Out-PL	Total PL[c]
HPL16	LB	4.7	0.02	0.04	0.06
	LB + Gau	4.8	0.11	0.54	0.66
	LB + Guu	4.8	0.07	0.45	0.52
YEJ16	LB	3.9	0.03	0.05	0.08
	LB + Gau	3.7	0.06	0.08	0.14
	LB + Guu	3.9	0.02	0.08	0.11

[a]HPL16 contains pelE and its regulatory regions. EcoRI-SalI fragment of pHPL16 was subcloned into the expression vector, pYEJOO1.

[b]Galacturonate (Gau) and glucuronate (Guu) was added to 0.1%.

[c] The supernatant after centrifugation was used for the assay of extracellular PL activity (Out-PL), and this precipitation was sonicated and used for the assay of intercellular activity (In-PL). Activity of PL was expressed as the units/OD_{600}. One unit of PL activity is defined as the amount of enzyme that produces an increase in absorbance of 1.0 in min.

mechanisms: (1) the induction by DGH and/or DTH at the early log phase growth and (2) the induction by KDG at the later growth phase.

This growth-phase dependent increase of PL specific activity, however, can be seen even when grown in the absence of these uronates (Fig. 3). Thus, the second induction by KDG may not be specific. Though there are no data to show that such growth-phase dependent increase is operated at the transcriptional level, the presence of putative operator-specific binding protein in *E. coli* may indicate such a possibility. The gene to express this phenotype was identified as *gpiR* (Condemine et al, 1986). Analysis of this mutant should make it possible to test this aspect.

4. CRP Binding Site

Since adenyl cyclase minus mutants cannot be induced to synthesize PL (Mount et al., 1979), the sensitivity of PL production to catabolite repression seems to be effected in a manner similar to the one in *E. coli*. In fact, the consensus CAP-binding site was found near the promoter region of *pel* genes (Keen et al., 1986). We have further confirmed this conclusion from the observation that a mutant, NP-CRP1, which was mutagenized at the CAP-binding site of *pel*E, by site-directed mutagenesis, produced a repressed level of PLE even in the absence of glucose (data not shown). This result strongly suggests that at least the basic regulatory mechanism may be operating at the transcriptional level.

5. Effectors of PL Activity

All purified PL require calcium ion for their activity. Many organic acids such as citrate inhibit pectinases in a noncompetitive manner (Tsuyumu, 1981). Since the induction of PL requires the breakdown and metabolism of pectate

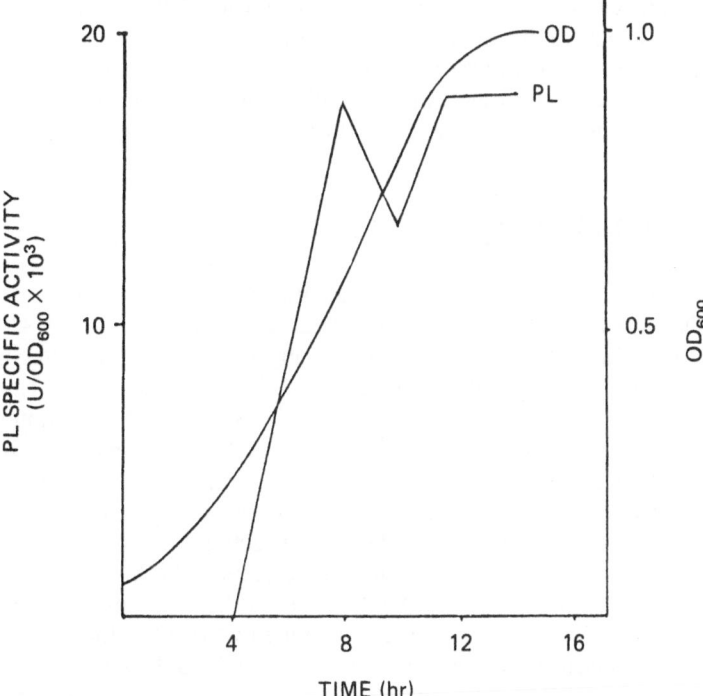

Figure 3. Increase of PLe specific activity in the absence of uronates. *E. coli* clone, HP16, was grown in L-broth. Time course samples were taken for measurement of absorbance at 600 nm and of PL activity.

according to the above-mentioned "product-induction mechanism," the effectors should influence the efficiency in PL induction by increasing or decreasing the concentration of degradative products from pectate. In fact, addition of EDTA, which chelates calcium ions, into the medium severely affected PL induction (Tsuyumu, 1977). Of course, the presence of these effectors would also control the activities of the already excreted pectinases.

6. Excretion of PL

The excretion of PL into the surrounding space may be another important aspect of the disease development by soft-rotting *Erwinia*. Although transport through the inner membrane by signal peptides seems to be common in these bacteria, the mechanism of transport through the outer membrane has not been elucidated. The latter which is also responsible for the transport of other proteins such as protease and cellulase is a unique characteristic of these pathogens, as this mechanism is absent in most of the gram-negative bacteria including *E. coli*. The studies using the mutants (OUT⁻) which were deficient in excretion (Thurn et al., 1985) will reveal the nature of this mechanism. Although most of the *E. coli* clones of PL isozymes of different *Erwinia* strains are unable to excrete PL, the clone HPL16 could excrete PL at the later growth phase as shown in Table 1. The efficiency of excretion varied depending on the choice of the recipient *E. coli* strain (data not shown). This excretion seemed to be not due to its leakiness, since a periplasmic protein, alkaline phosphatase, was retained in the periplasmic region under the same conditions (data not shown). Thus, the 2 kb fragment in this clone seems to contain the additional information necessary for the transport of PL through the outer membrane.

Regulation of the Synthesis of Endopectin Lyase

The synthesis of PNL is induced by the addition of DNA-damaging agents such as nalidixic acid, mitomycin C, and UV (Kamimiya et al., 1974). A survey by Tsuyumu et al., (1984) of extracts of mitomycin C-induced cultures of soft-rotting Erwinia revealed that induction of PNL by these agents is a shared trait among these organisms. Purified PNL from Ech strain EC183 had a molecular weight of 41,000 and isoelectric point at around 10. This PNL preferentially cleaved Link pectin (chemically esterified pectate) in an endo-type manner and macerated plant tissues (Tsuyumu et al., 1985). Furthermore, when soft-rotting *Erwinia* was inoculated into plant disks, PNL activity in the macerated disks often reached the level of PL seen in induced cultures (Table 2). Thus, the in planta induction of PNL seems to be the result of the action of DNA-damaging agents in plants. This conclusion was supported by the result that the induction patterns of ß-galactosidase syntheses in *lac⁻* Eca HMgal9 strain containing the gene fusion of *pnl-lacZ* or of *umuC-lacZ* were very similar both in vitro and in planta (Tsuyumu et al., 1990). The presence of DNA-damaging agents in various plants was confirmed by the *rec*-assay (Tsuyumu et al., 1985). Since formation of free radicals was detected as the electron spin resonance absorption of the freeze-dried whole plant tissues (data not shown), it is highly possible that

Table 2. The activities of PL and PNL in the macerated plant tissues. Eca strains (EC1, T29) and Ech strains (D7, D12, EC183) were inoculated onto various plant tissues. The activity was expressed as the units/10^9 cells.

Strains	Enzyme	Potato	Carrot	Radish	Ch. cabbage	Cabbage	Lettuce
D7	PL	11	6.7	22	42	54	140
	PNL	0.5	0.5	2.4	4.1	-	4.5
D12	PL	41	63	12	150	71	280
	PNL	3.2	4.1	6.2	16	11	10
EC183	PL	66	47	66	13	140	23
	PNL	2.6	4.0	4.4	8.2	13	3.6
EC1	PL	25	14	36	3.3	8.9	120
	PNL	-	1.0	0.8	0.5	-	3.8
T29	PL	3.5	1.2	7.8	3.5	2.6	48
	PNL	0.5	4.2	1.4	-	-	3.2

- no detectable activity.

intact plants contain DNA-damaging agents. If this is the case, PNL may be induced by these agents at a very early stage of invasion of these soft-rotting *Erwinia* into their host plants.

Although the basic regulatory and biochemical natures of PNL are quite similar among soft-rotting *Erwinia* strains, there seem to be at least two immunologically distinct groups in soft-rotting *Erwinia*. The antibody raised against either Eca PL (Itoh et al., 1988) or Ech PL (Tsuyumu et al., 1984) reacted with the extracts from mitomycin C-induced cultures of the strains in the same species, but the antibody for Ech PL did not react with PL of Eca subsp. *carotovora*. It was surprising, however, that the antibody for Ech PL reacted with the extracts of *E carotovora* subsp. *atroseptica* which is classified in the Eca group on the basis of many taxonomical characteristics. Although the significance of this immunological specificity in the evolution of this enzyme is unknown, the differentiation of Ech vs. Eca subsp. *carotovora* and Eca subsp. *carotovora* vs. Eca subsp. *atroseptica* can be made on this basis. The difference may also be significant in the differentiation of their pathogenicity.

Finally, it should be emphasized that there were considerable variations in the extent of PNL induction among different strains of soft-rotting *Erwinia* both in vitro (Tsuyumu et al., 1984) and in planta (Table 1). This difference may be considered either as negative or positive data for the significance of PNL in the pathogenicity of soft-rotting *Erwinia*. If we evaluate the pathogenic nature rather than the saprophytic nature of soft-rotting *Erwinia*, the latter possibility has to be carefully considered.

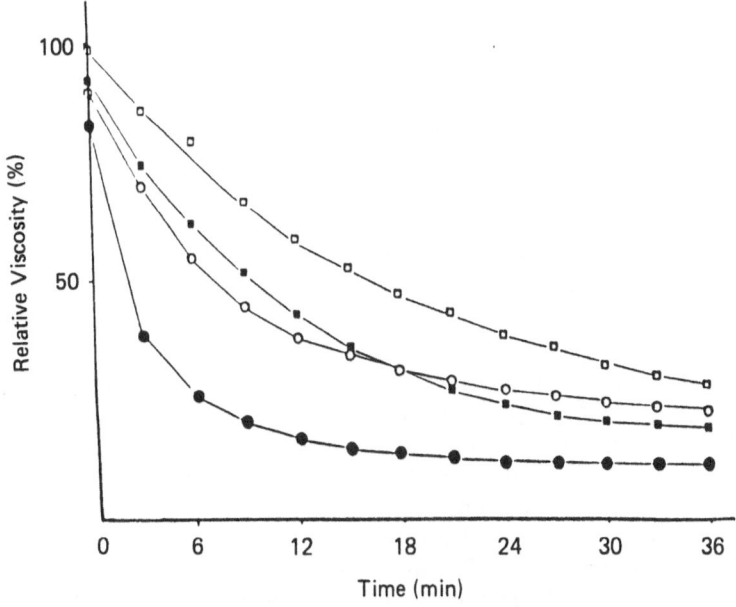

Figure 4. Cooperative reduction of the viscosity of pectin by simultaneous addition of PL and PNL. Time course measurements of the viscosity were done after addition of PL (O), PNL (□), PL and PNL (●), and one-half dilution PL and PNL (■).

Degradation of Pectin by Mixture of Pectinases

Pectic substances exist in a partially esterified state (i.e., pectin) in plants. Thus, PL alone should stop the degradation of pectin at the esterified portions. Addition of PME which desterifies these esterified portions should lead to complete degradation of pectin by PNL. Addition of PNL may also assist in the degradation of pectin by PL by cooperative degradation of the esterified and the deesterified portions in pectin, respecively. Such cooperative degradation of pectin can be seen in vitro, when the viscometric assay was conducted by simultaneous addition of these enzymes using pectin as the substrate (Fig. 4). Both of these enzymes degraded pectin to some extent, but the degradation gradually slowed down. When two enzymes were added together, however, the degradation proceeded further in addition to the apparent increase in the rate of initial degradation. Similarly, in the spectrophotometric assay, the activity of either PL or PNL alone slowed down, even if the amount of enzyme was increased. Addition of both enzymes simultaneously led not only to the rapid degradation of pectin at low concentration of enzymes but also to more extensive degradation of pectin at higher concentration of the enzymes (data not shown). These data demonstrate that PNL and PL alone cannot degrade pectin to smaller units such as the monomer and the dimer even at high concentration of enzymes and that the presence of these enzymes together makes such complete degradation possible. Maceration of plant tissues, however, can be caused by either PL or PNL alone. Thus, complete degradation of pectin to smaller units such as the dimer or the monomer may be unnecessary for maceration, while such degradation is required for soft-rotting *Erwinia* to metabolize them for energy and to induce PL synthesis.

Biochemical Nature of Polygalacturonase

Polygalacturonase (PG) degrades pectate hydrolytically. The necessity of exo-PG in the induction of PL by providing metabolizable forms was clearly demonstrated in Ech (Collmer and Bateman, 1981). While Ech strains are known to produce only exo-type of PG (see review by Collmer and Keen, 1986), Eca strains produce endo-PG also (Zink and Chatterjee, 1985). We have cloned about a 5.8 kb DNA fragment from Eca strain ECl in *E. coli*. Two types of endo-PG (alkaline and neutral) were purified from this clone. They, like other reported PGs of different origins, did not require Ca^{2+} for their activities and had pH optima in the acidic range. Considering that plant tissues have acidic pH, PG may be active in plants even at lesser amounts compared with PL whose optimum pH is in the alkaline range.

Like the *E. coli* clones of pel isozymes, the clone pEClOl produced more PG than its donor Eca strain ECl (data not shown), as observed in the clone of *peh* of Eca strain DB7l (Zink et al., 1985). It may be interesting to study this analogy in more detail to elucidate the regulatory mechanism for PG synthesis.

Elicitation of Soft-Rotting Disease

The symptoms caused by different soft-rotting *Erwinia* strains are indistinguishable from each other. Similar symptoms can be seen by the inoculation of a high concentration of a purified enzyme or of *E. coli* clone harboring a single pectinase gene. However, distinct host ranges do exist for these pathogens, although they are broader compared with those of other plant pathogenic bacteria. In the field, these pathogens are often found at high concentration in the rhizosphere without initiating the infection (Tsuyama, 1965). Thus, the maceration tests using wounded plant blocks may not reflect the pathogenicity of the organisms in the field. The finite controls on the elicitation of the infection step should be a clue to the understanding of their pathogenicity.

Based on the presently available information, we may be able to hypothesize how the distinct regulatory mechanisms of each pectinase may contribute to disease development. When soft-rotting *Erwinia* enter into plants through natural openings, none of pectinases are expected to be induced. The first available inducer of pectinases may be DNA damaging agents which induce the synthesis of PNL. PNL together with constitutively synthesized extracellular endoPL, exoPG and endoPG (in the case of Eca) would start to degrade the pectic substance in a cooperative manner. Only after metabolizable forms become available through degradation of the pectic substance, PL and possibly PG syntheses may be induced. These pathogens should then become capable of starting effective maceration and consequently disperse themselves throughout the plant tissues. Otherwise, these pathogens may be localized and severely affected by host defense mechanisms such as the accumulation of phytoalexins. The effectors in plant tissues would influence the rate of this process by their control of the extracellular pectinases. Once these pectinases are induced, the dispersed pathogens would repeat this cycle at new locations leading to the development of distinct symptoms of the disease.

Acknowledgements

This work was supported in part by Grants-in-Aid for Scientific Research Nos. 61560050 and 63560041 (1987-1989) from the Ministry of Education, Science and Culture of Japan. We thank Drs. N. Okabe, M. Goto, and Y. Takikawa for their encouragement and suggestions, Dr. A. K. Chatterjee for useful discussions and the gift of bacterial strains, and Drs. S. Ouchi and H. Kunoh for reading the manuscript.

References

Basham, H.G. and Bateman, D.F., 1975, Relationship of cell death in plant tissue treated with a homogeneous endopectate lyase to cell wall degradation, *J. Bacteriol.* **164**:831-835.

Collmer, A. and Bateman, D.F., 1981, Impaired induction and self-catabolite repression of extracellular pectate lyase in *Erwinia chrysanthemi* mutants deficient in oligogalacturonide lyase, *Proc. Natl. Acad. Sci. USA* **78**:3920-3924.

Collmer, A. and Keen, N.T., 1986, The role of pectic enzymes in plant pathogenesis, *Annu. Rev. Phytopathol.* **24**:383-409.

Condemine, G., Hugouvieus-Cotte-Pataat, N. and Robert-Baudouy, J. 1986, Isolation of *Erwinia chrysanthemi kduD* mutants altered in pectin degradation, *J. Bacteriol.* **165**:917-941.

Goto, M. and Okabe, N., 1962, Studies on pectin-methyl esterase secreted by *Erwinia carotovora* (Jones) Holland, with special reference to the cultural conditions relating to the enzyme production and the difference of enzyme activity due to the kind of strains, *Ann. Phytopathol. Soc. Japan* **27**:1-9.

Hugouvieux-Cotte-Pataat, N., et al., 1986, Regulatory mutations affecting the synthesis of pectate lyase in *Erwinia chrysanthemi*, *J. Gen. Microbiol.* **132**:2099-2106.

Itoh, K., et al., 1988, DNA structure of pectate lyase I gene cloned from *Erwinia carotovora*, *Agri. Biol. Chem.* **52**:479-487.

Kamimiya, S., et al., 1974. Purification and properties of a pectin trans eliminase in *Erwinia aroideae* formed in the presence of nalidixic acid, *Agri. Biol. Chem.* **38**:1071-1078.

Keen, N.T. and Tamaki, S., 1986, Structure of two pectate lyase genes from *Erwinia chrysanthemi* EC16 and their high-level expression in *Escherichia coli*, *J. Bacteriol.* **168**:595-606.

Kotoujansky, A., 1987, Molecular genetics of pathogenesis by soft-rot erwinias, *Annu. Rev. Phytopathol.* **25**:405-430.

Lei, S.P., et al., 1985, Evidence that polygalacturonase is a virulence determinant in *Erwinia carotovora*, *J. Bacteriol.* **164**:831-835.

Lei, S.P., et al., 1987, Characterization of the *Erwinia carotovora* pelB gene and its product pectate lyase, *J. Bacteriol.* **169**:4379-4383.

Miura, M., Tsuyumu, S. and Takikawa, Y., 1988, Effect of starvation on induction of pectate lyase in soft-rotting *Erwinia*, *Int. Cong. Plant Pathol., Kyoto* p.II-1-25 (Abstr.).

Mount, M.S., et al., 1979, Regulation of endopolygalacturonate transeliminase in an adenosine 3',5'-cyclic monophosphate-deficient mutant of *Erwinia carotovora*, *Phytopathology* **69**:117-120.

Thurn, K.K. and Chatterjee, A. K., 1985, Single-site chromosomal Tn5 insertions affect the export of pectolytic and cellulolytic enzymes in *Erwinia chrysanthemi* EC16, *Appl. Environ. Microbiol.* **50**:894-898.

Tsuyama, H., 1965, Ecology and control of soft-rot bacteria of vegetables, *Ann. Phytopathol. Soc. Japan* **31**:159-164.

Tsuyumu, S., 1977, Inducer of pectic acid lyase in *Erwinia carotovora*, *Nature* **269**:237-238.

Tsuyumu, S., 1979, "Self-catabolite repression" of pectate lyase in *Erwinia carotovora*, *J. Bacteriol.* **137**:1035-1036.

Tsuyumu, S., 1981, Non-toxic chemical control of soft rot diseases, *Proc. 5th. Int. Conf. Plant Pathog. Bact., Cali.* pp.461-471.

Tsuyumu, S. and Chatterjee, A.K., 1984, Pectin lyase production in *Erwinia chrysanthemi* and other soft-rot *Erwinia* species, *Physiol. Plant Pathol.* **24**:291-302.

Tsuyumu, S., Funakubo, T., and Chatterjee, A.K., 1985, Purification and partial characterization of a mitomycin C-induced pectin lyase of *Erwinia chrysanthemi* strain EC183, *Physiol. Plant Pathol.* **27**:119-130.

Tsuyumu, S., et al., 1985, Presence of DNA damaging agents in plants as the possible inducers of pectin lyases of soft-rot *Erwinia, Ann. Phytopathol. Soc. Japan* **51**:294-302.

Tsuyumu, S., et al., 1990, Induction of pectin lyase and a SOS (*umuC*) gene in various plants, *Proc. 7th. Int. Conf. Plant Pathog. Bact., Budapest* (in press).

Zink, R.T. and Chatterjee, A.K., 1985, Cloning and expression in *Escherichia coli* of pectinase genes of *Erwinia carotovora* subsp. *carotovora, Appl. Environ. Microbiol.* **49**:714-717.

Summary of Discussion of Tsuyumu's Paper

Chatterjee asked which *E. coli* strain was used in the experiment where pseudoinduction by glucuronate or galacturonate was observed. *Tsuyumu* replied that it was C600. *Chatterjee* further asked whether galacturonate deficient mutants were tested. *Tsuyumu* said no, and stated that the reason for using C600 was that they excreted the enzyme even though other enzymes like alkaline phosphatase remain in the periplasmic space; in the case of HB101, excretion was very low, he said. *Chatterjee* then asked about the conditions under which the EC16 bacteria were grown from which extracts were taken for the gel retardation experiment. Were they grown under inducing conditions, in LB, or were the extracts from PGA-grown cells? *Tsuyumu* said that they were grown in LB. He added that his priority was to set up a system knocking out not only one part but by making a double mutant in OP1 and OP2, before he does experiments with extracts.

Chatterjee then asked whether in studies with mutants, modified fragments (*pelE*, OP1 and OP2), were marker exchanged with the wild type genome. *Tsuyumu* said that he had not, and said that he was talking about the plasmid in *E. coli* and eventually he wants to put these modified fragments in *Erwinia*. *Chumley* asked whether *Tsuyumu* could say something more on the mechanism of the noncompetitive inhibition by carboxylic acids; perhaps chelation of divalent cations or maybe there is some effect on the substrate as well as the enzyme? *Tsuyumu* said he did not know the answer. He had expected the inhibition to be competitive, but it turned out to be noncompetitive. He speculated that the carboxylic acids may act directly on the substrate and somehow alter it. *Chumley* observed that divalent cations play an important role in holding pectate monomers together, and chelating calcium changes the structure of the substrate. *Tsuyumu* said that chelation may not be involved because they used sodium salts and these still had inhibitory activity. *Morris* inquired whether other polyanions such as polyacrylic acid or alginic acid act as competitive inhibitors. *Tsuyumu* answered that as the molecular weight increases a degree of specificity with respect to inhibition emerges. For instance, citrate has a higher inhibitory effect on pectate lyase but very low inhibitory effect on polygalacturonate.

Bushnell asked whether rust fungi would have to have pectin degrading capabilities to enter from the intercellular space. There was no consensus on this question among the participants. *Macko* asked whether there were other factors besides the pectic enzymes that are involved in pathogenicity. *Tsuyumu* said that *Handa* at Purdue has shown that mutants defective in pathogenicity still produced pectic enzymes indicating that something else was involved in pathogenicity. *Chatterjee* observed that in *E. chrysanthemi* there is good evidence that outer membrane proteins play a role in iron uptake and pathogenicity. He said he did not know whether this is true in the cases of *E. carotovora* and *E. atroseptica*.

Keen asked whether there was any indication of the presence of a *pel* box. *Tsuyumu* replied that they have looked for it but without success. *Keen* further observed that a KDG box has been recognized by *Condmine in Lyon*, and *Chatterjee* observed that *Hinton* has found putative KdgR-binding sequences in

E. carotovora. Keen then said that this palindrome is not found in any of the EC16 strains and that if this type of a box is involved in regulation of *pel*E then one wonders whether similar regulatory circuits may be involved with respect to other *pel* genes.

Mills asked whether there were any differences in the upstream regulatory sequences of *pel* genes of *E. carotovora* and *E. atroseptica*, which infect potatoes in hot and cold climates respectively. *Tsuyumu* answered that they did not have the disease caused by *E. atroseptica* in Japan, however, he did expect a difference in the finite control of regulation between the two organisms. *Chatterjee* added that Pel production is a "temperature sensitive" phenomenon; if one grows *E. chrysanthemi* at 32° C one gets a severe drop in activity levels, and he said that it is not known whether the drop is due to a transcriptional problem or due to some other reason. He further speculated that the effect of temperature in the alleged differential pathogenic responses of *E. carotovora* and *E. atroseptica* may not be at the level of production of pectate-degrading enzymes.   *Durbin* asked whether both pathogens are found in the same environment. *Chatterjee* said that they are but it is often difficult to distinguish one from the other. *Toyoda* asked what transposon-*lacZ* system was used in *Tsuyumu's* studies on *pnl*. *Tsuyumu* said that he used the same transposon that *Chatterjee* had used in his studies of *pnl* genes in *E. carotovora*.

Chapter 4
Regulation of the Production of Pectinases and Other Extracellular Enzymes in the Soft-Rotting *Erwinia* spp.

A.K. Chatterjee, J.L. McEvoy, H. Murata, and A. Collmer

Members of the enterobacterial genus *Erwinia* cause diseases in a wide variety of plant hosts producing an array of symptoms including necrosis, wilt, gall, and soft-rot. In addition to bona fide plant pathogens, this genus houses commensals such as *E. herbicola* that largely sustain an epiphytic lifestyle, but can occasionally infect immunocompromised humans and animals. During the past decade *Erwinia* spp. have been extensively studied and several reviews (Chatterjee et al., 1990; Chatterjee et al., 1986; Collmer et al., 1986; Kotoujansky, 1987) discuss contemporary developments. For that reason and for the sake of brevity in this report we focus primarily on our recent work on the regulation of the production of pectinases and other extracellular enzymes in the soft-rotting bacteria: *E. chrysanthemi* (Echr) strain EC16 and *E. carotovora* subsp. *carotovora* (Ecc) strain Ecc71. The evidence presented here allows the following general conclusions: (1) The *pel* (pectate lyase) genes of EC16 are regulated differently. (2) Extracellular enzyme production in Ecc71 is subject to a common control. (3) Pectin lyase production in Ecc71 in response to DNA-damaging agents requires a transcriptional activator.

Extracellular Enzymes

Many soft-rotting *Erwinia* spp. produce an assortment of degradative enzymes (pectate lyase isozymes, Pels; polygalacturonase, Peh; cellulase, Cel; protease, Prt; pectin methylesterase, Pme; pectin lyase, Pnl; phospholipase, Plc) that act on preformed plant structures. Most of these enzymes are true extracellular proteins since they are completely dissociated from the productive cells, and are actively exported across the cell envelope in the absence of lysis or a nonspecific release of cell-bound proteins. This conclusion is further supported by the isolation of export-defective (Out⁻) mutants of Echr and Ecc wherein the enzymes are confined to the periplasm instead of being secreted out of the cell (Andro et al., 1984; Hinton et al., 1987; Ji et al., 1987; Murata et al., 1990; Thurn et al., 1985). Both physiological and genetic evidence (Chatterjee et al.,

1990; Murata et al., 1990) indicates that enzyme export is tightly linked with tissue-macerating ability of these *Erwinia* spp. These findings taken along with the knowledge of the type of substrates cleaved by the exported proteins suggest that each of these enzymes has the potential to contribute to the pathogenicity of the soft-rotting erwinias. Moreover, studies with genetically modified bacteria or fractionated enzymes have established that Pels of Echr and Ecc or endo-Peh of Ecc are sufficient to elicit cell separation (i.e., tissue maceration) and changes in cell membrane permeability (Chatterjee et al., 1990; Collmer et al., 1986; Kotoujansky, 1987).

Studies with cloned DNAs and the enzymes produced in the *E. coli* clones as well as in *Erwinia* revealed that while EC16 contained at least four *pel* genes and an exo-*peh* gene, Ecc71 carried five *pel* genes, an endo-*peh* gene and an exo-*peh* gene (Barras et al., 1987; Collmer et al., 1986; Collmer et al., 1988; Keen et al., 1984; Willis et al., 1987; Zink et al., 1985). The EC16 *pel* genes and the corresponding Pel enzymes have been studied more thoroughly than the *Ecc*71 genes and proteins (Collmer et al., 1986; Keen et al., 1986; Roeder et al., 1985, 1987; Schoedel et al., 1986; Tamaki et al., 1988; Willis et al., 1987); the findings pertinent to this report are listed below. (1) The four *pel* genes occur in EC16 in two clusters, one consisting of *pel*A and *pel*E, and the other containing *pel*B and *pel*C. (2) While *pel*B and *pel*C share extensive homology, *pel*A and *pel*E share some, albeit dispersed, homology. (3) All four *pel* genes contain sequences for an N-terminal signal peptide presumed to direct polypeptide translocation across cell membrane(s). (4) Pel species differ with respect to their tissue-macerating ability: PelE has the highest activity followed by PelB and PelC; PelA does not cause tissue maceration. A dominant role of PelE in pathogenicity of EC16 has been confirmed by studying genetically modified bacteria (Payne et al., 1987; Ried et al., 1988; Roeder et al., 1985, 1987). Further evidence for a differential role of pectinases in pathogenicity comes from the studies with other systems. The PelA-deficient mutants of the Echr strain 3937 failed to cause systemic symptoms in the normally susceptible host, *Saintpaulia ionantha* (Boccara et al., 1988). Moreover, a Pme⁻ mutant (deficient in pectin methylesterase; the corresponding gene locus = *pem*) was noninvasive when inoculated into *S. ionantha* (Boccara et al, 1989). Thus, it would appear that the activities of both Pme and PelA are required in systemic invasion of *S. ionantha* by *E. chrysanthemi* strain 3937. In light of that, it would be of interest to assess the role of PelA and Pme in Echr strains infecting other hosts.

Pel Regulation

In view of the importance of pectinases, specifically the Pels, in pathogenicity, considerable attention has been directed to the analysis of the regulation of enzyme production. Early studies revealed induction by pectate catabolic products and susceptibility to catabolite repression (Chatterjee et al., 1986; Kotoujansky, 1987; Tsuyumu, 1977, 1979). Various mutants have been isolated that are altered in Pel production (Chatterjee et al., 1990; Kotoujansky, 1987). We should note that most of the early studies did not take into account the possibility that each Pel species could be regulated differently. The use of reporter gene systems has, however, facilitated exploration of the differential expression of individual *pel* genes and the identification of regulatory loci. Using *lacZ* transcriptional fusions, Robert-Baudouy and associates (Condemine et al., 1987; Hugouviex-Cotte-Pattat et al., 1987, 1989; Reverchon et al., 1987) have found that KdgR (= regulator of several genes in the ketodeoxygluconate pathway) represses *pel, pem, ogl* (oligogalacturonate lyase) and other pectate catabolic genes in Echr strain 3937. Nucleotide sequence data (Reverchon et al., 1989) suggested the presence of a putative KdgR binding motif at the 5-prime end of the *pel*E and *ogl* genes of this bacterial strain. In EC16, 3-deoxy-D-glycero-2,5-hexodiulosonate functions as an inducer and activates the expression of *pel, ogl, peh*, and *kdu*D (3-deoxy-D-glycero-2,5-hexodiulosonate dehydrogenase). Moreover, the expression of some of those genes is affected in Ogl⁻ and KduD⁻ mutants (Chatterjee et al., 1985). These findings indicate that the genes in EC16 are subject to a common regulation.

To test differential expression of *pel* genes, we examined Pel production in *E. coli* strain HB101 and Echr strain UM1005 (devoid of endo-Pel activity; Ried et al., 1988) harboring Pel⁺ plasmids (Barras et al., 1987). Our data revealed three classes of Pels in EC16: Class I represented by PelE, has a low basal level that is efficiently induced with pectate catabolic products. Class II, represented by Pels B and C, has a very high basal level that is subject to slight induction. Class III, represented by PelA, has a very low basal level that is not significantly altered by the pectate catabolic products.

The differential expression of *pel*A and *pel*E genes in Echr (EC16) can be explained by invoking the presence of a transcriptional activator(s) responding to specific signals. The observation that the *pel*A gene of the Echr strain 3937 is stimulated by plant extracts (Beaulieu et al., 1990) further supports this hypothesis. We should note that findings of Reverchon et al., (1987) with the strain 3937 are also consistent with a positive regulation of some of the *pel* genes.

To further analyze *pel*E regulation, we isolated by EMS mutagenesis of UMJ1008 (*pel*E⁺, Δ*pel*A, Δ[*pel*B-*pel*C]) strains that apparently were altered in Pel production. One class of mutants, designated as Pecl, was noninducible by pectate. Furthermore, Pel production was not restored by *out*⁺ or *pel*E⁺ DNA. To determine if the noninducible phenotype was exerted *in trans* we examined Pel production in the Pec1 mutant, AC4232, and its parent, UMJ1008, carrying *pel*A-E⁺ or *pel*B-C⁺ plasmids. Pel production in both strains carrying the *pel*B⁻C⁺ plasmid was constitutive although levels were two- to four fold higher in UMJ1008. In contrast, basal and induced levels of Pel produced by AC4232 carrying a *pel*A-E⁺ plasmid were 2.2% and 0.3%, respectively, of those produced by UMJ1008 carrying the plasmid. Since *pel*A is at most only marginally inducible with pectate or its catabolic products (see above), the lack of Pel induction in AC4232 may result from a mutation in a gene encoding a *trans*-acting factor that specifically stimulates the expression of *pel*E. However, confirmation of this hypothesis would require isolation of the "activator" gene.

Coregulation of Extracellular Enzyme Production

Aside from specific regulation of *pel* genes in the Echr strains discussed above, recent evidence both with Ecc and Echr indicates that Pel production may be coregulated along with other extracellular enzymes such as Peh, Cel, and Prt. By Tn5, TnphoA and Tn10-lacZ mutagenesis of Ecc71 we have obtained strains that are pleiotropically defective in the production of Pel, Peh, Cel, and Prt. These mutants were also reduced in their ability to macerate potato tubers and celery petioles. A mutant (AC4231) of a similar phenotype was obtained from a derivative of the Echr strain EC16 by EMS mutagenesis; however, most of our genetic and physiological studies described below were done with Ecc71 and its derivatives.

Following mobilization of a gene library into one of the pleiotropic mutants, we identified a cosmid (pAKC264) that restored extracellular Pel, Peh, Cel, and Prt production. pAKC264 and its subclone (pAKC602) carrying a 6.5 kb DNA segment restored production of these enzymes in all of our Tn-generated mutants. The following evidence indicates that the pleiotropic phenotype did not result from insertions in genes specifying enzyme export (*out*), adenyl cyclase (*cya*), or cAMP receptor protein (*crp*). (1) The mutants utilize as a carbon source compounds that are not catabolized by Cya$^-$ or Crp$^-$ mutants (Mount et al., 1979; Pastan et al., 1976). (2) The cloned DNA that restores enzyme production in the pleiotropic mutants does not complement *cya* or *crp* mutations in *E. coli*. (3) In the mutants, the levels of Pel, Peh, and Cel, as well as Prt are considerably reduced. This contrasts with the bona fide Out$^-$ mutants wherein the localization of enzymes and not their levels are affected. Moreover, there are at least two pathways directing enzyme export in Ecc and Echr (see for example Chatterjee et al., 1987, Murata et al., 1990): one mediates the export of Prt and the other mediates export of Cel, Pel, and Peh. Out$^-$ mutants, defective in the export of Pel, Peh, and Cel, are not affected in the production of protease. (4) The pleiotropic defect is rectified by plasmids that do not restore the Out phenotype in any of the export-defective mutants. (5) Conversely, none of the Out$^+$ plasmids restore enzyme production in the mutants deficient in extracellular enzymes. The gene responsible for the pleiotropic phenotype was designated *aep* for activation of extracellular protein production.

In addition to complementing mutations causing the pleiotropic phenotype, the Aep$^+$ plasmids stimulated production of Pel, Peh, Cel, and Prt in the wildtype strain Ecc71 by about threefold. The *aep*$^+$ DNA, however, had no apparent effect on the levels of periplasmic enzymes (cyclic phosphodiesterase, ß-lactamase) or a cytoplasmic enzyme (ß-galactosidase). These data demonstrated that the *aep* product specifically stimulates the production of extracellular enzymes.

To establish linkage between the *aep* gene and the pleiotropic phenotype, we inactivated the gene by insertion of the mini-Mu *lac* fusion element, Mu d11734 (Castilho et al., 1984). Two classes of insertions were recovered: Class I had a very low basal level of ß-galactosidase and this level remained unchanged under various cultural conditions. Class II produced a moderate level of ß-galactosidase readily detectable under various growth conditions. Restriction analysis revealed that while the mini-Mu element had inserted into the same general area of the cloned DNA, they were oriented differently. One representative construct of each class was transferred into a Lac$^-$ derivative (AC5006) of Ecc71, and the *aep-lacZ* DNAs were placed on the chromosome by marker exchange. In this manner two types of Aep$^-$ strains were derived:

AC5024 contained a *lacZ* insertion which responded to the *aep* promoter; AC5025, on the other hand, contained *lacZ* in the reverse orientation such that *lacZ* expression was independent of the *aep* promoter. In all our subsequent analysis the strain AC5025 served as a control. As might be expected, both AC5024 and AC5025 failed to produce extracellular Pel, Peh, Cel, and Prt. We should point out that the aep^+ allele confers a *trans*-dominant phenotype as indicated by restoration of enzyme production in aep^-/aep^+ merodiploid strains.

The availability of *aep-lacZ* transcriptional fusions allowed us to examine the regulation of *aep* expression under various cultural conditions known to stimulate extracellular enzyme production. In previous analyses (Murata et al., 1990; Murata and Chatterjee, unpublished) we had determined the following. Growth of Ecc71 with pectate did not result in high levels of Pel activity; in fact, the induction ratio ranged from 2 to 4. Compared with pectate, pectin allowed a better induction of Pel since induction ratios of 10 to 15 were obtained when Ecc71 or its derivatives were grown in media supplemented with pectin. In these media production of other extracellular enzymes (i.e., Peh, Cel, or Prt) was not adversely affected. These responses of Ecc71 with pectin and pectate contrast the response of Echr strain EC16 wherein maximal stimulation of Pel production occurs with pectate. These findings suggested that in these bacteria the production of pectate lyases (and other extracellular enzymes) was stimulated by different signals (=effectors). Further support for this hypothesis comes from our observations of differential activation of *pel* expression by plant (carrot root, celery petiole, and potato tuber) extracts in Ecc, Echr, and *E. carotovora* subsp. *atroseptica*. These observations raised the following two issues. (1) Is the *aep* expression activated by cultural conditions that stimulate Pel (extracellular enzyme) production? (2) If so, is there a temporal control in the expression of *aep* and *pel* genes? To resolve the first issue, we used an Aep$^+$ strain and an Aep-LacZ strain, both derived from a Lac$^-$ Ecc71 derivative. Pel production in the Aep$^+$ strain and ß-galactosidase production in the Aep-LacZ strain were higher in cultures containing citrus pectin and celery extract than in glycerol cultures. Induction ratios of ß-galactosidase in the Aep-LacZ strain were ca. 3 and 5 with citrus pectin and celery extract, respectively. Likewise, in the Aep$^+$ strain Pel activity was five-fold and 13- fold higher in citrus pectin and celery extract media, respectively, than in glycerol medium. These data show that transcription of *aep* is stimulated by substances that stimulate Pel production.

We next examined the issue of a temporal control of *aep, pel, cel*, and *prt* genes. For this study, we used a derivative of Ecc71 with the following genotype: *aep-lacZ, lacZ, pel$^+$, cel$^+$, peh$^+$, prt$^+$/aep$^+$*; the *aep$^+$* gene was on the low copy IncPI vector, pRK404 (Ditta et al., 1985). The strain with the same chromosomal markers but carrying pRK404 in place of the *aep$^+$* plasmid served as the control. The kinetic data revealed that in the strain carrying the *aep$^+$* plasmid high rates (=induced levels) of Pel, Cel, and Prt production commenced after the induction of ß-galactosidase. The *aep$^+$* DNA did not affect the expression of the *aep-lacZ* fusion. These data taken along with the genetic evidence demonstrate that (1) *aep* gene transcription is not subject to autogenous regulation and (2) *aep* transcription is required for the production of extracellular enzymes in Ecc71.

Pectin Lyase Production in Response to DNA-Damaging Agents

Pectin lyases (Pnls) produced by *Erwinia* spp. differ from the Pels in many respects (for details see Chatterjee et al., 1990). A significant difference lies in the mode of regulation of these enzymes. While Pels are induced by pectin or its catabolic products (see above), Pnls of *E. carotovora* are produced in response to DNA-damaging agents such as nalidixic acid, mitomycin C, and UV light (Itoh et al., 1980; Itoh et al., 1982; Kamimiya et al., 1972; Tomizawa et al., 1971; Zink et al., 1985). We have extended these findings by showing that Pnl induction by DNA-damaging agents occurs in strains of *E. carotovora* subsp. *atroseptica*, *E. c.* subsp. *carotovora* (Ecc), *E. chrysanthemi* and *E. rhapontici* (McEvoy et al., 1990; Tsuyumu et al., 1984). In most strains Pnl production is accompanied by cellular lysis and production of a bacteriophage or bacteriocin. In addition, Pnl production by Ecc71 requires a functional *recA* gene (Zink et al., 1985). This *recA*-dependent production of Pnl was noted in potato tubers as well as in culture. Furthermore, Tsuyumu et al. (1985) have shown activation of Pnl production by extracts of various plant tissues.

The *recA*-mediated damage-inducible phenotypes of Ecc71 are reminiscent of the SOS system of *E. coli* wherein a signal generated by DNA damage activates the product of the *recA* gene which is then involved with cleaving LexA, the repressor of a number of unlinked genes (for details see Walker, 1987). We have found that while the *E. coli lexA* represses an SOS-like system in Ecc71 (McEvoy et al., 1987), the basal and induced levels of Pnl and Ctv were not adversely affected by the *E. coli lexA*. Although these findings would argue against an involvement of LexA in Pnl production, our assays may not have been sensitive enough to discern an incomplete repression operating in a heterologous system.

Molecular Cloning of *pnl*A

Several Ecc71 cosmid libraries in RecA$^+$ or RecA$^-$ *E. coli* strains were screened for clones producing Pnl. No Pnl$^+$ clones were found. Subsequently, by mobilization of an Ecc71 cosmid library into Ecc193 and screening approximately 200 transconjugants for Pnl activity, two clones were detected which produced high levels of Pnl. The cosmids responsible for this hyperproduction of Pnl restored a wildtype phenotype to five previously isolated *pnl*::Tn*5* mutants of Ecc71 but did not restore production of the bacteriocin carotovoricin (Ctv) to *ctv*::Tn*5* mutants. These cosmids were found to be virtually identical by restriction analysis. The gene, localized to approximately a 1.4-kb segment of DNA, was determined to be the structural gene, *pnl*A, since *E. coli* strains harboring this gene on high copy number plasmids produced detectable levels of Pnl activity. However, Pnl was not inducible by DNA-damaging agents in RecA$^+$ or RecA$^-$ *E. coli* strains harboring *pnl*A.

Having isolated the *pnl*A gene of Ecc71, we examined the homology between *pnl* genes of various Pnl producing *Erwinia* spp. (McEvoy et al., 1990). A 1.4-kb probe, containing the *pnl*A sequence, hybridized under stringent conditions with various size *EcoRI* fragments of three Ecc and four Eca strains tested. No homology was detected between the Ecc71 *pnl*A and genomic digests of six Echr strains and three of four *E. rhapontici* strains, even under low stringency conditions. One *E. rhapontici* strain was exceptional in that a 4.3-kb *EcoRI* fragment weakly hybridized with the Ecc71 *pnl*A, as detected following low stringency washes. Since all of the strains included in the Southern analysis produce Pnl upon exposure to DNA-damaging agents, these data reflect a divergence in the *pnl* sequences of these *Erwinia* spp.

From the nucleotide sequence of *pnl*A of Ecc71 it was deduced that the protein has a molecular mass of 32,100 daltons with an isoelectric point of 9.92. This corresponds well with earlier studies (Itoh et al., 1982) indicating that Pnls of *E. carotovora* subsp. *carotovora* have a mass of approximately 28,000 daltons and isoelectric points of >9.55. The first 20 amino acid residues deduced from the nucleotide sequence matched perfectly with those of the N-terminus of the purified Ecc71 protein. There is no indication of processing of the N-terminus (i.e., no evidence for a signal sequence) which contrasts the findings with Pels of Echr and Ecc (Hinton et al., 1989; Keen et al., 1986; Lei et al., 1987; Tamaki et al., 1988; Trollinger et al., 1989). A presumed ribosome binding site (AAGGAA) was found at -6 bp to the translational start site. Within 600 bp upstream of the *pnl*A translational start site no sequences resembling a LexA binding site (SOS-box) were found, although at -175 bp a near perfect 24-bp palindromic sequence occurs which is a candidate for binding of a regulatory molecule. In addition, at -85 bp there is another smaller palindromic sequence, parts of which resemble the consensus sequence of the KdgR binding site (Hinton et al., 1989; Reverchon et al., 1989). The significance of these palindromic sequences in the regulation of *pnl*A transcription is currently being tested.

Expression of *pnl*A and *pnl*A-*lacZ*

*pnl*A-*lacZ* transcriptional fusions were constructed using the mini Mu-lac element Mu *d*I1734 (Castilho et al., 1984). These fusions were inducible by mitomycin C or nalidixic acid in RecA[+] Ecc71 derivatives. A representative fusion, when present as a single copy on the chromosome of the Lac[-] derivative of Ecc71, was inducible 70-fold in the presence of mitomycin C. *pnl*A-*lacZ* was not inducible in a RecA[-] Ecc71 derivative or in RecA[+] or RecA[-] *E. coli* strains. These findings indicated that Pnl regulation involves transcriptional activation of *pnl*A and that this activation requires RecA and another factor which is absent in *E. coli*.

Isolation of *pnl*R mutants

We had previously isolated a Tn*5* mutant of Ecc71, AC5130, which failed to produce Pnl and Ctv and to lyse following treatment with DNA-damaging agents. The induction of Pnl or Ctv production in this strain did not occur following the introduction of a *recA*[+] or a *pnl*A[+] plasmid. These results suggested that AC5130 may harbor a mutation in a regulatory gene. Subsequent to the isolation of AC5130, we obtained Pnl regulatory mutants of the Ecc71 derivative, AC5022, wherein *lacZ* served as the reporter gene for *pnl*A. In mutants obtained by EMS mutagenesis, ß-galactosidase and cellular lysis were not inducible by mitomycin C. These phenotypes could not be restored in these mutants by the introduction of a *rec*A$^+$ plasmid. Furthermore, the mutants were not inducible for Pnl production when carrying a *pnl*A$^+$ plasmid. These findings suggested that the mutants had a defect(s) in *pnl*R which encodes a transcriptional activator of *pnl*A as well as other genes.

Molecular Cloning of *pnl*R

In order to clone the activator gene, an Ecc71 gene library in the cosmid pSF6 was mobilized into the strain AC5130. Out of approximately 500 transconjugants screened, one was found which produced Pnl and Ctv and lysed upon exposure to mitomycin C or nalidixic acid. The cosmid, carrying about 40 kb of Ecc71 DNA, restored damage-inducible ß-galactosidase production in the PnlR⁻ mutants of AC5022. The gene responsible was localized on a 6.7-kb *Eco*RI fragment. A RecA$^+$ *E. coli* strain harboring this segment of DNA and the *pnl*A$^+$ gene or *pnl*A-*lacZ* was inducible for Pnl or ß-galactosidase by addition of mitomycin C. Thus *pnl*R encodes a product required in the RecA-dependent activation of *pnl*A transcription.

Recently, *pnl*R-*lacZ* transcriptional fusions were constructed using the mini Mu *lac* element Mu *d*I1734 (Castilho et al., 1984). These fusions were inducible in a RecA$^+$ LexA$^+$ *E. coli* as well as the PnlR⁻ RecA$^+$ (LexA$^+$?) Ecc mutant, AC5130. These findings, albeit preliminary, suggest that (1) *pnl*R is not autoregulated, (2) *pnl*R expression may be regulated by the *lex*A product, and (3) the stimulation of Pnl production by DNA-damaging agents results from an increased pool of the *pnl*R product required for transcription of *pnl*A.

Acknowledgements

Research in our laboratories is supported by the National Science Foundation (grant DMB-8796262), the Science and Education Administration of the U.S. Department of Agriculture grant 25846, and a grant from the Food for the 21st Century program of the University of Missouri-Columbia. We gratefully acknowledge Jean-Pierre Chambost, Asita Chatterjee, Wesley Chun, and Judy Engwall for the analysis of some of the cloned DNA and Pnl proteins and Eva L. Chatterjee for assistance with preparation of the manuscript.

This manuscript is journal series number 11,175 of the Missouri Agricultural Experiment Station.

References

Andro, T., et al., 1984, Mutants of *Erwinia chrysanthemi* defective in secretion of pectinase and cellulase, *J. Bacteriol.* **160**:1199-1203.

Barras, F., Thurn, K.K., and Chatterjee, A. K. 1987, Resolution of four pectate lyase structural genes of *Erwinia chrysanthemi* (EC16) and characterization of the enzymes produced in *Escherichia coli, Mol. Gen. Genet.* **209**:319-325.

Beaulieu, C., and Van Gijsegem, F., 1990, Identification of plant-inducible genes in *Erwinia chrysanthemi* 3937, *J. Bacteriol.* **172**:1569-1575.

Boccara, M., and Chatain, V., 1989, Regulation and role in pathogenicity of *Erwinia chrysanthemi* 3937 pectin methylesterase, *J. Bacteriol.* **171**:4085-4087.

Boccara, M., et al., 1988, The role of individual pectate lyases of *Erwinia chrysanthemi* strain 3937 in pathogenicity on saintpaulia plants, *Physiol. Mol. Plant Pathol.* **33**:95-104.

Castilho, B.A., Olfson, P. and Casadaban, M.J., 1984, Plasmid insertion mutagenesis and *lac* gene fusion with mini-Mu bacteriophage transposons, *J. Bacteriol.* **158**:488-495.

Chatterjee, A.K., Thurn, K.K. and Tyrell, D.J., 1985, Isolation and characterization of Tn5 insertion mutants of *Erwinia chrysanthemi* that are deficient in polygalacturonate catabolic enzymes oligogalacturonate lyase and 3-deoxy-D-glycero-2,5-hexodiulosonate dehydrogenase, *J. Bacteriol.* **162**:708-714.

Chatterjee, A.K., and Vidaver, A.K., 1986, Genetics of pathogenicity factors: application to phytopathogenic bacteria. *In* Advances in Plant Pathology, Ingram, D.S. and Williams, P.H., eds. vol 4, pp. 1-224 Academic Press, London.

Chatterjee, A.K., et al., 1987, Molecular genetics of soft-rot *Erwinia* and their plant paghogenic determinants. *In* Biotechnology in Agriculture, Natesh, S., Chopra, V.L., and Ramachandra, S., eds., Oxford IBH, New Delhi, pp. 261-273.

Chatterjee, A.K., et al., 1990, Molecular genetics of regulation and export of *Erwinia* pectinases. *In* Proceedings of the International Symposium on Biochemistry and Molecular Biology of Plant Pathogen Interactions. Phytochemical Society of Europe (in press).

Collmer, A., and Keen, N.T., 1986, The role of pectic enzymes in plant pathogenesis. *Annu. Rev. Phytopathol.* **24**:383-409.

Collmer, A., et al., 1988, Construction and characterization of *Erwinia chrysanthemi* mutants containing mutations in genes encoding extracellular pectic enzymes. *In* Molecular Genetics of Plant-Microbe Interactions, Palacios, R., and Verma, D.P.S. eds., pp. 356-361. APS Press, St. Paul.

Condemine, G., and Robert-Baudouy, J., 1987, Tn5 insertion in *kdgR*, a regulatory gene of the polygalacturonate pathway in *Erwinia chrysanthemi, FEMS Microbiol. Lett.* **42**:39-46.

Ditta, G., et al., 1985, Plasmids related to the broad host range vector, pRK290, useful for gene cloning and for monitoring gene expression, *Plasmid* **13**:149-153.

Hinton, J.C.D., and Salmond, G.P.C., 1987, Use of Tn*phoA* to enrich for extracellular enzyme mutants of *Erwinia carotovora*, subspecies *carotovora, Mol. Microbiol.* **1**:381-386.

Hinton, J.C.D., et al., 1989, Extracellular and periplasmic isoenzymes of pectate lyase from *Erwinia carotovora* subspecies *carotovora* belong to different gene families, *Mol. Microbiol.* **3**:1785-1795.

Hugouvieux-Cotte-Pattat, N., and Robert-Baudouy, J. 1987, Hexouronate catabolism in *Erwinia chrysanthemi, J. Bacteriol.* **169**:1223-1231.

Hugouvieux-Cotte-Pattat, N., and Robert-Baudouy, J., 1989, Isolation of *Erwinia chrysanthemi* mutants altered in pectinolytic enzyme production, *Mol. Microbiol.* **3**:1587-1597.

Itoh, Y., Izaski, K., and Takahashi, H., 1980, Simultaneous synthesis of pectin lyase and cartovoricin induced by mitomycin C, nalidixic acid or ultraviolet light irradiation in *Erwinia carotovora*, *Agric. Biol. Chem.* **44**:1135-1140.

Itoh, Y., et al., 1982, Enzymological and immunological properties of pectin lyases from bacteriocinogenic strains of *Erwinia carotovora*, *Agric. Biol. Chem.* **46**:199-205.

Ji, J., Hugouvieux-Cotte-Pattat, N., and Robert-Baudouy, J., 1987, Use of Mu-lac insertions to study the secretion of pectate lyases by *Erwinia chrysanthemi*, *J. Gen. Microbiol.* **133**:798-802.

Kamimiya, S., Izaki, K., and Takahashi, H., 1972, A new pectolytic enzyme in *Erwinia aroideae* formed in the presence of nalidixic acid, *Agric. Biol. Chem.* **36**:2367-2372.

Keen, N.T., and Tamaki, S., 1986, Structure of two pectate lyase genes from *Erwinia chrysanthemi* EC16 and their high-level expression in *Escherichia coli*, *J. Bacteriol.* **168**:595-606.

Keen, N.T., et al., 1984, Molecular cloning of pectate lyase genes from *Erwinia chrysanthemi* and their expression in *Escherichia coli*, *J. Bacteriol.* **159**:825-831.

Kotoujansky, A., 1987, Molecular genetics of pathogenesis by soft-rot *Erwinia*, *Annu. Rev. Phytopathol.* **25**:405-430.

Lei, S., et al., 1987, Characterization of the *Erwinia carotovora pelB* gene and its product pectate lyase, *J. Bacteriol.* **169**:4379-4383.

McEvoy, J.L., Murata, H., and Chatterjee, A.K., 1990, Molecular cloning and characterization of a gene for pectin lyase of *Erwinia carotovora* subsp. *carotovora* that responds to DNA damaging agents, *J. Bacteriol.* **172**: 3284-3289.

McEvoy, J.L., Thurn, K.K., and Chatterjee, A.K., 1987, Expression of the *E. coli lexA$^+$* gene in *Erwinia carotovora* subsp. *carotovora* and its effect on production of pectin lyase and carotovoricin, *FEMS Microbiol. Lett.* **42**:205-208.

Mount, M.S., et al., 1979, Regulation of endopolygalacturonate transeliminase in an adenosine 3',5'-cyclic monophosphate-deficient mutant of *Erwinia carotovora*, *Phytopathology* **69**:117-120.

Murata, H., et al., 1990, Characterization of transposon insertion Out⁻ mutants of *Erwinia carotovora* subsp. *carotovora* defective in enzyme export and of a DNA segment that complements out mutations in *E. carotovora* subsp. *carotovora*, *E. carotovora* subsp. *atroseptica* and *E. chrysanthemi*, *J. Bacteriol.* **172**:2970-2978.

Pastan, I., and Adhya, S., 1976, Cyclic adenosine 5'-monophosphate in *Escherichia coli*, *Bacteriological. Rev.* **40**:527-551.

Payne, J.H., et al., 1987, Multiplication and virulence in plant tissues of *Escherichia coli* clones producing pectate lyase isozymes PLb and PLe at high levels and of an *Erwinia chrysanthemi* mutant deficient in PLe, *Appl. Environ. Microbiol.* **53**:2315-2320.

Reverchon, S., and Robert-Baudouy, J., 1987, Regulation of expression of pectate lyase genes *pelA*, *pelD*, and *pelE* in *Erwinia chrysanthemi*, *J. Bacteriol.* **169**:2417-2423.

Reverchon, S., et al., 1989, Nucleotide sequence of the *Erwinia chrysanthemi ogl* and *pelE* genes negatively regulated by *kdgR* gene product, *Gene* **85**:125-134.

Ried, J.L., and Collmer, A., 1988, Construction and characterization of an *Erwinia chrysanthemi* mutant with directed deletions in all of the pectate lyase structural genes, *Mol. Plant-Microbe Interact.* **1**:32-38.

Roeder, D.L., and Collmer, A., 1985, Marker-exchange mutagenesis of a pectate lyase isozyme gene in *Erwinia chrysanthemi*, *J. Bacteriol.* **164**:51-56.

Roeder, D.L., and Collmer, A., 1987, Marker-exchange mutagenesis of the *pel*B gene in *Erwinia chrysanthemi* CUCPB 1237. *In* Plant Pathogenic Bacteria Civerolo, E.L., Collmer, A., Davis, R.E., and Gillaspie, A.G., (eds.) pp. 218-223, Martinus Nijhoff Publishers, Dordecht.

Schoedel, C., and Collmer, A., 1986, Evidence of homology between the pectate lyase-encoding *pel*B and *pel*C genes in *Erwinia chrysanthemi*, *J. Bacteriol.* **167**:117-123.

Tamaki, S.J., et al., 1988, Structure and organization of the pel genes from *Erwinia chrysanthemi* EC16, *J. Bacteriol.* **170**:3468-3478.

Thurn, K.K., and Chatterjee, A.K., 1985, Single-site chromosomal Tn5 insertions affect the export of pectolytic and cellulolytic enzymes in *Erwinia chrysanthemi* EC16, *Appl. Environ. Microbiol.* **50**:894-898.

Tomizawa, H., and Takahashi, H., 1971, Stimulation of pectolytic enzyme formation of *Erwinia aroideae* by nalidixic acid, mitomycin C and bleomycin, *Agric. Biol. Chem.* **35**:191-200.

Trollinger, D., et al., 1989, Cloning and characterization of a pectate lyase gene from *Erwinia carotovora* EC153, *Mol. Plant-Microbe Interact.* **2**:17-25.

Tsuyumu, S., 1977, Inducer of pectic acid lyase in *Erwinia carotovora*, *Nature* **269**:237-38.

Tsuyumu, S., 1979, "Self-catabolite repression" of pectate lyase in *Erwinia carotovora*, *J. Bacteriol.* **137**:1035-1036.

Tsuyumu, S., and Chatterjee, A.K., 1984, Pectin lyase production in *Erwinia chrysanthemi* and other soft-rot *Erwinia* species, *Physiol. Plant Pathol.* **24**:291-302.

Tsuyumu, S., et al., 1985, Presence of DNA damaging agents in plants as the possible inducers of pectin lyase of soft-rot *Erwinia*, *Annu. Phytopathol. Soc. Japan* **51**:294-302.

Walker, G.C., 1987, The SOS response of *Escherichia coli*, p. 1346-1357. *In Escherichia coli* and *Salmonella typhimurium*: cellular and molecular biology Neidhardt, F.C., Ingraham, J.L., Low, K.B., Magasanik, B., et al. (eds.), vol. 2. pp.1346-1357 American Society for Microbiology, Washington, D.C.

Willis, J.W., Engwall, J.K., and Chatterjee, A.K., 1987, Cloning of genes for *Erwinia carotovora* subsp. *carotovora* pectolytic enzymes and further characterization of the polygalacturonases, *Phytopathology* **77**:1199-1205.

Zink, R.T., and Chatterjee, A.K., 1985, Cloning and expression in *Escherichia coli* of pectinase genes of *Erwinia carotovora* subsp. carotovora, *Appl. Environ. Microbiol.* **49**:714-717.

Zink, R.T., et al., 1985, recA function of *Erwinia carotovora* subspecies carotovora is required in the induction of pectin lyase and carotovoricin by DNA-damaging agents, *J. Bacteriol.* **164**:390-396.

Summary of Discussion of Chatterjee's Paper

Tsuyumu initiated the discussion by asking whether *Chatterjee* had any evidence that the *pnlR* product acts at the transcriptional level. *Chatterjee* said it does and added that they have demonstrated this using *pnl*A-*lacZ* fusions although they have not yet done transcript assays.

Bennetzen asked as to why *Erwinia* might have put another layer between LexA and gene activation and inquired whether there were other genes that might require *PnlR* as a positive regulator. *Chatterjee* responded that was indeed the case. When the bacteria are exposed to DNA damaging agents, bacteriocin production is induced along with Pnl; in addition, depending on the dosage of the DNA damaging agent, the cells may also start lysing. So, there is a subset of genes that are controlled by the DNA damaging agents and in *Erwinia* all these phenotypes are concomitantly controlled by *pnlR* (now redesignated as *digR*, for the regulator of several damage inducible genes). He speculated that this subset of genes is of the SOS type which is further regulated by the bacterium by using a positive transcriptional activator as opposed to being directly regulated by *LexA* which is known to control the SOS genes. This allows the bacterium some degree of flexibility in terms of controlling the expression of damage inducible genes. *Bennetzen* then asked whether this meant that *PnlR* (= DigR) would be controlling genes specifically designed to deal with the stress associated with infection. *Chatterjee* concurred.

Keen asked whether potato glycoalkaloids do anything. *Chatterjee* said they had not studied this. *Keen* then asked whether all the *E. carotovora pel* genes are coordinately regulated by the *aep* gene product. *Chatterjee* said as far as they knew all those encoding the extracellular ones are but they did not know about those specifying the periplasmic species. He said they see a very low level of cell-bound activity and consequently they are not sure about the periplasmic enzymes. *Keen* asked whether *Chatterjee* had any Lux insertions. *Chatterjee* replied that they did not yet have them.

Essenberg asked whether there was any relationship between the global regulatory gene, *aep*, and the global export gene of *Xanthomanas* studied by *Daniels*. *Chatterjee* said that there are two systems that affect extracellular proteins: one that mediates transport of proteins across the cell envelope and the other, involving the *aep* gene, that regulates the expression of the genes that code for exported proteins. Therefore, *Daniels* is looking at genes involved in the export of proteins whereas *Chatterjee* is looking at these as well as the gene that regulates the expression of the structural genes specifying exported proteins. *Essenberg* further asked whether *Chatterjee* knew the nature of the active component in the celery extract and whether it was stable. *Chatterjee* replied that they did not know the nature of the active component, and that autoclaving did not eliminate the activity of the extracts.

Keen asked whether the *aep* gene of *E. carotovora* had homology with *E. chrysanthemi*. *Chatterjee* said that they were in the process of doing the Southerns and will subsequently sequence the gene(s).

Nester asked whether there was anything unique about the celery extract and whether other plant extracts behaved similarly. *Chatterjee* said that carrot and potato extracts show some effect in *E. carotovora* subsp. *atroseptica*, less so in *E. carotovora* subsp. *carotovora*. He said that celery extracts have given them consistent results. He continued that they have not examined any other plant extracts besides those mentioned before. *Nester* then asked if *Chatterjee* had sequenced the *pnlR* (= *digR*) gene and whether they had identified a two component regulatory system yet. *Chatterjee* answered negatively to both questions.

Tsuyumu asked why the specific activity of Pel produced by *E. chrysanthemi* incresed in the minimal medium containing glycerol *Chatterjee* said that it (i.e., the strain EC16) produces high basal activity. He continued that most of their work had been with *E. carotovora* subsp. *carotovora* and *E. carotovora* subsp. *atroseptica* and not with *E. chrysanthemi* which showed high activity with glycerol The reason for this, they believe, may be due to PelB and PelC activities whose genes are expressed strongly because of their native promoters, whereas the genes encoding PelA and PelE are poorly expressed under non-inducing conditions. However, *pelE* expression can be stimulated about 200-fold when one grows the organism in the presence of polygalacturonate whereas *pelA* cannot be induced by pectate.

Mills asked whether *E. carotovora* subsp. *atroseptica* and *E. carotovora* subsp. *carotovora* were pathogenic to both potato and celery. *Chatterjee* said yes; in addition the *E. chrysanthemi* isolate they work with (EC16), which was isloated from chrysanthemum, causes disease in chrysanthemum and macerates a variety of tissues. *Chatterjee* continued that *Collmer* has found that when EC16 derivatives are grown in the presence of chrysanthemum cell walls new Pel species appear indicating that normally silent genes are activated by plant cell wall components.

Yoshikawa then asked how *Chatterjee* evaluated claims in the literature that pectate lyase (by itself) from *Erwinia* causes a hypersensitive reaction in plants. *Chatterjee* said that his lab had no evidence on that but that based on work with *E. rubrifaciens*, *Kado's* group had suggested something of this nature, *Chatterjee* said Collmer and coworkers have found that pretreatment of leaf tissue with Pel prevents subsequent elicitation of HR by *Pseudomonas syringae* pv. *syringae*. Further discussion on reports in the literature on this subject ended with the conclusion that pectolytic enzymes may induce HR in plants indirectly be releasing oligouronides which in turn induce the HR response in affected plant tissues, and the oligouronides might act as global regulators affecting many plant processes.

Essenberg asked whether it was the pectic enzyme quantity or quality or both that is involved in HR and phytoalexin production by the products of maceration by these enzymes. *Chatterjee* replied that it could be both, because a mixture of different types of enzymes may produce different products. He said that some enzymes may be involved in systemicity as the French workers believe whereas others may be involved in maceration, and both the quality and quantity may be important in the dynamics of HR, infection and disease.

Vance then asked whether a report in the literature showing homology between *Erwinia* pectin-degrading enzyme genes and those of *Rhizobium* had been further elaborated upon. *Chatterjee* replied that from the Southern data there appears to be some homology with *pelE* DNA but to his knowledge no enzyme activity in *Rhizobium* has been demonstrated. *Nester* said that they did not find homology in *Agrobacterium* with pectin-degrading enzyme genes.

Yokoyama asked if *Chatterjee* knew of other examples of activation of enzymes by DNA damaging agents. *Chatterjee* said that a chitinase in *Serratia marcesans* was activated in this way. *Shiraishi* asked if some of the pectin-degrading enzymes are toxic to cells. *Chatterjee* answered that these enzymes were not toxins in the classical sense of the term. *Yamada* observed that DNA damaging agents might damage genes that encode pectin lyase or other enzymes which would eventually require the cells to induce exonuclease and DNA polymerase activities to repair the damage, and asked whether *Chatterjee* knew the temporal sequence of events in the cell. *Chatterjee* said he did not know the answer to that question, however, he said that various events such as DNA repair, *rec*A activation, and *pnl*R (= *dig*R) activation may happen concomitantly. On the other hand, he said that *Walker's* work with *din-lac*Z fusions in *E. coli* indicates that there is a hierarchy of expression of different *din* genes. He concluded the discussion by saying that a similar hierarchy may exist in his system.

Chapter 5
Characterization and Function of Bacterial Avirulence Genes

Noel T. Keen

Considerable progress has occurred in the cloning and characterization of avirulence genes from viral and bacterial pathogens. All of the avirulence genes thus far cloned behave as single dominant alleles and encode unique protein products. Nevertheless, with one noteworthy exception, the functions of avirulence genes in the pathogens that harbor them are not known. The hypersensitive reaction (HR) in plants carrying the complementary avirulence gene, however, appears to be determined by elicitors-- either primary avirulence gene protein products or metabolites arising from their catalytic activity. Since this topic has already been reviewed in depth (Keen et al., 1990; Keen, 1990), in this paper, I will concentrate only on the most recent information.

Resistance Genes and Avirulence Genes

The interactions between plant disease resistance genes and pathogen avirulence genes that initiate the plant HR have been thoroughly documented genetically (Ellingboe, 1984; Flor, 1942). However, the biochemical mechanisms that confer this genetic complementarity are poorly understood. Several models have been proposed to mechanistically explain gene-for-gene specificity (Bushnell et al., 1981; Doke et al., 1979; Gabriel et al., 1990; Keen et al., 1977; Lamb et al., 1989). Although other mechanisms may occur, the available evidence from biochemical experiments as well as with cloned pathogen avirulence genes supports the elicitor-receptor model (Keen et al., 1990). This hypothesis proposes that specific elicitors are the primary or secondary products of avirulence alleles in the pathogen and are recognized by receptors encoded by the complementary plant disease resistance genes.

Unfortunately, disease resistance genes have not yet been cloned from any higher plant, but avirulence genes have been characterized from viral and bacterial pathogens. The first avirulence gene was identified from race 6 of *Pseudomonas syringae* pv. *glycinea* (Staskawicz et al., 1984). Subsequently, several additional avirulence genes were cloned from bacterial pathogens (Gabriel et al., 1986; Hitchin et al., 1989; Kelemu et al., 1990; Minsavage et al., 1990; Shintaku et al., 1989; Vivian et al., 1989; and see further references in Keen et al., 1990). Recent research has also raised the possibility that avirulence genes and their complementary plant disease resistance genes may account for higher level specificities, for example, pathovar level specificity in bacterial pathogens (Kobayashi et al., 1989; Whalen et al., 1988). The functions of avirulence genes in the pathogen have generally not been deduced. Recently, however, an avirulence gene from *Xanthomonas campestris* pv. *oryzae* was shown to possess high homology with the *phoS* gene of *Escherichia coli* (Kelemu et al., 1990; J. Leach, personal communication). It is not yet clear whether this avirulence gene exhibits the *phoS* phenotype or how it leads to the plant HR. There is now substantial evidence, however, that certain avirulence genes lead to the production of specific elicitors.

Specific Elicitors

I reported the detection of specific elicitors from *Phytophthora megasperma* f.sp. *glycinea* at this meeting several years ago (Keen, 1976). It was not possible, however, to isolate these elicitors and prove that they resulted from the activity of defined avirulence genes in the pathogen. In subsequent years, several other specific elicitors were reported [Bruegger and Keen, 1979, from *Pseudomonas syringae* pv. *glycinea*; De Wit et al., 1985, from *Cladosporium fulvum*; Tepper and Anderson, 1986, from *Colletotrichum lindemuthianum*; J.S. Huang from *Heterodera glycines* (personal communication); Keen et al., 1990 with avirulence gene D from *Pseudomonas syringae* pv. *tomato*; Mayama et al., 1986, with victorin]. Some of these elicitors have also been isolated and characterized. Noteworthy is the novel peptide host-selective toxin, victorin (Wolpert et al., 1985). Victorin was shown by Mayama et al. (1986) to behave as a specific elicitor of the HR only in oat plants carrying the *Pc-2* allele for crown rust resistance. A specific receptor has also recently been detected in *Pc-2* but not *pc-2* oat cells (Wolpert et al., 1989). These data all indicate that victorin is in fact an elicitor and not a toxin in the classic sense. Although victorin is produced by *Helminthosporium victoriae*, it is probable that *Puccinia coronata* races carrying the avirulence gene complementary to *Pc-2* also elaborate an analogous elicitor during infection.

Another characterized specific elicitor is the linear peptide produced by *Cladosporium fulvum* races carrying the *ACf9* avirulence gene (DeWit et al., 1985; Scholtens-Toma et al., 1988). The elicitor appears to be produced in quantity by the fungus only during plant infection. This is of particular interest, because, as we shall see later, certain bacterial avirulence genes are only expressed at high level in the plant.

Recent research in the laboratories of N. Okada and W. Dawson has shown that the coat protein gene of tobacco mosaic virus (TMV) is a specific elicitor of the HR in tobacco plants carrying the N' disease resistance gene. As reviewed in more detail by Keen et al. (1990), genetic experiments have proven that mutation of amino acids occurring on the exposed surface of TMV coat protein aggregates may alter their ability to elicit the HR (Culver et al., 1989 and personal communication). Transformation of the coat protein gene only from HR-inducing strains into N' tobacco resulted in transgenic plants that exhibited autogenic necrotic symptoms. However, introduction of the coat protein gene from compatible strains had no detectable effect on N' transgenic plants. Thus, the work with TMV establishes that the coat protein is a specific elicitor and the gene encoding it is an avirulence gene. This avirulence gene is therefore the first for which its function in the pathogen as well as the mechanism by which it interacts with the complementary plant disease resistance gene are known.

Although relatively few specific elicitors have yet been detected and even fewer chemically characterized, it appears that these metabolites convey the genetic specificity of the avirulence genes from which they emanate. Specific elicitors are chemically diverse, leading to the speculation that they are the plant equivalents of antigens in vertebrates. We have recently studied a specific elicitor resulting from the activity of a well-characterized avirulence gene, *avr*D. The molecular genetic data strongly indicate that this elicitor is the signal which initiates the HR in incompatible soybean cultivars.

Avirulence Gene D from *Pseudomonas syringae* pv. *tomato*

Avirulence gene D (*avr*D) was cloned from *Pseudomonas syringae* pv. *tomato* and shown to function in *P.s. glycinea* to elicit the HR in some but not all soybean cultivars (Kobayashi et al., 1989; 1990a). The characterization of *avr*D has resulted in considerable insight into the question of how bacteria harboring the gene interact with soybean plants carrying the complementary disease resistance gene, called *Rpg*4 (Keen et al., 1990). Since much of the work with *avr*D has been reviewed recently (Keen, 1990), I will present only the overall outline here and discuss the general significance of *avr*D to plant recognition of pathogens leading to the HR.

Kobayashi et al. (1990a) sequenced *avr*D and showed that it encoded a 34 kDa protein product that had no leader peptide sequence and appeared to be a soluble cytoplasmic protein. The protein isolated from *E. coli* cells overexpressing *avr*D did not function itself as an elicitor of the HR, but *E. coli* cells expressing *avr*D produced a low molecular weight substance that behaved as a specific elicitor of the soybean HR (Keen et al., 1990). This elicitor has recently been isolated (M. Stayton, S. Tamaki, and N. Keen, unpublished observations), but its structure has not yet been deduced. However, the elicitor is thermostable, soluble in 95% ethanol and behaves as an anion at neutral pH. The most noteworthy characteristic of the *avr*D elicitor is that it initiates the HR in only those soybean cultivars which are resistant to *P.s. glycinea* cells carrying the cloned *avr*D gene (Keen et al., 1990).

The specificity of the elicitor was confirmed when segregating progeny of a cross of soybean cultivars sensitive (cultivar Flambeau) to the elicitor or insensitive (Merit) were analyzed (Keen et al., 1990). It is important to note that it was previously impossible to screen soybean plants for segregation of a resistance gene complementing *avr*D because no known isolate of *P.s. glycinea* contains a functional *avr*D gene. This raised the question of whether such a resistance gene occurred and, if so, why it was present in soybean when *P.s. glycinea* did not express a functional *avr*D gene. Nevertheless, analysis of segregating cross-progeny established that a dominant Mendelian resistance gene which complements *avr*D occurred in certain cultivars of soybean (Keen et al., 1990). This gene, called *Rpg*4, appears to confer both disease resistance to *P.s. glycinea* race 4 cells carrying *avr*D and sensitivity to the isolated *avr*D elicitor.

Several bacteria which do not normally produce detectable amounts of the *avr*D elicitor all produced significant quantities when the cloned *avr*D gene was introduced on a broad host range plasmid. These bacteria are *Escherichia coli, Erwinia chrysanthemi, Pseudomonas syringae* pv. *glycinea* race 4, *Xanthomonas campestris* pv. *glycines* and *Rhizobium fredii P.s.* pv. *glycinea* race 4 and *X.c.* pv. *glycines* are normally pathogenic on all known cultivars of soybean. However, both bacteria yielded a hypersensitive reaction on incompatible soybean cultivars when the cloned *avr*D gene was introduced into them on plasmids. Thus, *avr*D caused the bacteria to produce *avr*D elicitor activity as well as to elicit a hypersensitive reaction on those soybean cultivars carrying the *Rpg*4 resistance gene. It therefore appears that several Gram-negative bacteria can produce the *avr*D elicitor, providing that the *avr*D gene is present. The evidence also strongly suggests that the *avr*D elicitor is the signal molecule which is detected by soybean plants carrying the *Rpg*4 resistance gene to initiate the HR. This represents the first case, therefore, in which the biochemical basis has been established for hypersensitive resistance conferred by a defined resistance gene.

No tested isolates of *P. syringae* pv. *glycinea* expressed the *avr*D phenotype, but all isolates of this bacterium contained DNA that hybridized to *avr*D (Kobayashi et al., 1990a). Kobayashi et al. (1990b) cloned and sequenced the homologue gene to *avr*D from race 4 of *P.s. glycinea* and observed that it encoded a colinear protein with no deletions or insertions. The homologue gene protein product contained 86% identical amino acids to the *P.s. tomato avr*D protein. However, the two downstream proteins contained 98 and 99 percent identical amino acid sequences. These data suggest that *P.s.* pv. *glycinea* has accumulated several mutations in the *avr*D gene in order to eliminate the avirulence phenotype. However, the fact that *P.s. glycinea* contains a homolgoue gene with no deletions or frame shift mutations indicates that its function in the bacterium may be retained.

*Avr*D constituted the first of five tandem open reading frames (Kobayashi et al., 1990a), raising the possibility that it encodes one enzyme in a secondary metabolic pathway. This speculation was supported by the finding that *avr*D and associated genes occur on a 75 kb plasmid present in all tested isolates of *P. syringae* pv. *tomato*. Bender et al., (1986) discovered this plasmid and designated it the "B plasmid," but no phenotypes had previously been assigned. We have thus far failed in attempts to cure *P. syringae* pv. *tomato* of the B plasmid, raising the possibility that essential genes occur on it.

Expression of the cloned *avr*D gene in *P. syringae* pathovars *glycinea* and *tomato* is very low when the cells are grown on several laboratory culture media but increases by ca. 100x when the bacteria are inoculated into plant leaves (Shen, Kobayashi, and Keen, unpublished observations). Plant cell suspension cultures also facilitate high expression of the *avr*D gene. Huynh et al. (1989) reported that *avr*B from *P. syringae* pv. *glycinea* is also expressed at high levels in plant leaves. Similarly, Tamaki and Keen (unpublished data) showed that expression of *avr*C was much greater when the bacteria were inoculated into leaves or grown on non-repressing carbon sources. However, unlike *avr*D, the expression of *avr*B and *avr*C appear to be mediated by catablite repression. *Avr*D, on the other hand, was poorly expressed in either rich or poor culture media, raising the possibility that its increased expression in plant tissues is due to specific induction. We are currently attempting to determine if plant tissues contain specific inducers of *avr*D expression.

Certain *P. syringae* pathovars in addition to *P.s. tomato* contained DNA that hybridized to *avr*D (Kobayashi et al., 1990a). Significantly, all strains that contained DNA homologous to *avr*D (isolates of pathovars *mori*, *phaseolicola*, *maculicola* and *lachrymans*) all produced *avr*D elicitor activity when grown in liquid medium (Keen et al., 1990). On the other hand, all isolates except pv. *glycinea* that lacked hybridizing DNA also failed to produce detectable *avr*D elicitor activity (isolates of pathovars *atropurpurea*, *syringae*, *pisi*, *morsprunorum*, *tabaci*, and *savastanoi*). This indicates that production of the *avr*D elicitor is strictly dependent on the presence of *avr*D.

While discovery of the *avr*D elicitor has considerably increased our insight concerning avirulence gene activity and signalling to elicit the HR, many questions remain. For instance, what is the chemical nature of the *avr*D elicitor and what is the presumed substrate molecule in Gram-negative bacteria upon which it acts? The elicitor-receptor model proposes that the *Rpg4* resistance gene in soybean should encode a protein product which is part of a specific receptor for the *avr*D elicitor. However, attempts have not yet been initiated to test this possibility. Also, other cloned bacterial avirulence genes have not yet been shown to direct production of specific elicitors. We have examined *E. coli* and *P. syringae* cells expressing the cloned *avr*B and *avr*C genes from *P.s. glycinea* (Tamaki et al., 1988), but have not detected elicitor activity in the culture fluids. The *avr*B and *avr*C proteins themselves also have not yielded elicitor activity. Thus, in these cases, the signalling mechanisms between bacteria and plant cells are not known.

Acknowledgements

The research on *avr*D was supported by a grant from the National Science Foundation. I also acknowledge my long-time collaborators at UC Berkeley, Brian Staskawicz and Doug Dahlbeck, as well as several graduate students and post-doctoral fellows at UC Riverside who contributed to the project. Particularly among the latter, I acknowledge Donald Kobayashi and Stanley Tamaki, whose contributions were invaluable.

References

Bender, C.L., and Cooksey, D.A., 1986, Indigenous plasmids in *Pseudomonas syringae* pv. *tomato*: conjugative transfer and role in copper resistance, *J. Bacteriol.* **165**:534-541.

Bruegger, B.B. and Keen, N.T., 1979, Specific elicitors of glyceollin accumulation in the *Pseudomonas glycinea*-soybean host-parasite system, *Physiol. Plant Pathol.* **15**:43-51.

Bushnell, W.R. and Rowell, J.B., 1981, Suppressors of defense reactions: a model for roles in specificity. *Phytopathology* **71**:1012-1014.

Culver, J.N., and Dawson, W.O., 1989, Tobacco mosaic virus coat protein: an elicitor of the hypersensitive reaction but not required for the development of mosaic symptoms in *Nicotiana sylvestris*, *Virology* 173:755-758.

De Wit, P.J.G.M., Hofman, A.E., Velthuis, G.C.M., and Kuc, J.A., 1985, Isolation and characterization of an elicitor of necrosis isolated from intercellular fluids of compatible interactions of *Cladosporium fulvum* (syn. *Fulvia fulva*) and tomato, *Plant Physiol.* 77:642-647.

Doke, N., Garas, N.A., and Kuc, J., 1979, Partial characterization and aspects of the mode of action of a hypersensitivity-inhibiting factor (HIF) isolated from *Phytophthora infestans*, *Physiol. Plant Pathol.* 15:127-140.

Ellingboe, A.H., 1984, Genetics of host-parasite relations: an essay, *Adv. Plant Pathol.* 2:131-151.

Flor, H.H., 1942, Inheritance of pathogenicity in *Melampsora lini*, *Phytopathology* 32:653-669.

Gabriel, D.W., Burges, A., and Lazo, G.R., 1986, Gene-for-gene interactions of five cloned avirulence genes from *Xanthomonas campestris* pv. *malvacearum* with specific resistance genes in cotton, *Proc. Natl. Acad. Sci., USA* 83:6415-6419.

Gabriel, D.W., and Rolfe, B.G., 1990, Working models of specific recognition in plant-microbe interactions, *Annu. Rev. Phytopathol.* (in press).

Hitchin, F.E., Jenner, C.E., Harper, S., Mansfield, J.W., Barber, C.E., and Daniels, M.J., 1989, Determinant of cultivar specific avirulence cloned from *Pseudomonas syringae* pv. *phaseolicola* race 3. *Physiol. Molec. Plant Pathol.* 34:309-322.

Huynh, T.V., Dahlbeck, D., and Staskawicz, B.J., 1989, Bacterial blight of soybean: regulation of a pathogen gene determining host cultivar specificity. *Science* 245:1374-1377.

Keen, N.T., 1976, Specific elicitors of phytoalexin production: determinants of race specificity? *In Biochemistry and Cytology of Plant-Parasite Interaction*, pp. 84-93, (Tomiyama, K. et al., eds.), Kodansha Ltd., Tokyo.

Keen, N.T., 1990, Gene-for-gene complementarity, *Annu. Rev. Genetics* 24:447-463.

Keen, N.T., and Bruegger, B., 1977, Phytoalexins and chemicals that elicit their production in plants, *ACS Symp. Ser.* 62:1-26.

Keen, N.T., and Buzzell, R.I., 1990, New disease resistance genes in soybean against *Pseudomonas syringae* pv. *glycinea*: evidence that one of them interacts with a bacterial elicitor. *Theor. Appl. Genet.* (in press).

Keen, N.T., and Dawson, W.O., 1990, Pathogen avirulence genes and elicitors of plant defense. *In Genes Involved in Plant Defense* (Boller, W. and Meins, F. eds.), , Springer-Verlag, New York. (in press).

Keen, N.T., Tamaki, S., Kobayashi, D., Gerhold, D., Stayton, M., Shen, H.. Gold, S., Lorang, J., Thordal-Christensen, H., Dahlbeck, D., and Staskawicz, B., 1990, Bacteria expressing avirulence gene D produce a specific elicitor of the soybean hypersensitive reaction, *Molec. Plant-Microbe Interact.* 3:112-121.

Kelemu, S., and Leach, J.E., 1990, Cloning and characterization of an avirulence gene from *Xanthomonas campestris* pv. *oryzae*, *Molec. Plant-Microbe Int.* 3:59-65.

Kobayashi, D.Y., Tamaki, S., and Keen, N.T., 1989, Cloned avirulence genes from the tomato pathogen *Pseudomonas syringae* pv. *tomato* confer cultivar specificity on soybean. *Proc. Natl. Acad. Sci. USA* 86:157-161.

Kobayashi, D.Y., Tamaki, S.J., and Keen, N.T., 1990a, Characterization of avirulence gene D from *Pseudomonas syringae* pv. *tomato*, *Molec. Plant-Microbe Interact.* 3:94-102.

Kobayashi, D.Y., Tamaki, S.J., Trollinger, D.J., Gold, S., and Keen, N.T., 1990b, A gene from *Pseudomonas syringae* pv. *glycinea* with homology to avirulence gene D from *P.s.* pv. *tomato* but devoid of the avirulence phenotype, *Molec. Plant-Microbe Interact.* **3**:103-111.

Lamb, C.J., Lawton, M.A., Dron, M., and Dixon, R.A., 1989, Signals and transduction mechansism for activation of plant defenses against microbial attack. *Cell* **56**:215-224.

Mayama, S., Tani, T. Ueno, T., Midland, S.L., Sims, J.J. and Keen, N.T., 1986, The purification of victorin and its phytoalexin elicitor activity in oat leaves, *Physiol. Molec. Plant Pathol.* **29**:1-18.

Minsavage, G.V., Dahlbeck, D., Whalen, M.C., Kearney, B., Bonas, U., Staskawicz, B.J., and Stall, R.E., 1990, Gene-for-gene relationships specifying disease resistance in *Xanthomonas campestris* pv. *vesicatoria*-pepper interactions. *Molec. Plant-Microbe Interact.* **3**:41-47.

Scholtens-Toma, I.M.J. and De Wit, P.J.G.M., 1988, Purification and primary structure of a necrosis-inducing peptide from the apoplastic fluids of tomato infected with *Cladosporium fulvum* (syn. *Fulvia fulva*), *Physiol. Molec. Plant Pathol.* **33**:59-67.

Shintaku, M.H., Kluepfel, D.A., Yacoub, A., and Patil, S.S., 1989, Cloning and partial characterization of an avirulence determinant from race 1 of *Pseudomonas syringae* pv. *phaseolicola*, *Physiol. Molec. Plant Pathol.* **35**:313-322.

Staskawicz, B.J., Dahlbeck, D., and Keen, N.T., 1984, Cloned avirulence gene of *Pseudomonas syringae* pv. *glycinea* determines race-specific incompatibility on *Glycine max* (L) Merr., *Proc. Natl. Acad. Sci. USA* **81**:6024-6028.

Tamaki, S., Dahlbeck, D., Staskawicz, B., and Keen, N.T., 1988, Characterization and expression of two avirulence genes cloned from *Pseudomonas syringae* pv. *glycinea*. *J. Bacteriol.* **170**:4846-4854.

Tepper, C.S. and Anderson, A.J., 1986, Two cultivars of bean display a differential response to extracellular components from *Colletotrichum lindemuthianum*, *Physiol. Molec. Plant Pathol.* **29**:411-420.

Vivian, A., Atherton, G.T., Bevan, J.R., Crute, I.R., Mur, L.A.J., and Taylor, J.D., 1989, Isolation and characterization of cloned DNA conferring specific avirulence in *Pseudomonas syringae* pv. *pisi* to pea (*Pisum sativum*) cultivars, which possess the resistance allele, R2, *Physiol. Molec. Plant Pathol.*, **34**:335-344.

Whalen, M.C., Stall, R.E., and Staskawicz, B.J., 1988, Characterization of a gene from a tomato pathogen determining hypersensitive resistance in non-host species and genetic analysis of this resistance in bean. *Proc. Natl. Acad. Sci. USA* **85**:6743-6747.

Wolpert, T.J., Macko, V., Acklin, W., Juan, B., Seibl, J., Meili, J., and Arigoni, D., 1985, Structure of victorin C, the major host-selective toxin from *Cochliobolus victoriae*, *Experientia* **41**:1524-1529.

Wolpert, T.J., and Macko, V., 1989, Specific binding of victorin to a 100-kDa protein from oats. *Proc. Natl. Acad. Sci. USA* **86**:4092-4096.

Summary of Discussion of Keen's Paper

Chumley opened the discussion by asking whether *Keen* had noticed any synergism or interaction between the other four reading frames and the *avr*D all of which were involved in some biosynthetic pathway. *Keen* said that they had not investigated this extensively, but that there was a suggestion that they get less elicitor production when all of them are present. He said that their suspicion was that these ORFs along with *avr*D produce products with enzymatic activities that are part of a biosynthetic pathway. *Chumley* then asked how the results on *avr*D reflect on the possible modes of action of *avr*A, B, and C. *Keen* replied that they had closely examined *avr*C to determine whether it produces an elicitor but were unable to find it. At the same time the *avr*C protein product did not, by itself, act as elicitor either. Thus, the mechanisms of induction of HR by *avr*C and *avr*D obviously have little in common, and there probably is no universal mechanism by which the different *avr* genes cause HR. *Keen* said that this assumption could be wrong because *avr* A, B, and C might produce unstable elicitors which would be difficult to detect.

Patil asked whether ORFs 2, 3, and 4 were required in HR induction. *Keen* speculated that these downstream ORFs may possibly encode enzymes that are part of a common pathway, *avr*D being the first step in that hypothetical pathway. *Patil* then asked whether *avr*D works in any of the tomato cultivars. *Keen* replied that they have tested the elicitor in many tomato cultivars but none showed the HR, and that extensive tests in other species have shown that it works only in certain cultivars of chrysanthemum and petunia. He said that he and his colleagues do not attach much significance to these observations. *Patil* concluded his questioning by asking whether the substance X in *Keen's* model comes from the bacterium. *Keen* replied that their assumption is that X is a common metabolite of Gram-negative bacteria but they have not shown this yet. *Mills* commented that if the *avr*D protein and substance X are both required then one should be able to mutate either of the genes that encode them which would prevent HR form occurring. *Keen* replied that *Tamaki*, in his laboratory, screened over 3000 *E. coli* random Tn5 mutants precisely with this question in mind but was unable to find the appropriate mutant. He continued that it was possible that X was a housekeeping metabolite in which case one would not find the mutant unless supplemented media were used to isolate the appropriate auxotrophic mutant. *Mills* then asked whether *Keen* had any idea of the nature of *avr*E. *Keen* replied that *Lorang* in his laboratory had been working on it and she had found that *avr*E lies at one end of the hrp cluster of *p.s.* pv. *tomato* described by *Boucher*. *Tsuyumu* asked what would happen if one inoculates *avr*D minus mutants in tobacco. *Keen* replied that marker exchange mutants of *p.s.* pv. *tomato* deficient in *avr*D remain HR proficient not only in tobacco but also in soybean. So functions other than *avr*D are involved in these HRs. *Tsuyumu* further asked whether HR produced by *avr*D in tobacco and in soybean are mechanistically different from each other. *Keen* said they are different from a recognitional point of view, and he assumed that the elicitors are different. *Lamb* asked whether the hosts of pathovars which produce traces of the *avr*D elicitor respond to the *avr*D elicitor. He said that they had examined this in six *Phaseolus vulgaris*, and four cucumber cultivars and in both species the results were negative. *Essenberg* asked that since *Keen* had hints that the plasmid carrying the *avr*D gene cannot be cured had he mutated the *avr*D gene in the plasmid and what effect did this have on the pathogenicity of pv. *tomato*. *Keen* said they have examined this issue but the results are inconclusive. *Morris* asked whether the two elicitor peaks have comparable biological activity and whether they synergize. *Keen* said that they appear to have comparable activity and do not appear to synergize. The nmr analysis of these peaks suggests that they differ by only two methylene groups. *Durbin* asked whether *Keen* had found a similar elicitor in *Xanthomonas*. *Keen* said that it had not been done. *Durbin* further asked whether in light of the work on *avr* genes it was safe to speculate that there are a finite number of resistance genes in plants similar to what appears to be the case with respect to the *avr* genes in pathovars. *Keen* said he believed that this may turn out to be the case. He further said that if one takes the gene-for-gene hypothesis literally and if an *avr* gene induces HR in two different plant species then by definition these plant species must share the same resistance gene.

Chapter 6
Organization and Function of Pathogenicity Genes of *Pseudomonas syringae* Pathovars *phaseolicola* and *syringae*

Dallice Mills, Pradip Mukhopadhyay, Yuqi Zhao, and Martin Romantschuk

The ability of a pathogen to multiply and incite disease in its host plant is determined by genes present both in the pathogen and the host. In recent years, the application of molecular genetic tools has resulted in major advances towards the identification and characterization of pathogeneticity-related genes from numerous phytopathogenic prokaryotes. Although biological functions for many of these genes remain generally ill-defined, there is growing optimism based on recent experimental evidence that a function for some of these genes will soon be determined. Concomitant with the rapid advances in this general area have been a number of timely review articles covering various aspects of the molecular genetics of plant pathogenic bacteria (Chatterjee et al., 1986; Daniels et al., 1988; Keen et al., 1988; Kotoujansky, 1987; Mills, 1985; and Panopoulos et al., 1985). This chapter briefly reviews recent published work on the molecular characterization of pathogenicity determinants of two bean pathogens, *Pseudomonas syringae* pvs. *phaseolicola* and *syringae*, as well as unpublished results from our laboratory that suggest a possible function of a pathogenicity gene common to both organisms.

Etiology

Pathogenic strains of *P. syringae* pv. *phaseolicola* and of an ecotype of *P. syringae* pv. *syringae* incite halo blight and brown spot disease, respectively, in common bean. Upon entry either through natural wounds, stomata or hydathodes of leaves of susceptible plants, the bacteria multiply in the intercellular spaces over a period of 4 to 7 days to concentrations that approach 10^7 cells/cm^2 (Bertoni et al., 1987). Halo blight symptoms usually appear within a week postinoculation and are initially manifested as water-soaked lesions that subsequently become necrotic. Associated with the lesion is a chlorotic halo resulting from the production of phaseolotoxin by the multiplying bacteria, which inhibits the plant enzyme ornithine carbamoyltransferase (see

Patil, this volume). Brown spot disease symptoms on leaves are characterized by necrotic, brown lesions that occasionally have chlorotic margins (Patel et al., 1964) and the leaves frequently pucker at the point of inoculation. Chlorosis associated with brown spot disease results from the synthesis of syringomycin, which is produced in a number of strains regardless of their host range.

The disease symptoms incited by each pathogen in susceptible bean cultivars differ somewhat, and their biochemical properties as well as their genetic variability, as measured by genomic sequence homology, clearly indicate that these are different organisms. However, sufficient similarities exist to suggest that they may have evolved pathogenicity genes with common functions that are essential for growth *in planta* and, ultimately, the elicitation of disease in susceptible bean plants.

Daniels et al. (1988) have suggested that genes already identified by their synthesis of known determinants of pathogenicity may represent only a small fraction of the total number of genes that are needed for pathogenesis. In recent years, the identification of genes that ostensibly function in pathogenesis has been made easier through the technique of transposon tagging (reviewed in Mills, 1985). This approach has been used to obtain random mutations in a wide spectrum of bacteria including pathovars *phaseolicola* and *syringae* (Anderson et al., 1985; Lindgren et al., 1986; Willis et al., 1990). Of particular interest to plant pathology has been the identification of a group of mutants that manifest an inability to produce disease in susceptible host plants. As the selection of these mutants routinely is made on a minimal growth medium, mutations to auxotrophy or resulting in poor growth are eliminated. It is surmised, therefore, that these mutations identify pathogenicity genes.

There has been recent discussion among plant pathologists regarding the question of the definition of a pathogenicity gene. Daniels et al. (1988) have argued that pathogens have adapted to the physicochemical environment within the host, and mutations that result in an attenuation or loss of disease symptoms in susceptible hosts necessarily involve pathogenicity genes. This position clearly establishes that a potentially large number of gene functions is essential for pathogenicity. Indeed, this is becoming quite apparent as a growing number of mutants are being analyzed. The spectrum of genes required for full pathogenicity include, but are not limited to, those that enable bacteria to 1) attach to leaf surfaces, 2) gain entry and multiply in the intercellular spaces, 3) overcome host defense mechanisms, and 4) produce substances that cause disease symptoms. Through mutational analysis an increasing number of genes are being demonstrated to be indespensable for pathogenesis.

Mutational Analysis of Pathogenesis

1. Mutations Affecting the Attachment of P. syringae pv. phaseolicola to Plant Leaf Surfaces

The initiation of infection depends on the ability of these pathogens to recognize and adhere to bean leaf surfaces. Pathovar *phaseolicola* expresses pili which function as primary receptors for a lipid-containing bacteriophage φ6

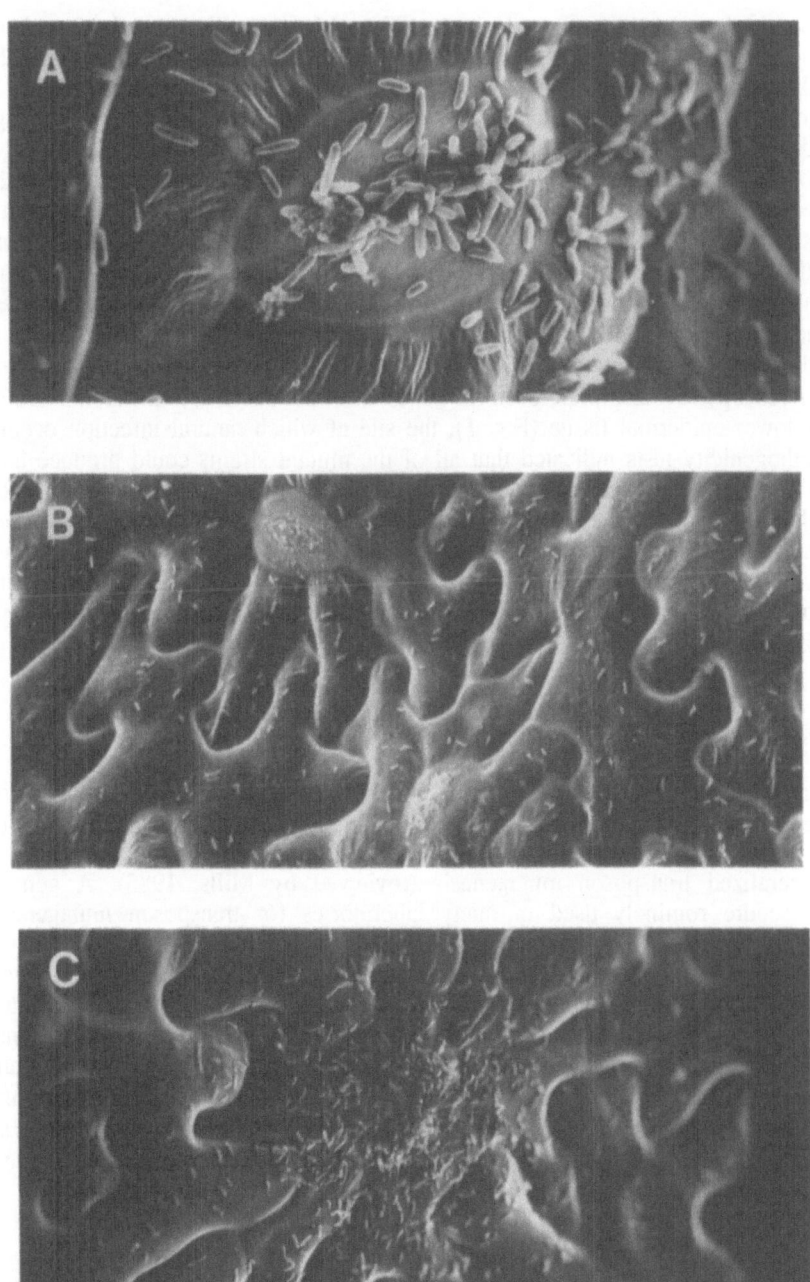

Figure 1. Adhesion of *Pseudomonas syringae* pv. *phaseolicola* cells to the surface of bean leaves. A. Wild type strain. B. Non-piliated mutant. C. Super-piliated mutant.

(Bamford et al., 1976; Cuppels et al., 1979). Mutant strains have been isolated and placed into three groups according to the level of piliation, as measured by adsorption. These include 1) non-piliated strains that do not adsorb φ6, 2) strains with reduced pilus expression that also exhibit a reduced level of phage adsorption, and 3) super-piliated strains that have an elevated level of binding of φ6 (Romantschuk et al., 1985). The ability of the mutant strains to adsorb to the bean leaf surfaces was also directly correlated with the level of piliation (Romantschuk et al., 1986). Relative to the wild type strain, the binding of non-piliated strains was reduced four- to fivefold, whereas the super-piliated strains showed a twofold increase in the number of bound cells. Of interest also was the observation that Red Kidney leaves were about twice as efficient in adsorbing bacteria as the leaves of the cultivar Prelude. Scanning electron microscopy revealed preference for adhesion of the bacteria to the stomatal area of lower epidermal tissue (Fig. 1), the site at which natural infection occurs. Pathogenicity tests indicated that all of the mutant strains could produce halo blight symptoms when the bacteria were forced into the leaf tissue. However, when the bacteria were spray-inoculated to more closely mimic the conditions leading to natural infection, the non-piliated strains failed to produce disease symptoms. These results clearly indicate that pili-mediated adsorbtion of this bacterium to bean leaf surfaces is a crucial early step leading to halo blight disease.

2. Identification and Cloning of Pathogenicity Genes by Transposon Tagging

Rapid advances in the identification and cloning of a variety of pathogenicity genes were made possible because of the development of suitable suicide plasmids that carry antibiotic resistance-encoding transposable elements for generalized transposon mutagenesis (reviewed by Mills, 1985). A general procedure routinely used in many laboratories for transposon mutagenesis within the *P. syringae* taxon has been previously outlined (Mills et al., 1990). Among the pathogenicity genes identified and cloned have been those involved in toxin production (see Patil, this volume), genes that are essential for primary symptom development in susceptible hosts, referred to as disease specific genes (Dsg), as well as genes that are involved in primary symptom development and the elicitation of the hypersensitive response (HR) in resistant bean cultivars and non-host plants. These latter genes have been designated *hrp* by Lindgren et al. (1986) to describe the phenotypic change in the hypersensitive reaction and pathogenicity.

3. Disease-Specific Pathogenicity Genes (Dsg)

Strain PS9024::Tn5 of *P. syringae* pv. *syringae* is mutated at a locus that affects only the ability to produce disease symptoms in previously susceptible bean cultivars, Eagle and Red Mexican (Mills et al., 1990). During the initial 3 days after hypodermic injection of approximately 10^3 cells into either bean cultivar, PS9024::Tn5 multiplies at a rate similar to the wild type parental strain (Bertoni et al., 1987). Thereafter, the number of viable cells rapidly decreases and only very minor or no symptoms appear at the site of inoculation. The *Eco*RI fragment that contains Tn5 is approximately 10 kb in size and was previously shown to have only weak or no homology with other strains in the *P. syringae* taxon including pv. *phaseolicola*, and very unexpectedly, some strains of pv. *syringae* that produce brown spot disease (Mills et al., 1987). Tn*5* was mapped to within 1.5 kb of one end of this fragment. Recently, an adjacent *Sal*I fragment located approximately 300 base pairs internal to Tn*5* was determined to have homology with other internal fragments extending to the other end of this *Eco*RI fragment but not with the 1.5 kb region containing Tn5 (Zhao et al., unpublished results). At least one other *Eco*RI fragment (ca. 8 kb) in PS9024::Tn*5* has homology with this repeated DNA sequence. Failure to detect the homologous gene in other strains of pv. *syringae* may have been due simply to the large amount of heterologous DNA present in the probe, which consisted of pBR322, Tn*5*, and repetitive DNA in the *Eco*RI insert. Several cosmids that contain this *Eco*RI fragment and adjacent genomic sequences are being subjected to molecular genetic analysis to define the limits of this pathogenicity locus and to characterize the genes.

Transposon mutagenesis of other pv. *syringae* genes affecting only pathogenicity have been reported by Rich et al. (1989) and Hrabak et al. (1989). A mutant designated *lem*A, results in the loss of lesion formation. Cosmid clones that complement the *lem*A mutation have been isolated from a genomic library and the genes are being characterized (Willis et al., 1990).

4. The hrp *Genes of* P. syringae

Transposon mutagenesis of pathovars *phaseolicola* and *syringae* has produced phenotypically similar pathogenicity mutants deficient in *hrp* gene functions (Niepold et al., 1985; Lindgren et al., 1986; Huang et al., 1988). The location of several Tn*5* insertions reveal a cluster of *hrp* genes within a 20 kb region of the *phaseolicola* genome. The organization and functions of these genes have been extensively studied by genetic complementation, gene replacement in other pathovars, promoter analysis, and comparative sequence analysis (Rahme, et al., 1989; Mindrinos et al., 1990). Another *hrp* locus designated *hrp*M (Mukhopadhyay et al., 1988a,b), was cloned from pathovar *syringae* and, also independently from pathovar *phaseolicola* (see Mindrinos et al., 1990). Apparently these genes are not limited to these two pathovars, but are present with varying degrees of homology in other pathovars of *P. syringae* as determined from the strength of the hybridization signals when used as probes in Southern blot analyses with other DNA (Lindgren et al., 1988). Furthermore, *hrp* mutants have been obtained in closely related pathovars *tabaci* and *glycinea* by gene replacement but not in pv. *tomato*, when the source of the mutant gene was from the 20 kb *hrp* gene cluster. Attempts to obtain a *hrp* mutant in pv. *phaseolicola* using *hrp*M as the mutant source DNA failed, although, unexpectedly, a *dsg* mutant was obtained (Mukhopadhyay et al., 1988b). However, gene replacement was achieved in other strains of pv. *syringae* using mutant *hrp*M as the source of DNA. These combined results clearly demonstrate that homologous *hrp* gene functions in other pathovars are also essential for pathogenesis.

Organization and Function of *hrp* Genes

1. The <u>hrp</u> Gene Cluster of pv. <u>phaseolicola</u>

This 20 kb *hrp* cluster is comprised of six genes designated *hrp*A, *hrp*B, *hrp*C, *hrp*D, *hrp*S, and *hrp*R, which extend throughout the entire length of this region (Mindrinos et al., 1990). A 9 kb segment contains three complementation groups, *hrp*AB, which comprises two functional units that appear to be expressed from a single promoter, and *hrp*C and *hrp*D, which are expressed from separate promoters. At least four other complementation groups appear to make up the remainder of this *hrp* cluster. Among these, *hrp*S was shown to be a regulatory gene. The *hrp*AB, *hrp*C and *hrp*D loci are inducible in the plant and in vitro through nutritional downshift. The nucleotide sequences of *hrp*S and *hrp*R reveal that they encode putative proteins with significant amino acid homology (ca. 40 to 46 percent homology) with a highly conserved domain of several of the NtrC family of prokaryotic regulatory proteins. That the *hrp*S gene function is required for *hrp*AB, *hrp*C and *hrp*D promoter activity was demonstrated in nutritional downshift experiments in which the promoter of each of these genes was fused to a reporter gene. Similarly, it was determined that *hrp*D promoter activity is under the control of *hrp*S in plants. Mutations within *hrp*R apparently have a limited effect on expression of reporter genes controlled by promoters from *hrp*AB, *hrp*C and *hrp*D. The complete and partial sequence of homologous *hrp*S and *hrp*R, respectively, from pv. *savastanoi* reveal greater than 99 percent homology with corresponding genes from pv. *phaseolicola*. Moreover, *hrp*S and *hrp*M have approximately 60 percent homology with each other.

Of additional significance is the observation that expression of these *hrp* genes is greatly reduced or repressed by raising the osmolyte concentration (Mindrinos et al., 1990). Apparently the electrolyte leakage that occurs during the HR could produce an osmolyte concentration that would prevent *hrp* gene expression which is known to be essential for pathogenesis. These recent preliminary findings suggest that testable hypotheses for the molecular bases of pathogenesis may be formulated that involve the direct role for control by *hrp* gene functions.

2. The <u>hrpM</u> locus

The *hrp*M locus of pv. *syringae* is comprised of two open reading frames, ORF1 and ORF2, that are expressed from *hrp*M1 and *hrp*M2 (Fig. 2). They encode putative polypeptides of 40 and 83 kilodaltons (kDa), respectively (Mukhopadhyay et al., 1988a,b). Tn5 insertions in either ORF result in the mutant *hrp* phenotype. The nucleotide sequence of this locus and promoter probe analysis suggest that these genes are expressed from a single promoter upstream of ORF1. The -10 region of this promoter overlaps a putative translation initiation codon ATG for the 40 kDa polypeptide, suggesting that a second ATG codon located 405 nucleotides downstream and in-frame with the translational stop codon may be the correct translation initiation codon. Initiation at this ATG would yield a 28 kDa polypeptide. The initiation codon of

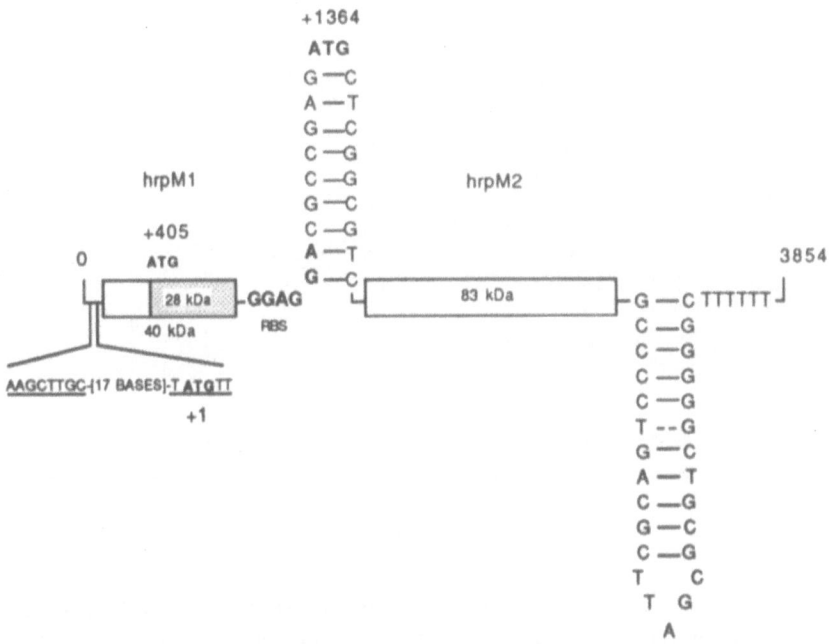

Figure 2. Organization of the *hrp*M locus of *Pseudomonas syringae* pv. *syringae*. Features of this locus are discussed in the text. Adapted from Mills et al., 1990.

ORF2 is located 205 nucleotides downstream of the stop codon of ORF1, and immediately preceding it is a consensus ribosome-binding site, GGAGGA. The last two bases, GA, are the first two nucleotides of a nine base pair repeat that forms a perfect palindrome with the ATG codon at its apex (Fig. 2). This configuration suggests that expression of ORF2 is controlled at the level of translation. A consensus transcription terminator is located 175 nucleotides downstream of a putative stop codon for ORF2. The nucleotide sequences of homologous genes cloned from pv. *phaseolicola* are reported by Marinos et al., (1990) to be highly conserved, with codon variation occurring primarily in the third position. They report, however, variance at nucleotides 3456 to 3460 in ORF2 that could suggest a sequencing error by Mukhopadhyay et al. (1988), which if confirmed would lead to an increase in the size of the polypeptide from 708 to 760 amino acids.

3. Putative Function of hrpM

A search of the GenBank and European Molecular Biology Laboratory data bases (Mukhopadhyay et al., 1988) revealed that the putative polypeptide encoded by ORF1 had 40 to 45 percent identity with histone H1 of rainbow trout, chicken, and *Xenopus laevis*, intimating that it functions as a DNA

A   B

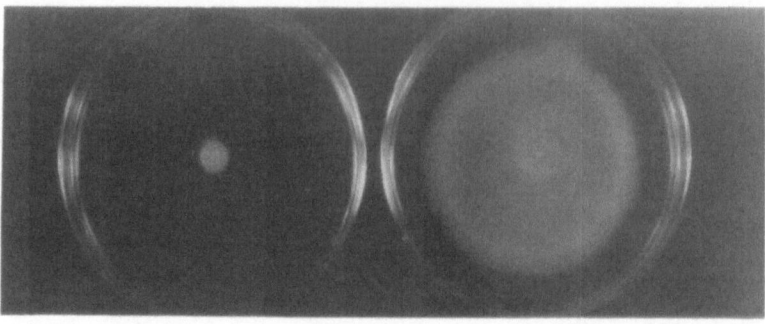

Figure 3. Growth of *Pseudomonas syringae* pv. *syringae* strains on 20-fold diluted King's medium B. A. Growth on 20-fold diluted medium; B. 20-fold diluted medium amended with ammonium chloride.

binding protein. A hyrophobicity plot of the putative product of ORF2 revealed that it was similar to many transmembrane proteins, but that it had no extensive amino acid identity with any protein in the data bases. The demonstration of a function for any gene which has no sequence identity with other proteins has become a challenge that has somewhat limited progress in ascertaining the functions of various pathogenicity determinants.

It was initially observed that strains mutated in *hrp*M have a mucoidal morphology (hence the designation *hrp*M) which could have resulted from an altered cell surface protein. Because the product of *hrp*M2 (ORF2) was likely a transmembrane protein, we adopted a position that its function could involve one of several cell processes, including cell motility and nutrient acquisition. However, *hrp*M mutants grow on a minimal salts medium in vitro indicating that loss of function does not produce an obvious shift towards auxotrophy. Unexpectedly, it was determined by a series of biochemical and physiological tests that *hrp*M mutants unlike the parental strain, could not grow on either 20-fold diluted King's medium B or minimal salts medium (Fig. 3). It was further shown that growth resumes if only the nitrogen source, ammonium chloride, is amended to normal concentrations. Moreover, several nitrogen sources including amino acids and organic nitrogen sources known to be synthesized by the bean plant could be substituted for the ammonium ion. However, glutamine which is present in bean xylem tissue and known to repress nitrogen uptake in *Escherichia coli*, inhibited growth of the wild type and *hrp*M mutant when used as the sole nitrogen source or when included with various other nitrogen sources (Fig. 4). Although the growth kinetics of the wild type strain is reduced by 10 μM glutamine in media containing either ammonium, allantoin or urea, *hrp*M mutants fail to grow at all under these conditions (Fig. 4).

These results suggest that the *hrp*M gene product functions either in the uptake or assimilation of nitrogenous compounds in environments where

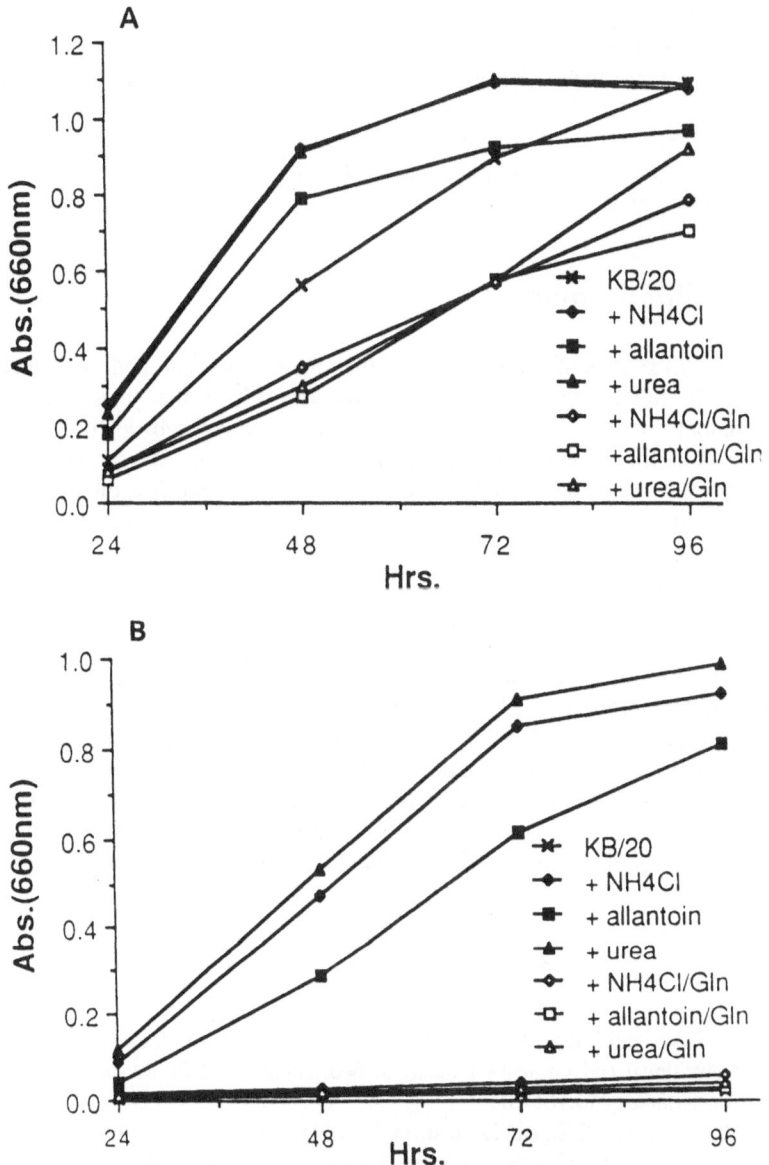

Figure 4. Kinetics of growth of *Pseudomonas syringae* pv. *syringae* strains on various nitrogen sources. A. Wild type strain PS9020. B. PS9021::Tn5 (*hrp*M mutant). Concentration of nitrogen sources, 10 m*M*; glutamine, 10 μ*M*.

glutamine is present. It is surmised that the uptake of ammonium ion, either through diffusion or by a general permease system, is sufficient to support growth of *hrp*M mutants in vitro. However, in environments where glutamine is present, growth of cells is dependent upon functional *hrp*M1 and *hrp*M2 gene products. The data support a role for *hrp*M in the uptake of nitrogenous compounds *in planta*. It could be speculated that the product of *hrp*M2, which has properties of transmembrane proteins, is a permease and that its synthesis is regulated by the product of *hrp*M1, which is thought to be a DNA binding protein. These genes are highly conserved in pathovars *phaseolicola* and *syringae*, and we have identified homologous sequences in over 30 strains representing 13 pathovars of *P. syringae* (Zhao and Mills, unpublished data). Although speculative, it is possible that homologous *hrp*M functions are required for growth of these pathovars in hosts that have significant levels of glutamine in their tissues.

Concluding Remarks

The results of studies of *hrp* genes strengthen the position adopted by Daniels et al., (1988) that as plant pathogens have adapted to the physicochemical environments of their hosts, certain genes have evolved whose functions are indispensible for growth and, ultimately, the production of compounds that incite disease. One can only speculate that *hrp*M and the *hrp* cluster encode only a few of the gene functions that are needed for full pathogenicity.

Acknowledgements

The published and unpublished results from the laboratory of D.M. were supported by U.S. Department of Agriculture Science and Education Administration grant 85-CRCR-1-1771 from the Competitive Research Grants Office and grant MB8315689 from the National Science Founudation and the Oregon State Agricultural Experiment Station, from which this is technical paper no. 9334. Acknowledged also is support from the Academy of Finland to Dennis Bamford, University of Helsinki, in whose lab the studies of piliation mutants were performed by M.R., who was also supported by the Alfred Kordelin Foundation.

References

Anderson, D.M., and Mills, D., 1985, The use of transposon mutagenesis in the isolation of nutritional and virulence mutants in two pathovars of *Pseudomonas syringae*. *Phytopathology* **75**:104-108.

Bamford, D.H., Palva, E.T., and Lounatmaa, K., 1976, Ultrastructure and the life cycle of the lipid-containing bacteriophage φ6, *J. Gen. Virol.* **32**:249-259.

Bertoni, G., and Mills, D., 1987, A simple method to monitor growth of bacterial populations in leaf tissue, *Phytopathology* **77**:832-835.

Chatterjee, A.K., and Vidaver A.K., 1986, Genetics of pathogenicity factors: application to phytopathogenic bacteria, *Adv. Plant Pathol.* **4**:1-218.

Cuppels, D.A., Vidaver, A.K., and Van Etten, J.L., 1979, Resistance to bacteriophage φ6 by *Pseudomonas phaseolicola*, *J. Gen. Virol.* **44**:493-504.

Daniels, M.J., Dow, J.M. and Osbourn, A.E., 1988, Molecular genetics of pathogenicity in phytopathogenic bacteria, *Annu. Rev. Phytopathol.* **26**:285-312.

Hrabak, E.M. et al., 1989, Characterization of a locus from *Pseudomonas syringae* pv. *syringae* involved in lesion formation. Abstr. Pseudomonas 89 Biotransformations, Pathogenesis, and Evolving Biotechnology, Chicago, Ill., July 9-13, 1989, p. 23.

Huang, H.C. et al., 1988, Molecular cloning of a *Pseudomonas syringae* gene cluster that enables *Pseudomonas fluorescens* to elicit a hypersensitive response in tobacco, *J. Bacteriol.* 170:4748-4756.

Keen, N.T. and Staskawicz, B., 1988, Host range determinants in plant pathogens and symbionts, *Annu. Rev. Microbiol.* 42:421-440.

Kotoujansky, D.Y., 1987, Molecular genetics of pathogenesis by soft-rot erwinias. *Annu. Rev. Phytopathol.* 25:405-430.

Lindgren, P.B., Peet, R.C., and Panopoulos, N. J., 1986, Gene cluster of *Pseudomonas syringae* pv. "*phaseolicola*" control pathogenicity of bean plants and hypersensitivity on nonhost plants, *J. Bacteriol.* 168:512-522.

Lindgren, P.B. et al., 1988, Genes required for pathogenicity and hypersensitivity are conserved and interchangeable among pathovars of *Pseudomonas syringae*, *Mol. Gen. Genet.* 211:499-506.

Mills, D., 1985, Transposon mutagenesis and its potential for studying virulence genes in plant pathogens, *Annu. Rev. Phytopathol.* 23:381-419.14.

Mills, D. and Mukhopadhyay P., 1990, Organization of the *hrpM* locus of *Pseudomonas syringae* pv. *syringae* and its potential function in pathogenesis. *In Pseudomonas* Biotransformations, Pathogenesis, and Evolving Biotechnology (Silver, S., Chakrabarty, A.M., Iglewski, B., and Kaplan, S., eds.) ASM Press, Washington, D.C. pp. 47-57.

Mills, D. and Niepold, F., 1987, Molecular analysis of pathogenesis of *Pseudomonas syringae* pv. *syringae*. *In* Molecular Determinants of Plant Diseases (Nishimura, S., Vance, C.P. and Doke, N., eds.) Japan Scientific Societies Press, Springer-Verlag, Tokyo, pp. 185-198.

Mindrinos, M.N. et al., 1990, Structure, function, regulation, and evolution of genes involved in pathogenicity, the hypersensitive response, and phaseolotoxin immunity in the bean halo blight pathogen. *In* Pseudomonas Biotransformations, Pathogenesis, and Evolving Biotechnology (Silver, S., Chakrabarty A.M., Iglewski, B. and Kaplan, S.) ASM Press, Washington D.C., pp. 74-81.

Mukhopadhyay, P. et al., 1988a, Pathogenicity genes of *Pseudomonas syringae*. *In* Molecular Genetics of Plant-Microbe Interactions, (Palacios, R., and Verma, D.P.S. eds.) APS Press, St. Paul, MN, pp. 247-252.

Mukhopadhyay, P., Williams, J., and Mills, D., 1988b, Molecular analysis of a pathogenicity locus in *Pseudomonas syringae*, pv. *syringae*. *J. Bacteriol.* 170:5479-5488.

Niepold, F., Anderson, D., and Mills, D., 1985, Cloning determinants of pathogenesis from *Pseudomonas syringae* pathovar *syringae*, *Proc. Natl. Acad. Sci. USA* 82:406-410.

Panopoulos, N., and Peet, R., 1985, The molecular genetics of plant pathogenic bacteria and their plasmids, *Annu. Rev. Phytopathol.* 23:381-419.

Patel, P. N. et al., 1964, Bacterial brown spot of bean in central Wisconsin, *Plant. Dis. Rep.* 48:335-337.

Rahme, L. et al., 1989, Organization and expression of the *hrp* gene cluster of *Pseudomonas syringae* pv. *phaseolicola. In* Vascular Wilt Diseases of Plants: Basic Studies and Control, (Tjamos, P., and Beckman, C. eds.) Springer-Verlag, Berlin. pp. 303-314.

Rich, J.J., Hrabak, E.M., and Willis, D.K., 1989, A mutation affecting lesion formation in *Pseudomonas syringae* pv. *syringae* is restored by a clone from *Pseudomonas syringae* pv. *phaseolicola.* Abstr. Pseudomonas 89 Biotransformations, Pathogenesis, and Evolving Biotechnology, Chicago, Ill. July 9-13, 1989, pg. 23).

Romantschuk, M., and Bamford, D. H., 1985, Function of pili in bacteriophage φ6 penetration, *J. Gen. Virol.* **66**:2461-2469.

Romantschuk, M., and Bamford, D.H., 1986, The causal agent of halo blight in bean, *Pseudomonas syringae* pv. *phaseolicola*, attaches to stomata via its pili, *Microbial Pathogenesis* **1**:139-148.

Willis, D.K. et al., 1990, Isolation and characterization of a *Pseudomonas syringae* pv. *syringae* mutant deficient in lesion formation on bean, *Mol. Plant-Microbe Interact.* **3**:149-156.

Summary of Discussion of Mills's Paper

Chatterjee asked whether the mutant was capsulated, or mucoid. *Mills* stated that it was much more mucoid than the wild type. *Chatterjee* then asked whether it was possible that a negative regulator of a gene for EPS production might have been inactivated. *Mills* responded that it could be; although in other backgrounds this mucoid effect was not as pronounced. *Chatterjee* asked what happens if one complements this mutant by putting the plasmid containing the wild type sequence into the mutant. Are all the phenotypes restored? Mills indicated that they were. *Mayama* asked about the product of the pathogenicity gene and what causes resistance in *Mills'* system. *Mills* responded that nothing is known about the gene product, other than the fact that from sequence data it looks like it is a transmembrane protein. Mutations in an open reading frame upstream of that gene also block disease expression, which would argue that it is part of an operon. The fact that the mutant has a problem with nitrogen uptake or utilization is the only thing attributable to this phenotype. It is obvious that mutations in different kinds of genes will lead to the same *hrp* phenotype, so the *hrp* gene cluster of pathovar *phaseolicola* codes for something apparently different from this, even though mutations in those genes give the same phenotypic response. *Bushnell* wondered if there was any way to screen the HR genes apart from the P genes. If it turns out that the *hrp* genes do not produce HR because the bacterium is growing poorly, then we are not getting to the HR part of the system. *Mills* responded that he thought some of the *hrp* mutants that *Panopoulos* isolated were screened on the basis of the HR negative phenotype, which were subsequently determined to be also pathogenicity minus. He thought some of those mutants grow somewhat, but generally they do not appear to grow very much in the host or non-host. *Keen* stated that it appears not to make any difference if the *hrp* genes are different, as long as the bacteria gain *hrp* competency one way or another. *Mills* stated that it appears that many different kinds of *hrp* gene functions, if lost through mutation, will give a similar phenotype. *Macko* then asked whether it was significant that the electron micrograph shows preference for stomata by the superpiliated strain. *Mills* responded that bacteria enter through stomata or wounds and, although the normal strain also adheres preferentially at that site, the superpiliated strain shows even a higher degree of adhesion of bacteria to the leaf surface, and this could be one of the early steps in the infection process.

Chapter 7
The Role of Indoleacetic Acid Biosynthetic Genes in Tumorigenicity

Tetsuji Yamada, Tomoki Nishino, Tomonori Shiraishi, Tom Gaffney, Frank Roberto, Curt J. Palm, Hachiro Oku and Tsune Kosuge[1]

Various disease symptoms in plants are caused by infection with microorganisms. Among the most well-characterized, on a molecular basis, is hyperplasia. Plant tumors incited by bacterial pathogens include crown gall (*Agrobacterium tumefaciens*), olive knot (*Pseudomonas savastanoi*), bacterial witches' broom (*Corynebacterium fascians*), bacterial gall of Japanese wisteria (*Erwinia milletiae*), and others (Nester et al., 1981; Nester et al., 1984; Kemper et al., 1985; Kosuge et al., 1983; Kado, 1984; Okajima et al., 1974). It has been shown that these hyperplastic tissues arise because of hormone imbalances, mainly in indoleacetic acid (IAA) and cytokinin. These imbalances arise in the host plant after bacterial infection.

Taphrina wiesneri (Rath.) Mix, *T. deformans* (Berkeley) Tulasne, and *T. pruni* Tulasne, the causal agents of cherry witches's broom, peach leaf curl, and plum pocket, respectively, also accumulate large amounts of IAA in culture. Production of IAA and other indole compounds in culture filtrates of several *Taphrina* species has been reported (Hirata, 1957; Kern et al., 1975; Perley et al., 1966). Although convincing evidence is not yet available, it has been speculated that hypertrophy is due to high concentrations of auxin and/or cytokinin produced by these organisms and that these concentrations perturb the hormone balance in the host cells. Mechanisms regulating phytohormone biosynthesis in microorganisms can be related to the expression of pathogen virulence and ultimately to symptoms developed in the host. Hence elucidation of these mechanisms involved in the expression of virulence factors will help clarify the pathogen's fitness to compete and survive in nature. In this chapter, the regulation of IAA production in a gall-producing bacterium, *P. savastanoi*, the role of genes (*iaa*) specifying IAA biosynthesis in tumorigenicity, and their evolution in plant-associated bacteria will be discussed. In addition, the results of a study of IAA production in some fungal systems will be presented.

---

[1]Dedicated to the memory of Dr. Tsune Kosuge (who died on 13 March 1988).

Genes Encoding Indoleacetic Acid Biosynthesis in Tumorigenic Bacteria

1. Common Origin of the iaa Genes of Tumorigenic Bacteria

The association of the tumor-forming bacterium *P. savastanoi* and its hosts, olive and oleander plants, provides a system for studying the molecular basis of plant hyperplasia. Tumor formation is presumed to be a response to high concentrations of IAA introduced into infected tissues by the bacteria (Comai et al., 1982a, 1982b). The genes involved in IAA biosynthesis are borne on a plasmid pIAA, in oleander strains of the pathogen. In olive strains these genes are on the chromosome. Mutants cured of pIAA are weakly virulent on oleander; when transformed with pIAA, they are restored to full virulence (Comai et al., 1980).

Earlier, we found similarities between the *P. savastanoi* and *A. tumefaciens* genes for IAA biosynthesis. The nucleotide sequences of *iaa*M and *iaa*H from *P. savastanoi* show significant homology with the *tms*1 and *tms*2 genes of *A. tumefaciens* (Yamada et al., 1985). These genes in *A. tumefaciens* encode enzymes functionally identical to the tryptophan monooxygenase and indoleacetamide hydrolase encoded by *iaa*M and *iaa*H in *P. savastanoi* (Shroeder, 1987). Significant nucleotide sequence homology has also been observed between a transposable DNA element IS51 bordering the *iaa* genes of *P. savastanoi*, and the Tc region of the T-DNA of *A. tumefaciens* (Yamada et al., 1986). Further, sequences homologous to IS66 from *A. tumefaciens* (Machida et al., 1984) and to an IS-like sequence from *Rhizobium leguminosarum* (Yun et al., 1987) are located adjacent to each other at the IS51-homologous locus in the Tc region of the octopine type Ti plasmid of *A. tumefaciens* strain 15955 (Fig. 1) (Barker et al., 1983). These results strongly suggest that genes responsible for tumorigenicity have a common origin, namely the *iaa* transposon bordered by insertion sequences, and have arisen by duplication and DNA rearrangement.

2. Iaa Genes in Other Plant Pathogenic Bacteria

Isolates of *Pseudomonas syringae* pv. *pisi* and *P. syringae* pv. *syringae* also contain *iaa* genes nearly identical to those of *P. savastanoi* (Ziegler et al., 1987). Hybridization shows that sequences homologous to the *P. savastanoi iaa* genes are possibly contained in *Xanthomonas campestris* pv. *pruni* and *Erwinia herbicola* pv. *milletiae*. Genes encoding the same enzymes, tryptophan monooxygenase and indoleacetamide hydrolase, are also present in plant symbiotic bacteria including some *Rhizobium* and *Bradyrhizobium* species (Sekine et al., 1989; Badenoch-Jones et al., 1982; Ernstsen et al., 1987).

Although the function of the *iaa* genes in these plant-associated bacteria is unknown, it is predicted that their expression influences the plant defense mechanisms in order to optimize their association with the plants. As demonstrated by Smidt and Kosuge (1978), the expression of *iaa* genes is not only responsible for tumorigenicity but is also associated with the ability to

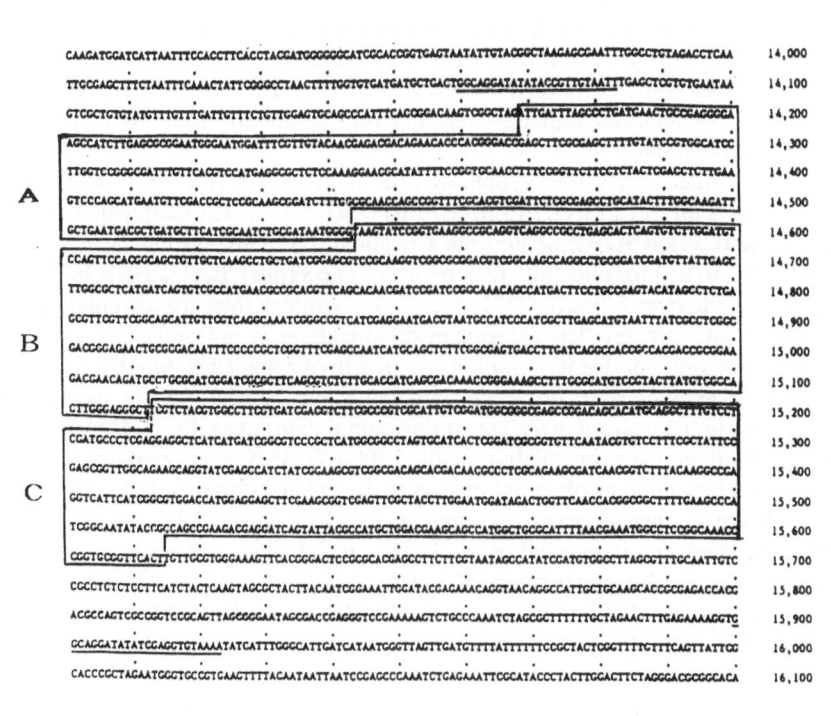

Figure 1. Nucleotide sequence of the T-DNA center region (Tc). Data from Barker et al., 1983. A. Homology to IS66 from *A. tumefaciens*, octopine type, A66 (Machida et al., 1984). B. Homology to IS51 from *P. savastanoi*. C. Homology to the putative IS element in *R. leguminosarum* 300 (Yun et al., 1987). Solid underlines indicate the 25 bp direct repeats located at the right end of the T$_L$-DNA and the left end of the T$_R$-DNA.

detoxify tryptophan analogues such as 5-methyltryptophan. It has been suggested that auxin and cytokinin concentrations in plant tissues invaded by pathogens can affect levels of hydrolytic enzymes such as chitinases and ß-1,3-glucanases which are thought to be involved in plant defense responses (Shinshi et al., 1987). One may also speculate that phytohormone production by the pathogens may alter the production of antimicrobial substances by the plant.

The presence of sequences homologous to *P. savastanoi* IS51 was also investigated in *P. syringae* pv. *pisi*, *X. campestris* pv. *pruni*, and *E. herbicola* pv. *milletiae*. Southern blot analysis indicated that homology was present in some species of *P. syringae* but not in the other organisms.

3. The role of iaa Genes in Tumorigenicity

It has been shown that when *P. savastanoi* strain 2009-3 is cured of its plasmid pIAA, it no longer exhibits virulence. The recombinant strain 2009-3(pIAA), in which pIAA has been reintroduced by co-transformation with RSF1010, regains IAA biosynthesis and virulence (Comai et al., 1980). To determine whether expression of the *iaa* operon is solely responsible for reacquisition of tumorigenicity, we constructed a recombinant plasmid pTET40 (Fig. 2) in which the *P. savastanoi iaa* operon was inserted into the wide host range vector pRK415 (Ditta et al., 1985). This construct was then conjugated into 2009-3. This recombinant did not regain virulence, suggesting that the *iaa* genes alone are not solely responsible for reacquisition of virulence. Some other genes on pIAA may be required, although loss of virulence due to instability of pTET40 cannot be ruled out.

To further verify this hypothesis, the hygromycin resistance gene (*hph*) was fused with *iaa* in pTET40 to give pTOM1 which was conjugated into *A. tumefaciens* strain A208, or into the pTi-cured avirulent *Agrobacterium* strain CBA4301, to test for alteration of virulence. The strain CBA4301(pTOM1) was avirulent, whereas A208(pTOM1) was more virulent than the parental strain. These results suggest that expression of *iaa* alone is not sufficient to incite tumor formation but that it does promote growth of established tumors.

Another attempt to prove the hypothesis was carried out by inoculating these recombinant *Agrobacterium* strains onto potato tuber discs. Hyperplasias which resulted from inoculation with A208(pTOM1) were characterized by hairy-root formation as observed in tobacco inoculated either with *A. rhizogenes* or with the *A. tumefaciens* rooty mutant, in which *tmr* had been insertionally inactivated by *Tn5* (Garfinkel et al., 1981). It is known that rooty type crown galls produced by inactivation of *tmr* contain high ratios of IAA/cytokinin (Akiyoshi et al. 1983; Nester et al., 1984; Schroeder, 1987). Since *A. tumefaciens* A208(pTOM1) accumulates large amounts of IAA in culture

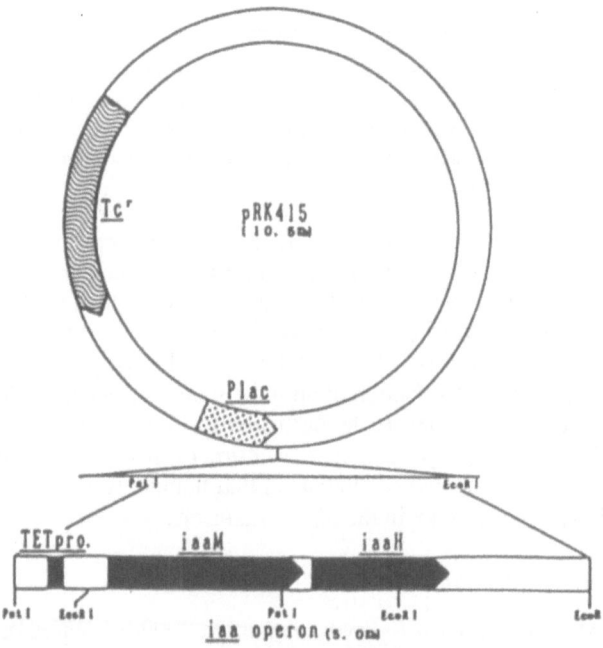

Figure 2. Schematic diagram of the plasmid pTET40. The *P. savastanoi iaa* operon was inserted into the wide host range vector pRK415.

media supplemented with tryptophan, it may have the capacity to create an environment in which there is a high ratio of IAA/cytokinin at the infection site. However, the exact mechanism of this phenomenon remains undetermined.

4. Regulation of the iaa Operon in P. savastanoi

Promotor activity of the 5'-flanking region of *iaa*M was monitored by constructing an *iaa*M-*lac*Z reporter fusion containing successive deletions in the promoter region. The results indicate that the promotor is located more than 400 bp upstream of the 5'-end of *iaa*M and that there is possibly a single 228 bp open reading frame between the promotor and *iaa*M. In *E. coli* the *iaa* promotor is one-fifth as active as in *P. savastanoi*. It contains sequences homologous to the -10 and -35 regions of the consensus (sigma 70) *E. coli* promotor (unpublished results).

Several other lines of evidence support the hypothesis that the *iaa* operon contains another transcriptional unit extending from the 3'-end of *iaa*M to the end of *iaa*H (unpublished results). Thus, IAM hydrolase activity is present in mutants in which the insertion sequences, IS51 or IS52, have been integrated into *iaa*M. The level of activity depends upon the position of IS integration; the closer the point of integration to *iaa*H, the lower the activity. Furthermore, possible *iaa*H transcripts can be seen in in vitro transcription/translation assays with a plasmid in which the kanamycin resistance gene (Km^r) is inserted into *iaa*M with opposite polarity to the CAT promotor (Palm et al., 1989). Finally, S_1 nuclease and primer extension mapping locate the putative *iaa*H transcriptional initiation site to the 3'-end of *iaa*M (unpublished results).

It has been shown that tryptophan 2-monooxygenase is product-inhibited by indoleacetamide ($K_i = 1$ mM) and feedback-inhibited by IAA ($Ki = 3$ mM) (Hutcheson et al., 1985). However, the substrate, tryptophan, induced neither tryptophan 2-monooxygenase activity nor the transcription of *iaa*. Moreover, expression of *iaa* does not seem to be subject to catabolite repression (unpublished results). Furthermore, even under tryptophan limitation, the level of *iaa*-mRNA accumulation is similar to that seen under tryptophan excess. A key question is how tryptophan secondary metabolism, namely IAA biosynthesis, is controlled via its primary metabolism, namely protein synthesis.

An IAA superproducing mutant of *P. savastanoi* has been reported which accumulates two to three times more IAA than its parent, and displays greater pathogenicity (Smidt et al., 1978). The superproducing phenotype is expressed only under the conditions of tryptophan limitation that one would expect to find in plant tissues, which are normally very low in tryptophan. The internal tryptophan pool size in the superproducer is 25- to 30-fold greater than that of the wild-type bacteria, but the mutant has levels of tryptophan 2-monooxygenase similar to its parent. These results indicate that the control of tryptophan pool size determines the fate of IAA biosynthesis as well as virulence (Kosuge et al., 1983; Kosuge et al., 1985; Comai et al., 1982a).

One might speculate that constitutive expression of *iaa* results in loss of energy to the cell economy. However, it seems apparent that relaxed control of tryptophan synthesis permits more IAA synthesis in the bacterium and that this is manifested as increased virulence.

Indoleacetic Acid Biosynthesis in *Taphrina*

1. Pathway of IAA Biosynthesis in <u>Taphrina wiesneri, T. deformans</u> and <u>T. pruni</u>

T. wiesneri, T. deformans, and *T. pruni,* the causal agents of cherry witches' broom, peach leaf curl, and plum pocket respectively, produce IAA from tryptophan (Trp) via indole-3-pyruvic acid (IPyA) and indole-3-acetaldehyde (IAAld) as intermediates (Fig. 3). They also convert indole-3-acetonitrile (IAN) to IAA. In *T. wiesneri,* the enzyme tryptophan aminotransferase, which converts Trp to IPyA, is not induced by Trp (Table 1). As discussed above for *P. savastanoi,* the enzyme catalyzing the first step of the conversion of tryptophan to IAA is not induced by the substrate. Again, it is likely that tryptophan aminotransferase is under relaxed control. Although it is common for the first step in a multienzyme reaction sequence to be modulated in order to provide maximum metabolite economy, this does not appear to be true for production of IAA by *Taphrina.*

2. Clofibric Acid Resistant Mutants of <u>Taphrina</u>

Mutants of *T. wiesneri* and *T. deformans* resistant to the antiauxin, clofibric acid, were selected. These mutants have a characteristic morphology in which the length of budding conidia is only one-third that of those of the wild-type. Mutants accumulate indole-3-ethanol (IEt) by reduction of IAAld instead of its oxidation to IAA. HPLC analysis showed that most tryptophan entering IAA secondary metabolism is converted into IAA. Very little was converted into IEt, suggesting that one of the main regulatory steps of IAA biosynthesis in *Taphrina* is the conversion of IAAld into IAA or IEt. This hypothesis is supported by a report of feedback inhibition of IEt oxidase by IAA in plants (Rivier et al., 1987).

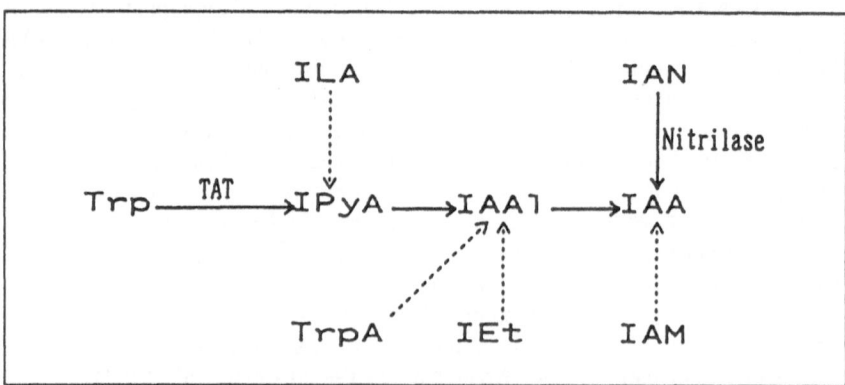

Figure 3. Biosynthetic pathway in *Taphrina* of IAA and IAA related indoles. Abbreviations are defined in the text.

Table 1. Changes in tryptophan aminotransferase activity and IAN nitrilase activity in cell-free extracts of *Taphrina* spp. supplemented with tryptophan or IAN

	Tryptophan aminotransferase					IAN nitrilase				
	Tryptophan (M)[a]					Indoleacetonitrile (mM)[b]				
	0	0.2	0.5	1.0		0	0.2	0.5	1.0	2.0
T. weisneri	7.86[c]	7.42	7.05	7.08		0	79	206	265	439
T. deformans	4.57	6.79	5.40	5.52		0	107	271	395	602
T. pruni	5.43	6.70	7.37	6.70		0	60	148	292	632

[a] Concentrations of Trp (mM) added to the culture.

[b] Concentration of IAN (mM) added to the culture.

[c] One unit of enzyme activity is defined as an increase of 1.0 A_{525} in the Salkowski reaction (Ehmann, 1977) per ng of protein.

The chromosomal DNA structure of wild-type and CFA resistant mutants was examined by pulsed field agarose gel electrophoresis. Extensive chromosomal DNA rearrangements were seen in CFA resistant mutants (Fig. 4).

Wild-type *T. wiesneri* carries at least three chromosomal DNAs whose sizes are estimated to be approximately 1430 (I), 1380 (II), and 1300 kb (III), whereas a CFA resistant (Iaa-) mutant carries at least four chromosome DNAs of 1430, 1380, 900, and 420 kb. It is assumed that in the mutant, chromosomal DNA (III) has been divided into two subchromosomal DNAs of 900 and 420 kb. Similarly, wild-type *T. deformans* carries at least three chromosomal DNAs whose sizes are 1530 (I), 1470 (II), and 1300 kb (III), whereas an Iaa⁻ mutant contains at least five chromosomal DNAs estimated to be 1470, 1380, 1120, 900, and 420 kb. It should be noted that the three chromosomal DNA bands corresponding to 1380, 900, and 420 kb are identical in Iaa⁻ mutants of *T. wiesneri* and *T. deformans*, although the band patterns of the parents are different. These results suggest that tumorigenic *Taphrina* species share extensive chromosomal homology and that mutations to the Iaa⁻ phenotype may result from similar chromosomal rearrangements at sites sharing common DNA.

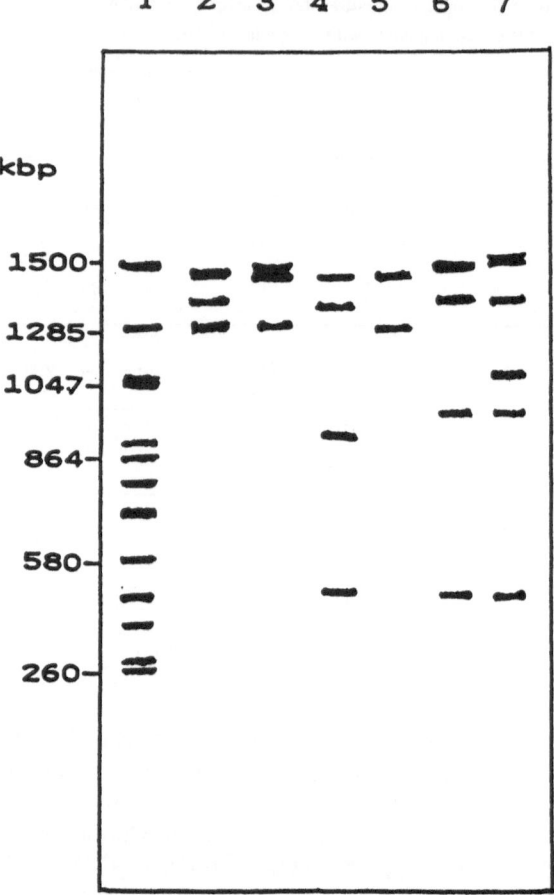

Figure 4. Pulse field gel electrophoresis of chromosomal DNA from wild type *Taphrina*, and CFA resistant mutants. Cells were grown for 2 days in YPD medium, harvested by centrifugation, washed in 0.2 *M* sorbitol, 20 m*M* EDTA, and 50 m*M* Tris-Cl (pH 7.5), and resuspended in 1.2 *M* sorbitol, 10 m*M* EDTA, and 100 m*M* citrate (pH 5.8) at a cell concentration of 150 mg/ml. Two percent melted agarose LGT (Beckman) was added to make 1% agarose and the mixture was solidified in a template. Agarose blocks containing *Taphrina* cells were incubated in osmotic stabilizing solution containing 1 mg/ml Novozyme 234 (Novo Industria), 1 mg/ml zymolyase 20T (Sigma), 4 μl/ml β-glucuronidase type H2 (Sigma) and 60 mg/ml of BSA, at 26°C for 24 hours. Electrophoresis was performed in Gene-Line[TM] LE agarose (Beckman) (1%) at 330 V/MA for 30 minutes with 4-second intervals followed by 220 V/MA for 18 hours with 60-second intervals. Lanes contained DNA from: *Saccharomyces cereviseae* DKD-5D (size marker), lane 1; wild-type *T. wiesneri*, lane 2; wild-type *T. deformans*, lane 3; wild-type *T. pruni*, lane 4; CFA resistant *T. wiesneri*, CFA-35, lane 5; CFA resistant *T. wiesneri*, CFA- 12, lane 6; CFA resistant *T. deformans* CFA-101, lane 7. Characteristic bands appeared in Iaa⁻ mutants of *T. wiesneri* and *T. deformans* as indicated by arrows.

Concluding Remarks

Phytohormone (and particularly IAA) production in tumorigenic bacteria and fungi plays a very important role in plant hyperplasia. Production of IAA is widespread in the plant kingdom although the level of production varies from one organism to the other. An important question is how the genes encoding the enzymes for IAA biosynthesis have evolved and diverged. The study of the structure and the regulation of *iaa* genes in *P. savastanoi* demonstrates that these genes are typically prokaryotic. On the other hand, comparable genes in the T-DNA of *A. tumefaciens* possess eukaryotic characteristics and are only expressed when the T-DNA is integrated into the plant genome. What is the biological significance of the fact that a prokaryote, *Agrobacterium*, carries eukaryotic-type genes necessary for virulence, and how were these genes originally acquired? Important clues reside in the sequences of the *iaa* genes of *P. savastanoi* and the comparable T-DNA genes in octopine *A. tumefaciens* strains. It is noteworthy that sequences homologous to insertion sequences of *P. savastanoi*, *A. tumefaciens*, and *R. leguminosarum* exist adjacent to each other in the T-DNA center (Tc) region of *A. tumefaciens*, which is not normally integrated into the plant genome. The data strongly support the hypothesis that tumorigenicity genes have a common origin, tentatively named the *iaa* transposon, and have undergone rearrangement in tumorigenic bacteria. It will be of interest to examine the structure of fungal *iaa* genes to further elucidate the evolution of hypertrophic virulence in plant pathogens.

References

Akiyoshi, D.E., et al., 1983, Cytokinin-auxin balance in crown gall tumors is regulated by specific loci in the T-DNA, *Proc. Natl. Acad. Sci. USA* **80**:407-411.

Badenoch-Jones, J., et al., 1982, Mass spectrometric quantification of indole-3-acetic acid in *Rhizobium* culture supernatants: relation to root curling and nodule initiation, *Appl. Environ. Microbiol.* **44**:275-280.

Barker, R.F., et al., 1983, Nucleotide sequence of the T-DNA region from the *Agrobacterium tumefaciens* octopine Ti plasmid pTi15955, *Plant Mol. Biol.* **2**:335-350.

Comai, L., and Kosuge, T., 1980, Involvement of plasmid deoxyribonucleic acid synthesis in *Pseudomonas savastanoi*, *J. Bacteriol.* **143**:950-957.

Comai, L., and Kosuge, T., 1982a, Metabolic regulation in plant pathogen interactions from the perspective of the pathogen. *In* Plant Infection: the Physiological and Biochemical Basis (Y. Asada et al., eds.), Springer-Verlag, Berlin, pp.175-186.

Comai, L., Kosuge, T., 1982b, Cloning and characterization of *iaaM*, a virulence determinant of *Pseudomonas savastanoi*, *J. Bacteriol.* **148**:40-46.

Ditta, G.T., et al., 1985, Plasmids related to the broad host range vector, pRK290, useful for gene cloning and for monitoring gene expression, *Plasmid* **13**:1249-1254.

Ehmann, A., 1977, The Van Urk-Salkowski reagent–a sensitive and specific chromatogenic reagent for silica gel thin-layer chromatographic detection and identification of indole derivatives, *J. Chromatogr.* **132**:267-276.

Ernstsen, A., et al., 1987, Endogenous indoles and the biosynthesis and metabolism of indole-3-acetic acid in cultures of *Rhizobium phaseoli*, *Planta*, **171**:422-428.

Garfinkel, D.J., et al., 1981, Genetic analysis of crown gall: Fine structure map of the T-DNA by site directed mutagenesis, *Cell* 27:143-153.

Hirata, S., 1957, Studies on the phytohormone in the malformed portion of the diseased plants. 3. Auxin formation on the culture grown *Exobasidium, Taphrina,* and *Ustilago* spp, *Phytopath. Soc. Japan,* 22:153-158.

Hutcheson, S.W. and Kosuge, T., 1985, Regulation of 3-indoleacetic acid production in *Pseudomonas savastanoi, J. Biol. Chem.* 260:6281-6287.

Kado, C.I., 1984, Phytohormone-mediated tumorigenesis by plant pathogenic bacteria. *In* Plant Gene Research; Genes Involved in Microbe-Plant Interactions (Verma, D.P.S. and Hohn, T., eds.), Springer-Verlag, Wein-New York pp. 311-336.

Kemper, E., et al., 1985, T-DNA-encoded auxin formation in crown-gall cells, *Planta* 165:257-262.

Kern, H. and Naef-Roth, S., 1975, Zur Bildung von Auxinen and Cytokininen durch *Taphrina-Arten. Phytopathol. Z.,* 83:193-222.

Kosuge, T., et al., 1983, Virulence Determinants in Plant-Pathogen Interactions. *In* Plant Molecular Biology (Goldberg, R.B., et al., eds.), Alan R. Liss, , New York, pp. 167-177.

Kosuge, T., et al., 1985, pIAA1, a virulence plasmid in *Pseudomonas savastanoi, In* Plasmids in Bacteria (Helinski, D.R., et al., eds.), Plenum Press, New York pp. 807-813.

Machida, Y., et al., 1984, Nucleotide sequence of the insertion sequence found in the T-DNA region of mutant Ti plasmid pTiA66 and distribution of its homologs in octopine Ti plasmid, *Proc. Natl. Acad. Sci. USA* 81:7495-7499.

Nester, E.W. and Kosuge, T., 1981, Plasmids specifying plant hyperplasias, *Annu. Rev. Microbiol.* 35:531-565.

Nester, E.W., et al., 1984, Crown gall; a molecular and physiological analysis, *Annu. Rev. Plant Physiol.* 35:387-413.

Okajima, T., et al., 1974, Histopathology of bacterial gall of Japanese wisteria (*Wisteria floribunda*), *Bot. Mag., Tokyo,* 85:177-185.

Palm, C.J., Gaffney, T. and Kosuge, T., 1989, Cotranscription of genes encoding indoleacetic acid production in *Pseudomonas savastanoi, J. Bacteriol.* 171:1002-1009.

Perley, J.E. and Stowe, B.B., 1966, On the ability of *Taphrina deformans* to produce indoleacetic acid from tryptophan by the way of tryptamine, *Plant Physiol.* 41:234-237.

Rivier, L. and Crozier, A., 1987, Principles and Practice of Plant Hormone Analysis, Academic Press, New York.

Sekine, M., et al., 1989, Molecular cloning of a gene for indole-3-acetamide hydrolase from *Bradyrhizobium japonicum, J. Bacteriol.* 29:867-874.

Shinshi, H., Mohnen, D., and Meins, F., 1987, Regulation of a plant pathogenesis-related enzyme: inhibition of chitinase and chitinase-mRNA accumulation in cultured tobacco tissues by auxin and cytokinin, *Pro. Natl. Acad. Sci. USA* 84:89-93.

Shroeder, J., 1987, Plant hormones in plant-microbe interactions. *In* Plant-Microbe Interaction, (Kosuge, T. and Nester, E.G., eds.), Macmillan, New York, pp. 40-63.

Smidt, M. and Kosuge, T., 1978, The role of indole-3-acetic acid accumulation by alpha methyl tryptophan-resistant mutants of *Pseudomonas savastanoi* in gall formation on oleanders, *Physiol. Plant Pathol.* 13:203-214.

Yamada, T., et al., 1985, Nucleotide sequences of the *Pseudomonas savastanoi* indoleacetic acid genes show homology with *Agrobacterium tumefaciens* T-DNA, *Proc. Natl. Acad. Sci. USA* **82**:6522-6526.

Yamada, T., Lee, P-D., and Kosuge, T., 1986, Insertion sequence elements of *Pseudomonas savastanoi*: nucleotide sequence and homology with *Agrobacterium tumefaciens* transfer DNA, *Proc. Natl. Acad. Sci. USA* **83**:8263-8267.

Yun, A.C., et al., 1987, A plasmid sequence from *Rhizobium leguminosarum* 300 contains homology to sequences near the octopine TL-DNA right border, *Mol. Gen. Genet.* **209**:580-584.

Ziegler, S.F., White, F.F., and Nester, E.W., 1987, Genes involved in indoleacetic acid production in plant pathogenic bacteria. *In* Plant Pathogenic Bacteria, (Collmer, A. et al., eds.), Martinius Nijhoff Publishers, Dordrecht, The Netherlands.

Summary of Discussion of Yamada's Paper

Keen initiated the discussion by asking whether the deletion of the first short ORF in the *iaa* operon results in the increased expression of the downstream ORFs. He said that if it did it would be consistent with the idea that the first ORF acts as an illicit start. *Yamada* replied that deletion of the first ORF does not result in higher expression of the downstream ORFs. *Nester* asked whether *Yamada* could speculate on the basis for the differences in host specificity of different strains (olive and oleander) and further asked about the nature of these differences. *Yamada* said that he could not answer that question because not enough data were available although in the strains they work on the structure of *iaa* genes in the chromosome and the plasmid is the same. *Nester* then asked whether there was a 20-fold increase in the tryptophan pool in the mutants which over-produced IAA. *Yamada* said yes, and added that the concentration of tryptophan controls the degree of virulence of the pathogen. The IAA superproducer mutants have a very high level of free tryptophan which is also highly virulent. Thus, what controls tryptophan biosynthesis controls IAA production and virulence. *Keen* then asked if the superproducer mutants have been characterized to determine whether the mutation was in the tryptophan pathway or the block was earlier in the pathway. *Nester* followed up on *Keen's* question by asking whether the mutants were selected on the basis of their resistance to a tryptophan analog. *Yamada* said that α-methyltryptophan was used for that purpose. *Nester* then said that the mutation may be in the anthranilate synthase rather than in earlier steps.

Mills asked whether anyone had looked at the organization of the chromosomal genes in a cosmid clone to determine whether or not they had come onto pIAA1 by recombination using the IS elements IS51 or IS52. *Yamada* said that to his knowledge no one has worked on that. *Van Alfen* inquired whether the biosynthetic pathway in *Taphrina* was the same as in the bacterium. *Yamada* said that he did not believe so. *Van Alfen* continued by asking whether the different spore sizes shown by *Yamada* had anything to do with the IAA mutations. *Yamada* said that they had complemented these mutants in the medium by providing extraneous IAA which did not change the spore sizes indicating that mutations in the *iaa* genes did not have anything to do with the spore size. *Bennetzen* asked that if IAA synthesis was constitutive in not only nonpathogens but even in the absence of plants by the pathogens as well, what was the metabolic fate of this IAA; why were these organisms making the IAA if there were no plants around? *Yamada* replied that he did not know the answer and speculated that constitutive production of IAA may be essential for the initial interaction between the pathogen and the plant. *Tsuyumu* mentioned that someone had claimed that in *E. coli* that IAA acts like cyclic AMP. *Yamada* replied that he was aware of that report but that the work was never elaborated upon.

Chapter 8
Molecular Analysis of Phaseolotoxin Production in *Pseudomonas syringae* pv. *phaseolicola*

Suresh S. Patil, Karla B. Rowley, H.V. Kamdar,
David Clements, Morton Mandel, and Tom Humphreys

Pseudomonas syringae pv. *phaseolicola* is the causal agent of halo blight of bean (*Phaseolus vulgaris* L.). The disease is manifested at cool temperatures when the bacteria enter the plant through stomata in the leaves. At the site of infection a green-yellow chlorotic zone or halo, ornithine accumulation and growth inhibition (Patil, 1974) occur. Systemic infection results in venetion of leaves and stunting of the growing point (Hoitink et al., 1966; Rudolph et al., 1966). The chlorosis, growth inhibion, and venetation are caused by an extracellular, nonspecific phytotoxin known as phaseolotoxin (Fig. 1) (Mitchell, 1984; Moore et al., 1984). It is not required for pathogenicity of *P.s.* pv. *phaseolicola* since naturally occurring toxin-deficient strains as well as Tox⁻ mutants are still capable of producing water-soaking but not the other symptoms. (Rudolph et al., 1966; Patil, et al., 1974; Russell, 1975; Quigley et al., 1985; Peet et al., 1986). During the past 15 years, extensive studies on the physiology, biochemistry, chemical structure and mode of action of phaseolotoxin have been reported (Mitchell, 1984). However, little is known about the biochemistry or genetics of its production. Phaseolotoxin is a potent inhibitor of the enzyme, ornithine carbamoyltransferase (OCT) of bean in vitro (Patil et al., 1970; Kwok et al., 1979). Octicidin, a peptidase degradation product of phaseolotoxin (Moore et al., 1984) which also inhibits OCT is responsible for symptom production in infected bean plants (Mitchell et al., 1977). Octicidin lacks the alanyl-homoarginine dipeptide moiety and is approximately 20 times more active against bean OCT than is phaseolotoxin (Kwok et al., 1982). Octicidin-induced chlorosis and growth inhibition in bean tissues can be reversed by L-citrulline and L-arginine (Patil et al., 1972).

Phaseolotoxin also inhibits OCT of *Escherichia coli* K-12, and this inhibition can be reversed by the addition of L-citrulline and L-arginine to the medium. This finding was used to develop a phaseolotoxin-specific microbial assay (Staskawicz et al., 1979).

Because phaseolotoxin inhibits a key enzyme in the arginine biosynthetic pathway, it is potentially autotoxic to *P.s.* pv. *phaseolicola*. However, it was found that the toxin is produced optimally at 18-20° C whereas no detectable

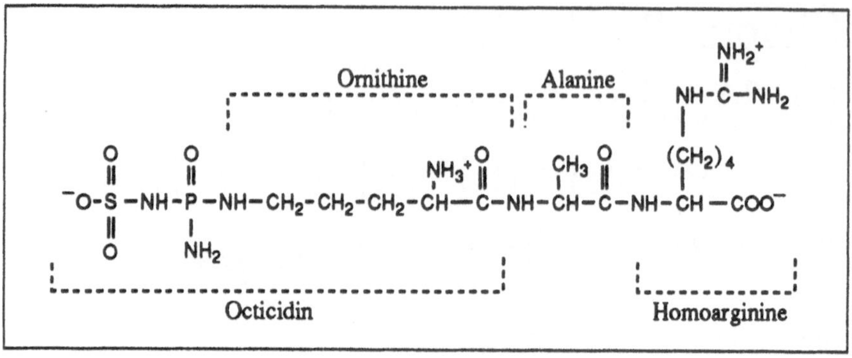

Figure 1. Structure of Phaseolotoxin

amounts are produced at 30^0 C (Goss, 1970; Mitchell, 1978; Staskawicz et al., 1979; Nuska et al., 1989), suggesting that an immunity mechanism exists in *P.s.* pv. *phaseolicola* at temperatures at which the toxin is produced. Subsequently, it was found that two physically distinct OCTs which differed in their sensitivity to phaseolotoxin were produced by *P.s.* pv. *phaseolicola* (Ferguson et al., 1980; Staskawicz et al., 1980; Jahn et al., 1985; Templeton et al., 1980). The phaseolotoxin-insensitive OCT was detectable only at toxin-producing temperatures (Staskawicz et al., 1980). It has also been suggested that resistant plant tissues may be involved in the regulation of toxin production (Oguchi et al., 1981).

Little is known about the number of genes involved in toxin biosynthesis, their genomic organization or their regulation. Recently, Peet et al., (1987) isolated the gene for the phaseolotoxin-insensitive OCT. They found that the ~22 kb *KpnI* clone also complemented Tn*5* Tox⁻ mutants, suggesting that at least some of the genes for toxin production are present in this fragement. We have been studying the genetics of phaseolotoxin production with a view to elucidating the number of genes and their organization in the biosynthetic pathway of this toxin, and to determine how it is regulated.

Isolation and Suppression of Tox⁻ Mutants

Our genetic studies of phaseolotoxin biosynthesis began with the production of toxin-deficient (Tox⁻) mutants of *P.s.* pv. *phaseolicola*, using Tn*5*, UV and EMS mutagenesis. The survivors were selected on a minimal medium (medium used to select Tn*5* mutants was supplemented with kanamycin). This procedure ensured against the selection of ornithine and alanine auxotrophs, two amino acids present in the tripeptide moiety of the toxin. Thus, mutations must be in homoarginine biosynthetic genes, in the genes that encode the inorganic moiety, or in regulatory genes. A total of 130 EMS and one UV mutant of strain G50-1 (Race 2) and five Tn*5* mutants from strain PDDC4612-1 (Race 2) were isolated.

Genomic DNA from two strains, PDDCC4612-1 and G50-1, of *P.s.* pv. *phaseolicola* was digested with appropriate restriction enzymes and cloned into the broad host range cosmids pLAFR1 and pLAFR3 (Staskawicz et al., 1987), respectively. Each library was mated en masse into 10 randomly selected independent EMS mutants or the UV mutant. Transconjugants showing the Tox$^+$ phenotype were selected using the *E. coli* growth inhibition assay and the cloned fragments were subjected to restriction endonuclease analysis. One clone (pDC938) was obtained by mating the PDDCC4612-1 cosmid genomic library en masse with a UV mutant (G50-1 Tox$^-$; Romeo et al., 1984) This clone also suppressed the Tox$^-$ phenotype of the 10 independent EMS Tox$^-$ mutants but failed to suppress any of the Tn5 Tox$^-$ mutants. In addition, seventeen other clones were obtained from the G50-1 cosmid genomic library by mating it with the 10 EMS mutants of G50-1. Analysis of the clones revealed that three (pHK31, pHK91, and pHK121) had identical restriction patterns but suppressed different mutants, G50-1A3, G50-1A9, and G50-1A12, respectively. Conversely, two clones (pHK81 and pHK91) with different restriction patterns complemented the same mutant, G50-1A9. This suggested that mutants used in this study, although independently obtained, may be similar in nature. A cross-complementation study of the clones using an additional 70 independent EMS mutants was undertaken. Four of the clones were found to be unstable and were not used in subsequent experiments. However, it was found that all of the remaining clones, regardless of the mutant-clone combination from which they were obtained suppressed every mutant. A comparison of the restriction patterns of the clones revealed that most were dissimilar and probably heterologous. This was confirmed by Southern blot analysis using as probes, three clones which did not share any restriction fragments. Based on these results, the clones were arranged into three distinct classes. Clones within a class share sequence homology, whereas no homology is detected between clones of different classes. Class I consists of clones pHK31, pHK91, pHK120, and pHK121, which harbor DNA inserts that hybridize to the probe DNA from pHK120. Class II consists of clones pHK10, pHK41, pHK42, pHK51, pHK81, pHK92, and pHK111 and hybridize to the probe DNA from clone pHK111. Class III consists of clones pDC938, pHK32, and pHK52 and hybridize to probe DNA from clone pDC938.

Clones belonging to these three classes were tested against five Tn5 Tox$^-$ mutants of strain PDDC4612-1. Clone pHK120 from class I suppressed all five Tn5 mutants and clones pHK31, pHK91, and pHK121 (also from Class I) suppressed all Tn5 Tox$^-$ mutants except 4612-1T11. Clones from Class II and Class III did not suppress any of the Tn5 mutants.

Molecular Analysis of Class III Clones

Class III genomic clone pDC938 was subcloned to determine the smallest biologically active fragment. pDC938 contains a 24.4 kb *Eco*R1 fragment from the genome of 4612-1 (Fig. 2). Subcloning revealed that the smallest fragment capable of restoring toxin production in the Tox$^-$ mutant was 1.0 kb and contained approximately 500 bp of DNA on either side of the unique *Bam*H1 site. Furthermore, deleting the 500 bp *Bg*lII-*Bam*H1 fragment resulted in loss of biological activity.

Tn3HoHo1 mutagenesis (Stachel et al., 1985) of the 4.9 *Bg*lII- *Eco*R1 fragment at the left end of the insert was done to further delineate the region of biological activity. We were not able to obtain any Tn5 insertions in the biologically active region of this fragment and since Tn3HoHo1 has a higher frequency of insertion it was chosen for mutagenizing the fragment. Twenty-one insertions were found in a 2.8 kb *SalI-Bg*lII fragment which encompasses the 1.0 kb fragment known to be biologically active. When these were mated into the G50-1UV Tox⁻ mutant, four insertions, located within 100 bp of the unique *Bam*H1 site, abolished biological activity. The four Tox⁻ insertions spanned a region of between 350 - 400 bp. Orientation of the four insertions which abolished biological activity was determined with respect to the promoterless β–galactosidase reporter gene. Results of β-galactosidase studies were inconclusive because neither orientation showed high levels of β-galactosidase activity, although there was a slight increase in activity when the Tn3HoHo1 was inserted so that transcription would be from the *Bg*lII toward the *Eco*RI site (right to the left, Fig. 2). If this region does contain a gene it must be quite small.

To determine whether this clone contains a primary structural gene required for the production of phaseolotoxin, an interposon was inserted into the unique *Bam*H1 site of the fragment to inactivate biological activity and marker exchanged into the genome of the wild-type strain. Marker exchange experiments using the Tn3HoHo1 insertional mutants were difficult to interpret because *Pseudomonas* is naturally resistant to ampicillin. However, Tn3HoHo1 mutagenesis and deletion cloning had suggested that the unique *Bam*H1 site in this region of pDC938 must remain intact for biological activity. Since *Pseudomonas* is sensitive to streptomycin and spectinomycin, the Ω fragment from pHP45Ω containing the Sm^{r}Sp^r resistance genes was inserted into the *Bam*H1 site of pDC938. Transconjugants in the UV Tox⁻ mutant failed to produce toxin. When transconjugants containing this plasmid were marker exchanged into the genome of the wild-type strain, the resulting homogenotes continued to produce the toxin, indicating that the mutated region in pDC938 is not directly involved in the production of toxin.

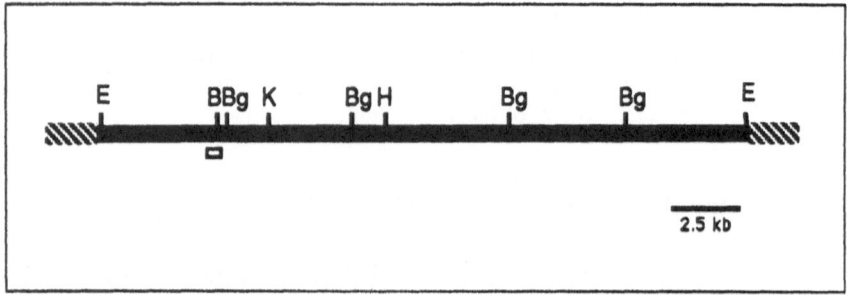

Figure 2. Restriction map of pDC938. The horizontal bar below the map represents the region of DNA involved in biological activity. E, *Eco*RI; B, *Bam*HI; Bg, *Bg*lII; K, *Kpn*I; H, *Hind*III.

Models Proposed to Explain Suppression of EMS Mutants by Class III Clones and Their Testing

1. "Multicopy Suppression" Model

According to this model, if enzyme A has overlapping substrate specificity with enzyme B, and the gene encoding enzyme A is mutated, such a mutation can be suppressed provided an amplified amount of enzyme B is available in the mutant. When the gene encoding enzyme B is cloned on a multicopy plasmid amplification of B occurrs in the A⁻ mutant cell which compensates for the missing enzyme A. Berg et al. (1988), have termed this phenomenon "multicopy suppression". They found that when heterologous sequences from *E. coli* are cloned on multicopy plasmids, they suppress *E. coli* K-12 alanine auxotrophs. They found that $avtA^+$, which encodes transaminase C, $tyrB^+$, which encodes transaminase D, and $alaB^+$, which encodes alanine-glutamate transaminase, suppress the same alanine auxotroph when cloned on a multicopy plasmid (Wang et al., 1987).

Since the identity of the enzymes involved in phaseolotoxin production is unknown, we cannot test this hypothesis in our system. Although we can rule out mutations in ornithine or alanine genes because all the Tox⁻ mutants suppressed by clones from classes II and III are prototrophic, it is possible that a mutation occurs in the homoarginine biosynthetic pathway since homoarginine, part of the tripeptide moiety (ornithinyl-alanyl-homoarginine) of phaseolotoxin (Fig. 1), is not essential for growth. Although little is known about this pathway in bacteria, in mammals, homoarginine appears to be synthesized from lysine, It has been proposed that a transamidinase is involved in the amidination of lysine to homoarginine (Ryan et al., 1969), and, like the transaminases, transamidinases may have overlapping substrate specificities so that when their genes are cloned onto multicopy plasmids and mated into the EMS and UV Tox⁻ mutants "multicopy suppression" occurs. Alternatively, the EMS and UV mutations may occur in the genes for the inorganic portion of the toxin and this mutation is suppressed by "multicopy suppression" when the cosmid clones are mobilized in the mutants.

2. Repressor Binding Site Model

According to this model, toxin production is regulated by a repressor. This model suggests that suppression of the EMS and UV Tox⁻ mutants by the heterologous clones occurs because of the presence of a DNA sequence which binds to and titrates a repressor binding protein. As discussed above, toxin is not produced at 30°C when the toxin-sensitive OCT is present in the cell (Staskawicz et al., 1980). It is possible that *Pseudomonas* may have evolved a regulatory mechanism to insure that toxin is not produced at higher temperatures. The fact that all the EMS and UV mutants were suppressed by all three classes of clones suggests that a single genetic function was mutated. We speculate that if this mutation results in constituitive production of the repressor, then fortuitous cloning of the repressor binding site would, in trans, titrate the repressor protein and allow production of toxin. This has been shown to be the case for the *gal* operon. Irani et al., (1983) found that when multiple copies of the *gal* operator are present in a cell the *gal* operon becomes constituitive because the repressor is completely titrated. If this hypothesis is correct, fragments from clones from all three classes must contain one or more binding sites and the Southern hybridization studies should have confirmed this. Since the repressor binding sequences on the different clones would most likely not be 100% homologous and would be quite small, the fact that it was not detected in the Southern hybridization analysis is not surprising given the stringency conditions used in our experiments.

3. Testing the Repressor Binding Site Model

We tested the repressor binding site model by doing the following experiment. Clone pDC938 was mated into *P.s.* pv. *phaseolicola* strain G50-1. The microbial assay was conducted to test the ability of the wild-type strain, the transconjugant containing pDC938 with the interposon (which insertionally inactivated the biologically active region), and the transconjugant containing pDC938 alone to produce toxin at 18^0 C and at 30^0 C. If the clone contains a sequence which binds to and "titrates" a repressor protein this clone should also "titrate" the repressor in the wild-type strain at 30^0 C when, according to this hypothesis, the repressor is being produced. Whereas all bacteria produced toxin at 18^0 C, only the transconjugant containing pDC938 produced toxin at 30^0 C. While this result does not rule out the first hypothesis, it strongly supports the repressor binding site hypothesis.

Molecular Analysis of Class I Clones

As stated above, Class I clones complemented Tn5 mutants as well as the EMS and UV mutants. Since the clone pHK120 complements all of the Tn5 mutants, this clone was chosen for molecular analysis. Tn5 mutagenesis (Ditta, 1986) of pHK120 was carried out in *E. coli* and mobilized en masse into wild-type *P.s.* pv. *phaseolicola* by triparental mating. Marker exchange of 55 Tn5 transposon insertions produced 31 Tox⁻ homogenotes, supporting our contention that transposon insertion had occurred in genes involved in toxin production. Preliminary mapping of the insertions revealed that they spanned the entire 25 kb region of pHK120, suggesting that the structural genes for toxin production are clustered in this fragment. Peet et al., (1987) have also isolated a clone that contains a cluster of genes that complements several of their Tn5 Tox⁻ mutants of *P.s.* pv. *phaseolicola*. Saturation mutagenisis of the insert in pHK120 and its marker exchange in the wild-type genome is currently in progress.

References

Berg, C.M., et al., 1988, Acquisition of new metabolic capabilities: Multicopy suppression by cloned transaminase genes in *Escherichia coli* K-12, *Gene* **65**:195-202.

Ditta, G., 1986, Tn5 mapping of *Rhizobium* nitrogen fixation genes, *Meth. in Enzymol.*, **118**:519-528.

Ferguson, A.R., Johnston, J.S., and Mitchell, R.E., 1980, Resistance of *Pseudomonas syringae* pv. *phaseolicola* to its own toxin, phaseolotoxin, *FEMS Microbiol. Lett.* 7:123-125.

Goss, R.W., 1970, The relation of temperature to common and halo blight of beans, *Phytopathology* 30:258-264.

Hoitink, H.A.S., Pelletier, R.C., and Coulson, J.G. 1966, Toxemia of halo blight of beans, *Phytopathology.* 56:1062-1065.

Irani, M.L., et al., 1973, Cyclic AMP-dependent constitutive expression of *gal* Operon: use of repressor titration to isolate operator mutations, *Proc. Natl. Aca. Sci.* 80:47754779.

John, O., Sauerstein, J., and Reuter, G., 1985, Detection of two ornithine carbomayltransferases in a phaseolotoxin-producing strain of *Pseudomonas syringae* pv. *phaseolicola,* *J. Basic Microbiol.* 25:543-546.

Kwok, O.C.H., Ako, H., and Patil, S.S., 1979, Inactivation of bean ornithine carbamoyltransferase by phaseolotoxin: effect of phosphate, *Biochem. Biophys. Res. Commun.* 89:1361-1368.

Kwok, O.C.H., and Patil, S.S., 1982, Activation of a chlorosis-inducing toxin of *Pseudomonas syringae* pv. *phaseolicola* by leucine aminopeptidase, *FEMS Microbiol. Lett.* 14:247-249.

Mitchell, R.E., and Bielski, R.L., 1977, Involvement of phaseolotoxin in halo blight of beans, *Plant Physiol.* 60:723-729.

Mitchell, R.E., 1978, Halo blight of beans: toxin production by several *Pseudomonas phaseolicola* isolates, *Physiol. Plant Pathol.* 13:37-49.

Mitchell, R.E., 1984, The relevance of non-host-specific toxins in the expression of virulence by pathogens, *Annu. Rev. Phytopathol.* 22:215-245.

Moore, R.E., et al., 1984, Inhibitors of ornithine carbamoyltransferase from *Pseudomonas syringae* pv. *phaseolicola,* Revised structure of phaseolotoxin, *Tetrahedron Lett.* 25:3931-3934.

Nüske, J., and Fritsche, W., 1989, Phaseolotoxin production by *Pseudomonas syringae* pv. *phaseolicola*: the influence of temperature, *J. Basic Microbiol.* 29:441-447.

Oguchi, T., et al., 1981, *Phaseolus Vulgaris - Pseudomonas syringae* pv. *phaseolicola* system: role of host tissue age and pathogenic toxin in establishment of pathogen in host, Proc. 5th Int. Conf. Plant Path. Bact., Cali, Central Internacional de Agriculture Tropical (CIAT), Cali, Columbia, pp. 288-294.

Patil, S.S., 1974, Toxins produced by pathogenic bacteria, *Annu. Rev. Phytopathol.* 12:259-279.

Patil, S.S., Kolattukudy, P.E., and Diamond, A.E., 1970, Inhibition of ornithine carbamoyltransferase from bean plant by the toxin of *Pseudomonas phaseolicola,* *Plant Physiol.* 46:752-753.

Patil, S.S., Tam, L.Q., and Sakai, W.S., 1972, Mode of action of the toxin from *Pseudomonas phaseolicola* I. Toxin specificity, chlorosis and ornithine accumulation, *Plant Physiol.* 49:803-807.

Patil, S.S., Hayward, A.C., and Emmons, R., 1974, An ultraviolet-induced nontoxigenic mutant of *Pseudomonas phaseolicola* of altered pathogenicity, *Phytopathology* 64:590-595.

Peet, R.C., et al., 1986, Identification and cloning of genes involved in phaseolotosin production by *Pseudomonas syringae* pv. *phaseolicola,* *J. Bacteriol.* 162:1096-1105.

Peet, R.C., and Panopoulos, N.J., 1987, Ornithine carbamoyltransferase genes and phaseolotoxin immunity in *Pseudomonas syringae* pv. *phaseolicola*. *EMBO Journal*. 6:3585-3591.

Quigley, N.B., Lane, D., and Bergquist, P., 1985, Genes for phaseolotoxin production are located on the chromosome of *Pseudomonas syringae* pv. *phaseolicola*, *Curr. Microbiol*. 12:295-300.

Romeo, C.J., and Patil, S.S., 1984, Abstr. no. 384, *Phytopathology* 74:837.

Rudolph, K., and Stahmann, M.S., 1966, The accumulation of L-ornithine in halo-blight infected bean plants (*Phaseolus-vulgaris* L.) induced by the toxin of the pathogen *Pseudomonas phaseolicola* (Burkh.), *Phyopathology* 57:29-46.

Russel, D.E., 1975, Observations on the *In Vivo* Growth and Symptom Production of *Pseudomonas phaseolicola* on *Phaseolus vulgaris*, *J. Appl. Bacteriol*. 43:167-170.

Ryan, W.L., Johnson, R.J., and Dimari, S., 1969, Homoarginine synthesis by rat kidney, *Arch. Biochem. Biophys* 131:521-526.

Stachel, S.E., An, G., Flores, C., and Nester, E.W., 1985, A Tn3 *lacZ* transposon for the random generation of β-galactosidase gene fusion: application to the analysis of gene expression in *Agrobacterium*, *EMBO* 4:891-898.

Staskawicz, B.J., and Panopoulos, N.J., 1979, A rapid and sensitive microbiological assay for phaseolotoxin, *Phytopathology* 69:663-666.

Staskawicz, B.J., Panopoulos, N.J., and Hoogenraad, N.J., 1980, Phaseolotoxin-insensitive ornithine carbamoyltransferase of *Pseudomonas syringae* pv. *phaseolicola*: basis for immunity to phaseolotoxin, *J. Bacteriol*. 142:720-723.

Staskawicz, B.J., et al., 1987, Molecular characterization of cloned avirulence genes from race 0 and race 1 of *Pseudomonas syringae* pv. *glycinea*, *J. Bacteriol*. 169:5789-5794.

Templeton, M.D., Sullivan, P.A., and Shepard, M.G., 1986, Phaseolotoxin- insensitive L-ornithine transcarbamoylase from *Pseudomonas syringae* pv. *phaseolicola*, *Physiol. Mol. Plant Pathol*. 29:393-403.

Wang, M.D., Buckley, L., and Berg, C.M., 1987, Cloning genes that suppress an *Escherichia coli* K-12 alanine auxotroph when present in multicopy plasmids, *J. Bacteriol*. 169:5610-5614.

Summary of Discussion of Patil's Paper

Chatterjee initiated the discussion by asking whether clones containing pDC10 with Tn3HoHo1 insertions were assayed for β- galaetosidase activity. *Patil* replied that it had been done but the basal level of β- galactosidase and that of cells harboring clones with Tn3HoHo1 insertions was not very high and therefore these experiments need to be repeated before drawing conclusions with regard to the direction of transcription of the putative gene in pDC10. *Chatterjee* said that the repressor model presented by *Patil* to explain the suppression of EMS Tox⁻ mutants by heterologous fragments probably makes more sense than the idea that heterologous genes produce products with the same biological function. He continued that there is precedence for repressor molecules being titrated out, and sited *Handa's* work on pectolytic enzymes where he had isolated DNA fragments from *E. caratovora* which seem to titrate out DNA binding proteins and results in changing the phenotype of the cell. *Chatterjee* further asked whether there was any evidence that genes are induced by plant constituents. *Patil* said that there was some indirect evidence for this. When plants of a resistant cultivar such as Red Mexican are inoculated with an incompatible race 1 cells, such as HB-33 one sees substantial growth of the organism before the occurrence of HR and restriction of growth. However, no concomitant toxin production occurs.

Keen asked whether saturation mutagenesis of pHK 120, which appears to contain the majority of toxin encoding genes, has been attempted, and suggested that transposon 1721 might be very useful in this because the *lacZ* fusions that result from the mutagenesis would be very useful. *Keen* further asked whether anything about the sequence of pDC10 was known. *Patil* replied that they have most of the sequence of the biologically active region of pDC10 and that if their assumption that a GTG is the start codon is correct, they have an open reading frame but that needs to be confirmed. *Keen* asked whether pHK120 containing transconjugants of other Gram-negative phytopathogenic bacteria produce toxin. *Patil* said they did not. He said that pHK120 may not contain all of the genes for toxin biosynthesis. *Durbin* asked whether only the phaseolotoxin molecule is exported from the bacterial cell. *Patil* said that in the culture medium they find both phaseolotoxin and octicidin but he did not know whether octicidin is exported by the bacteria or produced from phaseolotoxin in the medium. *Durbin* then commented that lack of ability to synthesize homoarginine may not be the explanation for the inability of some of the mutants to produce toxin because they could still produce octicidin. *Patil* replied that the bioassay used in their system was specific for phaseolotoxin since octicidin is not taken up by the *E. coli* test organism and that all the genetic work is based on this assay.

Bennetzen asked whether *Patil* believed that the reason why he was not able to complement the Tn5 mutants with the original wild-type library was because the Tn5 insertions produced polar effects and the DNA pieces in the library were not large enough to effect complementation. *Patil* concurred. *Chatterjee* then commented that if the repressor hypothesis is correct one should have been able to clone the regulatory gene on pHK120. *Patil* said they have not yet started working on that aspect but they expect to find this gene when they complete the saturation Tn5 mutagenesis of pHK120. *Essenberg* concluded the discussion by asking whether suppression of Tox⁻ mutants by these clones in multicopy plasmids explains both hypotheses. *Patil* agreed that it did.

Fungal Strategies

Chapter 9
Molecular Analysis of Pathogenesis in *Ustilago maydis*

Sally A. Leong, Jun Wang, James Kronstad,
David Holden, Allen Budde, Eunice Froeliger,
Thomas Kinscherf, Peilin Xu, William A. Russin,
Deborah Samac, Timothy Smith, Sara Covert,
Baigen Mei, and Christophe Voisard

We are isolating and studying genes required for pathogenicity of *Ustilago maydis*, the causative agent of corn smut disease (Christensen, 1963). This phytopathogenic basidiomycete offers an attractive system in which to gain a molecular understanding of host-parasite interactions. The organism grows as a haploid yeast on defined laboratory media, mutants are readily generated by UV or chemical mutagenesis, stable diploids can be constructed for mitotic recombination and complementation analysis, and the fungus is amenable to Mendelian genetic analysis (Holliday, 1974). These attributes are not found in combination in any other phytopathogenic fungus. Although this pathogen is no longer a production constraint in North America, where resistant hybrid corn is grown, it continues to be a problem in third world countries where susceptible varieties are still cultivated. Moreover, we hope that an understanding of mechanisms of pathogenesis in this host-parasite interaction will have application to more economically important and difficult to study fungal diseases such as the bunts and rusts. We have initiated a molecular genetic analysis of two gene systems thought to control pathogenic growth of *U. maydis* in maize. These include genes that program sexual development and genes involved in the high affinity, siderophore-mediated iron uptake system of the fungus. In order to conduct these analyses, one of our first goals has been to develop tools which enable us to clone, study, and transfer genes in *U. maydis*.

Gene Transfer in *Ustilago maydis*

To date only a few phytopathogenic fungi have been successfully transformed (reviewed in Wang et al., 1989). At the time research on transformation was started in our laboratory, no plant pathogenic fungus had been stably transformed. Our laboratory was the first to develop a routine and efficient gene transfer system for the phytopathogenic basidiomycetes. A method for the generation and regeneration of large numbers of protoplasts was established. A chimeric selectable marker was constructed by transcriptional fusion of an *U.*

maydis hsp70 (heat shock protein) gene promoter and the coding region of a gene for hygromycin B phosphotransferase. Using this system, we have stably transformed *U. maydis* (Wang et al., 1988), *Ustilago hordei* (Holden et al., 1988) and *Ustilago nigra* (Holden et al., 1988). The latter two species are pathogens of barley. The transforming DNA is integrated into chromosomal DNA and recombines homologously at the hsp70 locus in ~30% of *U. maydis* transformants. Linear DNA gives rise to higher numbers of transformants than circular DNA. A transformation frequency of ~1000 transformants per µg linearized vector was obtained. Copy number and position effects on expression of vector-associated genes were observed. Using one of our vectors, Holloman and coworkers isolated an *U. maydis* ARS (autonomously replicating sequence) and developed self-replicating vectors for *U. maydis* (Tsukuda et al., 1988). We have employed one of these ARS sequences to construct the self-replicating cosmid vector pJW42 (Fig. 1) and genomic DNA library for *U. maydis*. This library, which yields transformation frequencies of 2000-10,000 transformants per µg, has been employed to isolate the *pan1* gene and the *a* mating type locus of *U. maydis* (Froeliger et al., 1989). We have also constructed a self-replicating vector for *U. maydis* by incorporating a portion of a linear mitochondrial plasmid from *Fusarium solani* f.sp. *cucurbitae* into our integrative transformation vector (Samac et al., 1989). Transformation frequency is improved 20-fold, however, transformants grow slowly on selective medium and are mitotically unstable.

We (Kronstad et al., 1989; Holden et al., 1989) and others (Banks et al., 1988; Fotheringham et al., 1989) have established protocols for single step gene disruption and gene replacement in *U. maydis*. These events occur at high frequency in *U. maydis* when compared with other filamentous fungi. In our disruption studies, the coding sequences of the orotidine-5'-P-decarboxylase (pyr6) gene (Kronstad et al., 1989) as well as an hsp70 gene (Holden et al., 1989) of *U. maydis* were physically disrupted by insertion of a fragment carrying a gene for resistance to hygromycin B. Linear DNA fragments carrying the hygromycin B resistance gene were transformed into *U. maydis* haploid (pyr6::HygB^r) and diploid cells (hsp70::HygB^r). Approximately 70%

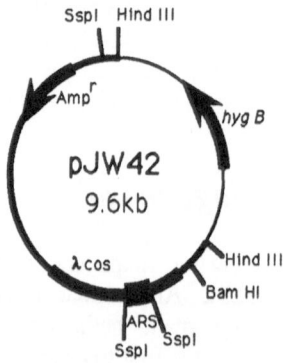

Figure 1. The self-replicating cosmid vector pJW42.

of the hygromycin resistant haploid transformants were found to carry the disrupted gene in place of the wild-type allele. In diploid transformants, only one allele was replaced. In the replacement experiment, a b2 mating type allele was substituted with a genomic DNA fragment encoding a b1 allele and carrying a linked gene for hygromycin B resistance (Kronstad et al., 1989a). Replacement of the b2 allele with the b1-HygB^r allele occurred in ~50% of transformants. These techniques permit isolation of genes from *Ustilago* by complementation, their manipulation in vitro, and their return to the *Ustilago* genome by homologous recombination. These tools are essential to the molecular genetic analysis of pathogenesis as well as other biochemical and developmental processes in *Ustilago*.

Using two approaches, we have isolated eight metabolic and stress response genes from *U. maydis* for use in construction of selectable markers and expression vectors and to gain an understanding of gene structure in *U. maydis*. Genomic DNA clones encoding the glyceraldehyde-3-P-dehydrogenase (GAPDH) gene (Smith et al., 1990) and an hsp70 (heat shock protein) gene family (Holden et al., 1989), were identified by heterologous hybridization with GAPDH and hsp70 gene probes from yeast. A cDNA library complementary to *U. maydis* mRNA was prepared in a yeast expression vector and used to complement mutations in the yeast genes URA3, LEU2, and TPI (Kronstad et al., 1989). The DNA sequence has been determined for the gene for GAPDH, the cDNA encoding pyr6, and the 5' portion of one hsp70 gene. These data confirm the functional assignments of these genes. The genomic copy of the pyr6 gene was isolated and employed to complement pyr6 mutants of *U. maydis*, thus providing a transformation system based on complementation for the fungus (Kronstad et al., 1989). Banks and Taylor (1988) have also developed a transformation system based on complementation of pyr3 mutants of *U. maydis*. The transcriptional regulation of the hsp70 gene family was analyzed (Holden et al., 1989). All four genes are expressed at normal growth temperatures; however, upon heat shock, transcription of two of the genes is up-modulated while that of one is down-modulated, and that of the last is unchanged. One of the genes was disrupted and shown to be essential to cell viability. Other than yeast, this is the only eucaryotic organism for which a heat shock gene has been shown by genetic mutation to be required for cell growth. The promoter of this gene was employed in the construction of a chimeric selectable marker for transformation of *Ustilago* (Wang et al., 1988).

Molecular Karyotype

As an aid to gene mapping in *U. maydis*, we have developed an electrophoretic karyotype for the fungus (Kinscherf et al., 1988). *U. maydis* has at least 20 distinct chromosome-sized DNAs ranging in size from 325 kbp to approximately 2000 kbp. Chromosome length variation was observed between isolates. A number of genes have been mapped on the karyotype by Southern hybridization (Kinscherf et al., 1988; Budde et al., 1990). This technique was used to document a translocation event in *U. maydis* (Kinscherf et al., 1988).

Mating Type Control

Successful infection of maize by *U. maydis* requires that haploid cells of the opposite mating type fuse to form a dikaryotic mycelium (reviewed in Froeliger and Kronstad, 1990). This mycelium is the parasitic phase of the life cycle; haploid cells are nonpathogenic. This parasitic mycelium colonizes host tissues, induces gall development, and eventually becomes sporogenous, giving rise to diploid teliospores. Two mating type loci control cell fusion and sexual development. The a locus is purported to control haploid cell fusion and may also regulate production of dikaryotic hyphae (Banuett et al., 1989). Two alleles a1 and a2 are present at this locus. The b locus controls production and maintenance of the dikaryotic mycelium as well as subsequent events in sexual development through meiosis (Day et al., 1971). At least 25 alleles have been identified at the b locus (Puhalla, 1968). Only mixtures of compatible haploid cells with different alleles at each mating type locus or diploid cells that are heterozygous at b can produce parasitic mycelium and elicit gall development. Thus, the b locus can be considered a master pathogenicity gene.

We have undertaken a molecular genetic analysis of the mating type system of *U. maydis*. Our goal is to understand how the products of the a and b loci, when heterozygous, function to control cell fusion, sexual development, and pathogencity of this fungus. We wish to compare the structure of the different a and b alleles, analyze their gene products, and determine how these loci are regulated and what genes are affected by their products. Toward this end, we have cloned two alleles of the a locus (Froeliger et al., 1989) and six alleles of the b locus (Kronstad et al., 1989a; Kronstad et al., unpublished findings). Four b alleles were also cloned by Schultz et al. (1990). These are among the first mating type control genes to be cloned from filamentous fungi. Other examples include Aα alleles of *Schizophyllum commune* (Giasson, 1989), Mata and MatA of *Neurospora crassa* (Glass, 1988) and MAT*1*-1 and MAT*1*-2 of *Cochliobolus heterostrophus* (Turgeon, 1989). The DNA sequence of seven different b alleles (Kronstad et al., 1989b; Schulz et al., 1990; Kronstad et al., unpublished findings) has been determined. These studies have identified an open reading frame of 410 amino acids in all seven alleles. Comparison of the predicted amino acid sequences of the alleles has revealed the presence of variable and constant domains. Sequences similar to a homeodomain motif (Schulz et al., 1990) and a nucleus localization sequence (Kronstad et al., 1989a; Kronstad et al., 1990) were also identified. To examine how the b locus controls sexual development and pathogenicity, a b*1* allele was disrupted (Kronstad et al., 1990). Mutation of the b1 allele did not effect cell growth; however, mutant cells were unable to form dikaryons with appropriate tester strains and were nonpathogenic either by themselves or in combination with compatible testers (Kronstad et al., 1990). These data indicate that the product of the b locus acts in a positive manner to control sexual development. We are also attempting to identify genes that are regulated by b. We have shown that haploid strains transformed with a different b allele become pathogenic (Kronstad et al., 1989). Thus, a dikaryotic or diploid nuclear state is not absolutely required for pathogenesis. We will exploit this finding by attempting to isolate nonpathogenic mutants from pathogenic haploid strains which carry two b alleles. Using this approach we anticipate identifying genes regulated by the b locus as well as general pathogenicity genes, i.e., genes required for phytohormone production and penetration of epidermal cell walls.

In contrast to the nearly identical alleles of the b locus, the two alleles of the a mating type locus of *U. maydis* are dissimilar (Fig. 2). A similar phenomenon has been observed for the MAT alleles of *N. crassa* (Glass, 1988) and yeast (Herskowitz, 1988), the Aα alleles of *S. commune* (Giasson, 1989), and MAT1 alleles of *C. heterostrophus* (Turgeon, 1989).

High Affinity Iron Transport

Colonization of host tissues requires that an invading microbial pathogen be capable of acquiring iron, as iron is an essential cofactor of many TCA cycle enzymes, cytochromes, ribotide reductase, among other enzymes. Iron is transported into most microorganisms by a high affinity iron uptake system which consists of siderophores--low molecular weight, ferric ion specific chelating agents--and their cognate membrane receptors (Leong et al., 1989). We are conducting a systematic analysis of the siderophore-mediatediron uptake system of *U. maydis* in order to assess its role in pathogenicity, spore physiology, and survival of the pathogen. We also wish to understand the structure and regulation of siderophore biosynthetic and transport genes of *U. maydis*.

U. maydis was confirmed to produce two siderophores, ferrichrome (Fig. 3) and ferrichrome A (Budde et al., 1989). Both siderophores were found intracellularly--in sporidia and teliospores--as well as extracellularly. Iron was found to repress biosynthesis of siderophores. Using novel screening bioassays, three classes of mutants defective in siderophore production have been isolated after chemical mutagenesis (Leong, 1987). Of these, one class (3 mutants) was found to produce siderophore constitutively. The other two classes are deficient in siderophore biogenesis. One class produces no siderophores (6 mutants) while the other is unable to produce ferrichrome (4 mutants). Mutants defective in siderophore transport were not identified. The class of mutants unable to

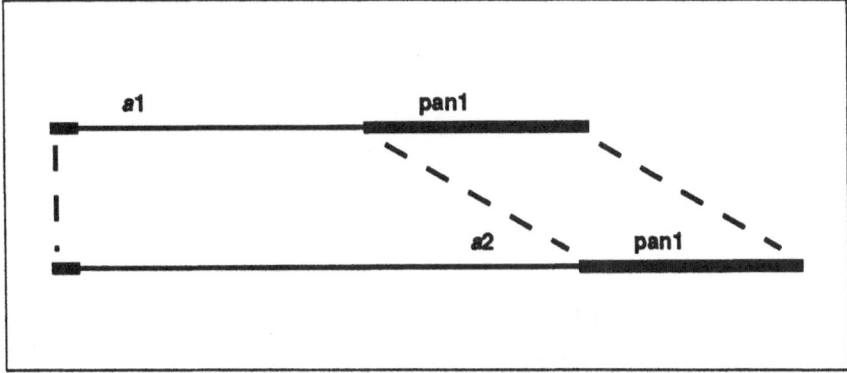

Figure 2. Comparison of the structure of the *a*1 and *a*2 alleles of *U. maydis*. Heavy bars denote regions of sequence identity; light bars indicate regions of nonidentity as determined by Southern hybridization analysis. The *a* locus is tightly linked to the *pan1* gene (Holliday, 1974).

produce either siderophore was examined further. Based on biochemical and genetic complementation tests and genetic segregation data, these mutants were surmised to be defective in ornithine-N[5]-oxygenase (Wang et al., 1989). This enzyme, which catalyzes the conversion of ornithine to δ-N-OH-ornithine, carries out the first committed step in the biosynthesis of ferrichrome and ferrichrome A. Preliminary studies of enzyme activity indicate that this enzyme is a flavoprotein which utilizes NADH or NADPH as a source of electrons. As expected, enzyme activity was absent in extracts prepared from an oxygenase mutant and from wild-type cells grown in the presence of repressing levels of iron. We have identified genomic DNA clones which complement this class of siderophore biosynthetic mutation (Wang et al., 1989) as well as mutations that deregulate siderophore production. The molecular structure and expression of these genes are being studied. The ornithine-N[5]-oxygenase gene was localized to a 8.1 kb *Hind*III fragment which maps to chromosome XIX (Wang et al., 1989). A cosmid clone capable of complementing two of the deregulated mutants was also identified. An 11 kb *Bam*HI subclone contains the complementing function. These are the first genes for siderophore biosynthesis and regulation to be cloned from a eucaryotic microorganism.

Phytopathogenicity of Siderophore Auxotrophs

Preliminary experiments were conducted to assess the phytopathogenicity of one siderophore-nonproducing mutant. To eliminate secondary mutations induced by the mutagenesis procedure, the siderophore mutation was backcrossed seven times to wild-type parental strains of opposite mating type. A mixture of haploid, mutant progenies of compatible mating type and carrying the siderophores lesion or teliospores of similar genotype were used to inoculate maize seedlings. The siderophore mutants caused disease, although disease ratings were generally lower than those obtained for infection with wild-type

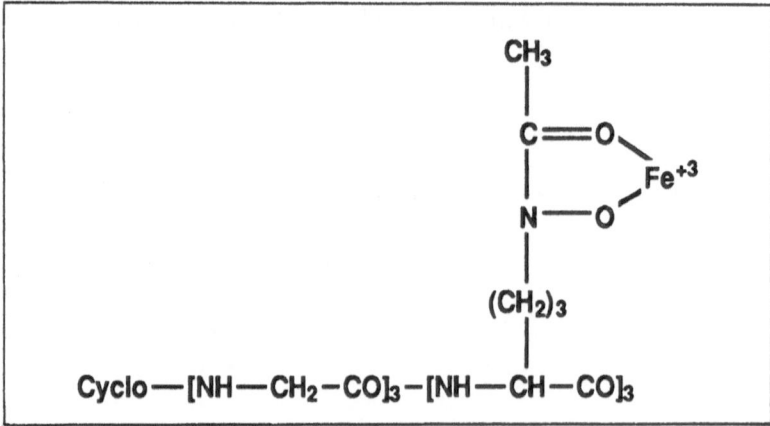

Figure 3. Ferrichrome.

strains (Leong, 1988). Exceptions were observed and standard deviations were large; therefore, the differences in disease rating for mutant and control inoculations may not be significant. Although the mutations were repeatedly backcrossed into a wild-type backround, closely linked secondary mutations may not have been eliminated. To insure that the siderophore mutation is causal to the lower disease rating, inoculations will be made in future with mutants constructed by molecular disruption of the oxygenase gene. A clear assessment of the role of siderophores in phytopathogenicity must await inoculation of maize seedlings with such mutants.

Disease Development in Maize Leaves Infected with *U. maydis*

To provide a visual framework for assessing the role of mating type genes and siderophores in disease development, cytological studies of infection and gall development were conducted with wild-type cells using scanning electron microscopy, transmission electron microscopy, and light microscopy. These studies confirm and extend those previously reported (Walter, 1933; Callow et

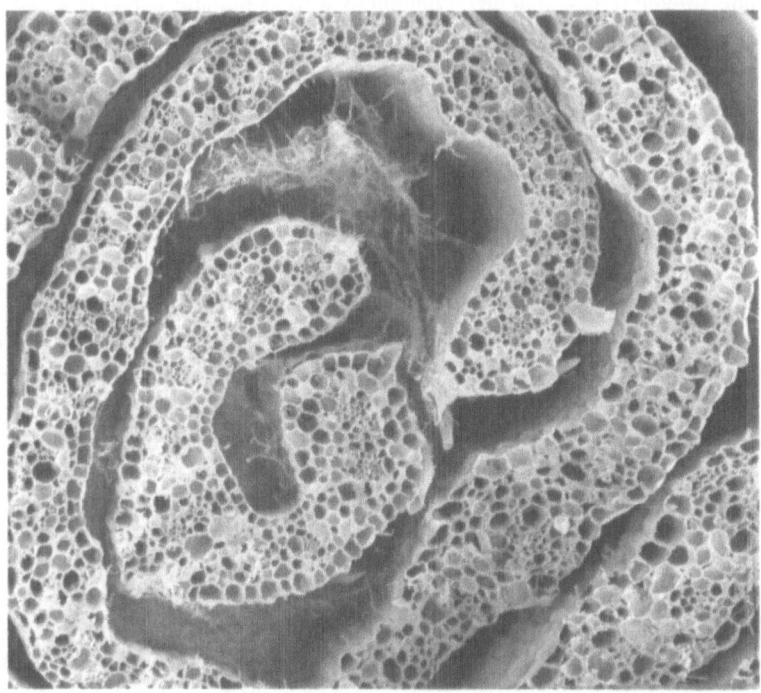

Figure 4. Freeze fracture of maize seedling whorl inoculated with *U. maydis* shows presence of a fibrillar matrix.

al., 1973: Mills et al., 1981). *U. maydis* haploid cells were embedded in an extensive fibrillar matrix within 24 hours after inoculation (Fig. 4). This material, which appears to emanate from the fungus, may promote attachment of haploid cells to host epidermis. Penetration of host cells by mixtures of germinated haploid sporidia of compatible mating type was found to occur through wounds and by direct penetration of epidermal cells of the leaf sheath. Penetration through natural openings such as stomates was not clearly documented. Cell fusion events (mating?) were observed prior to host infection. The fungus grows through host epidermal and mesophyll cell layers in order to reach the vascular tissue where it ramifies. All cell types of the vascular bundle were infected by 96 hours. Within 48 hours, neoplastic growth was evident in the vascular parenchyma. Visible gall development was observed at 5 days post-inoculation. Hyphae in the process of budding to form teliospores were documented in mature galls. Teliospores gave similar results to those obtained for compatible mixtures of haploid cells while solopathogenic diploids were found to be less aggressive and disease development was retarded. Haploid cells and mixtures of incompatible haploid cells were capable of limited growth in epidermal, mesophyll, and vascular tissue. However, no evidence of gall formation was observed. In future, we will evaluate disease development of mutants impaired in sexual development as well as mutants carrying a disrupted oxygenase gene.

Summary

Considerable progress has been made in recent years to develop *U. maydis* into a microbial phytopathogen that is amenable to molecular genetic analysis. The fungus is readily transformed, the fungal genome can be manipulated by gene disruption/replacement, and a molecular karyotype is available for gene mapping. Numerous genes have been isolated and their structure and regulation are being analyzed. These include genes controlling cell metabolism, stress response, siderophore biogenesis, sexual development and pathogenesis. With these tools and genetic materials in hand, no barriers remain to molecular studies of pathogenesis in the corn smut fungus.

Acknowledgements

This work was supported by Public Health Service grant 1 RO1 GM33716 from the National Institutes of Health to S.A.L., the U.S.D.A., the Graduate School of the University of Wisconsin, and an operating grant from the Natural Sciences and Engineering Research Council of Canada to J.W.K.

References

Banuett, F. and Herskowitz, I., 1989, Different *a* alleles of *Ustilago maydis* are necessary for maintenance of filamentous growth but not for meiosis, *Proc. Natl. Acad. Sci. USA* **86**:5878-5882.

Banks, G.R. and Taylor, S.Y., 1988, Cloning of the PYR3 gene of *Ustilago maydis* and its use in DNA transformation. *Mol. Cell. Biol,* **8**:5417-5424.

Budde, A.D. and Leong, S.A., 1989, Characterization of siderophores from *Ustilago maydis, Mycopathologia* **108**:125-133.

Budde, A.D. and Leong, S.A., 1990, *Ustilago maydis. In* Genetic Maps, vol. 5 (S.J. O'Brien, ed.), Cold Spring Harbor Laboratory, New York.

Callow, J.A. and Ling, I.T., 1973, Histology of neoplasms and chlorotic lesions in maize seedlings following the injection of sporidia of *Ustilago maydis* (DC) Corda. *Physiol. Plant Pathol.* **3**:489-494.

Christensen, J.J., 1963, Corn smut caused by *Ustilago maydis, Am. Phytopathol. Soc. Mono.*

Day, P.R., Anagnostakis, S.L. and Puhalla, J.E., 1971, Pathogenicity resulting from mutation at the *b* locus of *U. maydis, Proc. Natl. Acad. Sci. USA* **68**:533-535.

Fotheringham, S. and Holloman, W.K., 1989, Cloning and disruption of *Ustilago maydis* genes, *Mol. Cell. Biol.* **9**:4052-4055.

Froeliger, E.H. and Kronstad, J.W., 1990, Mating and pathogenicity in *Ustilago maydis. In* Seminars in Developmental Biology: Fungal Development and Mating Interactions (C.A. Raper and D.I. Johnson, eds.), Saunders Scientific Publications, London. (in press)

Froeliger, E.H., and Leong, S.A., 1989, Isolation of mating type genes from *Ustilago maydis.* Genetics and Cellular Biology of Basidiomycetes (Abstr.). University of Toronto Press, Toronto pp. 25.

Giasson, L. et al., 1989, Cloning and comparison of Aα mating-type alleles of the basidiomycete *Schizophyllum commune, Mol. Gen. Genet.* **218**:72-77.

Glass, N.L. et al., 1988, DNAs of the two mating type alleles of *Neurospora crassa* are highly dissimilar, *Science* 241:570-573.

Herskowitz, I., 1988, Life cycle of the budding yeast *Saccharomyces cerevisiae, Microbiol. Rev.* **52**:536-553.

Holden, D., Wang, J. and Leong, S.A., 1988, DNA-mediated transformation of *Ustilago hordei* and *Ustilago nigra, Physiol. Mol. Plant Pathol.* **33**:235-239.

Holden, D.W., Kronstad, J. and Leong, S.A., 1989, Mutation in a heat-regulated hsp70 gene of *Ustilago maydis, EMBO J.* **8**:1927-1934.

Holliday, R., 1974, *Ustilago maydis, In* Handbook of Genetics, vol. 1 (King, R.C., ed.) Plenum Press, New York. pp. 575-595.

Kinscherf, T., and Leong, S.A., 1988, Molecular analysis of the karyotype of *Ustilago maydis, Chromosoma* **96**:427-433.

Kronstad, J.W., et al., 1989, Isolation of metabolic genes and demonstration of gene disruption in *Ustilago maydis, Gene* **79**:97-106.

Kronstad, J.W. and Leong, S.A., 1989a, Isolation of two alleles of the *b* locus of *Ustilago maydis, Proc. Natl. Acad. Sci. USA* **86**:978-982.

Kronstad, J.W. and Leong, S.A., 1989b, Molecular characterization of the *b* locus of *Ustilago maydis. Fungal Genet. Newsl.* **36**:26.

Kronstad, J.W. and Leong, S.A., 1990, The *b* mating type forms of *Ustilago Maydis* contains variable and constant domains. *Genes Dev.* **4**:1384-1395.

Leong, S.A., 1987, Molecular strategies for the analysis of the interaction of *Ustilago maydis* and maize. *In* Molecular Strategies for Crop Protection (Arntzen, C., and Ryan, C. eds.), Alan R. Liss, New York, pp. 95-106.

Leong, S.A., 1988, Identification and molecular characterization of genes which control pathogenic growth of *Ustilago maydis* in maize. *In* Molecular Genetics of Plant-Microbe Interactions (Verma, D.P. and Palacios, R. eds.), APS Press. St. Paul. pp. 241-246.

Leong, S.A. and Expert, D., 1989, Siderophores in plant-pathogen interactions. *In* Plant-Microbe Interactions–A Molecular Genetic Perspective, vol. 3 (Nester, E. and Kosuge, T., eds.), McGraw-Hill, New York, pp. 62-83.

Mills, L.J., and Kotze, J.M., 1981, Scanning electron microscopy of the germination, growth and infection of *Ustilago maydis* on maize, *Phytopath. Z.* **102**:21-27.

Puhalla, J.E., 1968, Compatibility reactions on solid medium and interstrain inhibition in *Ustilago maydis*, *Genetics* **60**:461-474.

Samac, D.A., and Leong, S.A., 1989, Characterization of the termini of linear plasmids from *Nectria haematoccoca* and their use in construction of an autonomously replicating transformation vector, *Curr. Genet.* **16**:187-194.

Schulz, B., et al., 1990, The *b* alleles of *U. maydis*, whose combinations program pathogenic development, code for polypeptides containing a homeodomain-related motif, *Cell* **60**:295-306.

Smith, T., and Leong, S.A., 1990, Structure of a glyceraldehyde-3-phosphate dehydrogenase gene from *Ustilago maydis*, *Gene*. (in press)

Tsukuda, T., et al., 1988, Isolation and characterization of an automonously replicating sequence from *Ustilago maydis*, *Mol. Cell. Biol.* **8**:3703-3709.

Turgeon, G., 1989, Molecular Analyses of the mating type locus of *Cochliobolus heterostrophus*. *Fungal. Genet. Newsl.* #36 (Peter J. Russel, ed) p. 6.

Walter, J.M., 1933, The mode of entrance of *Ustilago zeae* into corn, *Phytopathology* **24**:1012-1020.

Wang, J., Holden, D., and Leong, S.A., 1988, Gene transfer system for the phytopathogenic fungus *Ustilago maydis*, *Proc. Natl. Acad. Sci. USA* **85**:865-869.

Wang, J., Budde, A., and Leong, S.A., 1989, Analysis of ferrichrome biosynthesis in the phytopathogenic fungus *Ustilago maydis*: cloning of an ornithine-N^5-oxygenase gene, *J. Bacteriol.* **171**:2811-2818.

Wang, J., and Leong, S.A., 1989, DNA-mediated transformation of phytopathogenic fungi. *In* Genetic Engineering, Principles and Practice, (Setlow, J. ed.), Plenum Press, New York, vol. 11 pp. 127-143.

Summary of Discussion of Leong's Paper

Chatterjee initiated the questioning regarding the siderophore genes by asking whether the oxygenase gene had been expressed in *E. coli*. *Leong* responded that they have not progressed that far yet, but that they would like to purify the protein at some stage. They may have to do some tailoring because they did not know, for example, if the gene has introns. She further stated that fungal genes did not generally function in prokaryotes, so they will probably have to go to cDNA cloning. If there are no introns they will probably have to put an *E. coli* promoter upstream of the gene so it is probably not going to be straightforward. *Chatterjee* then asked whether they had looked at other genes in the pathway. Leong responded that they had not, but they plan to. She said that the intent is to complement some of our other mutations and begin to acquire an understanding of the enzymology involved in the whole biosynthetic pathway of ferriochrome as well as the molecular biology of the genes and their regulation. *Yamada* then asked whether they have looked at IAA production in the haploid and diploid strains of *U. maydis*. *Leong* responded that they have not, but there were some old reports on phytohormone biosynthesis by *U. maydis*, and the auxins and cytokinins in culture. *Yamada* then asked about what was known about growth of the nonpathogenic strains in planta, and whether they grow as well as pathogens. *Leong* responded that they had looked at inoculations with haploids and incompatible combinations of haploid cells, in other words, ones that are unable to mate and cause disease. They found that both the haploids and the incompatible combinations were capable of limited growth in the host. They were also able to cause the chlorotic flecking but not galls. *Van Alfen* then asked whether disruption of the *b* locus allows the dikaryon to still retain the mycelial form. *Leong* responded that it did not, and that the mycelial form is definitely associated with virulence. *Van Alfen* then followed up by asking whether it was possible to force a dikaryon in culture by adding charcoal to the medium, as if to remove some component. *Leong* responded that the mating medium they use is a charcoal medium, but to her knowledge, nobody knew what it is about the medium that activates the dikaryon formation. *Chumley* then asked whether they have looked at expression of the *a* and *b* alleles and whether there might be some developmental regulation. Leong responded that it was not yet done. *Chumley* then asked *Leong* to say something more about the pathogenicity of haploids that are diploid for the *b* locus.

Leong responded that they did not know a lot about them. She said that they were not as aggressive as the haploid strains of opposite mating type coming together to form a dikaryon, and one sees small gall formations; there seemed to be a kind of hierarchy. The haploids are the weakest of the pathogens, diploids are kind of intermediate, and the dikaryons are the most pathogenic. *Bushnell* then asked whether there has been a cytological karyotype done of this fungus and how that might compare with the electorophetic karyotype. *Leong* indicated there had not been, and that these chromosomes were very small, in the 325 to about 2000 kilobase range. Her understanding was that fungal chromosomes that size are really hard to visualize by microscopic means. *Essenberg* then asked whether the b1 and b2 alleles simply control the ability to make the mycelial form which is the requirement for pathogenicity, or whether there is some other function for the *b* locus. *Leong* indicated that she thought that the *b* locus is a master regulatory gene which centrally controls many of these events throughout the sexual phase of the life cycle including teliospore formation and meiosis. She said that it is very hard to separate the various stages of pathogenesis because gall formation is closely tied to the formation of the teliospores.

Chapter 10
Molecular Analysis of Genes for Pathogenicity of *Alternaria alternata* Japanese Pear Pathotype, a Host-Specific Toxin Producer

Takashi Tsuge and Hirokazu Kobayashi

There are now at least seven known plant diseases caused by the fungus *Alternaria alternata*, in which host-specific toxins (HSTs) are produced as pathogenicity factors (Nishimura et al., 1983). That HSTs participate in establishment of plant disease in a host selective manner is one of the most clearly revealed mechanisms: capacity to infect and host specificity are most likely governed by the pathogen's ability to produce HSTs (Nishimura et al., 1983; Scheffer et al., 1984; Yoder, 1980).

Knowledge of pathways and genes for HST biosynthesis contributes not only to basic science but also to the development of alternative strategies for the protection of plants from attack by the pathogens. However, classical genetic analysis of HST production in *A. alternata* is impossible because the fungus has no known sexual cycle. Although the chemical structures of some *A. alternata* HSTs have already been elucidated (Nishimura et al., 1989), information about HST metabolism is limited only to the chemical structures of certain intermediates (Feng, et al., 1990; Nakatsuka et al., 1990). We have begun studies to elucidate the pathways and genes for HST synthesis in *A. alternata* pathogens, especially in the Japanese pear pathotype of *A. alternata*. In this paper, we present our recent progress in gene analysis of the pathogen.

Alternaria alternata Japanese Pear Pathotype

The Japanese pear pathotype of *A. alternata* (Fries) Keissler (formerly, *A. kikuchiana* Tanaka) is the pathogen of black spot of Japanese pear. Tanaka (1933) showed with foresight that the pathogen produced an unusual toxic metabolite responsible for its host-specific pathogenicity. Nishimura and his colleagues started studies on the toxic metabolites in the 1960s, including research on the Japanese pear pathotype as well as on other pathotypes of *A. alternata* (Nishimura, 1987; Nishimura et al., 1983; Nishimura et al., 1989).

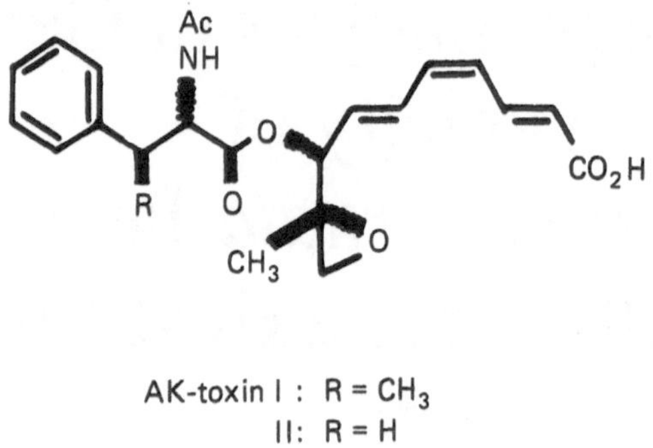

Figure 1. Chemical structures of AK-toxins I and II.

The phytotoxin of the Japanese pear pathotype, designated as AK-toxin, was isolated and characterized as two related molecular species, AK-toxins I and II (Nakashima et al., 1985) (Fig. 1). AK-toxin I is a major toxin both in yield and biological activity, and it functions as a host recognition factor at primary infection sites (Hayashi et al., 1989). The role of AK-toxins in pathogenesis and the action mode of the toxins are reviewed by Otani et al. in this volume.

Population Dynamics of the Japanese Pear Pathotype

A. alternata is a ubiquitous fungus surviving on various dead materials in the field. However, once the fungus acquired the ability to produce an HST, it simultaneously became a destructive pathogen in an agroecosystem containing highly susceptible host genotypes. Although work on the ecological significances of HSTs has been extremely limited (Scheffer, 1983; Scheffer et al., 1984), HST-producing pathogens seem to offer a convenient tool for epidemiological surveys of fungal populations in the field. As an extension of ecological studies of HST producers, we analyzed restriction fragment length polymorphism (RFLP) of the Japanese pear pathotype as a way to assess variability in population structure along with AK-toxin productivity.

More than 250 isolates collected from 27 pear fields in different areas in Japan were subjected to RFLP analysis. We chose ribosomal RNA genes (rDNA) as a target for analysis, because rDNA polymorphisms were shown to be characteristic of each isolate by the restriction enzyme, *Xba*I, in a preliminary experiment (Tanabe et al., 1989). Since an rDNA clone of Japanese pear pathotype strain 15A has been selected from a genomic library and characterized as containing an 8.15-kb repeating unit (Tsuge et al., 1989), the rDNA clone was used for hybridization to locate polymorphisms. We detected at least five classes of variation of rDNA in the pathogen population. They are distinguished from one another in polymorphisms of *Xba*I-recognition sites in regions non-coding mature rRNAs and/or in the length of a repeated unit. The individual pathogen populations in all fields contained two or more classes of rDNA variants. Although the same kinds of RFLPs were detected in population structures from different areas, distribution patterns of rDNA variants in each population structure were significantly correlated with geographic origins. Although the polymorphisms in rDNA repeating unit length and the RFLPs in the non-coding spacer DNA have been reported in other fungi (Garber et al., 1988; Petes et al., 1977; Specht et al., 1983; Wu et al., 1983), those of *A. alternata* rDNA were much higher and are thought to be unique among fungi. The saprophytic ability of *A. alternata* pathogens is thought to be potentially related to the higher rDNA polymorphism in the pathogen population. We are assessing genetic variations in population structures in various areas and fields which differ in cultivation history of the Japanese pear cultivar Nijisseiki, in order to elucidate the population dynamics of the Japanese pear pathotype.

Presence of Extrachromosomal Genetic Elements

The Japanese pear pathotype of *A. alternata* seems to be genetically unstable in AK-toxin productivity (Tsuge et al., 1986). Field isolates of AK-toxin producers often lost the ability to produce the toxin which resulted in the loss of pathogenicity during culture on media. Such toxin-deficient mutants cannot be distinguished from saprophytic *A. alternata* in morphological characteristics.

Possible participation of DNA plasmids in AK-toxin productivity was examined. Total DNA purified from many field isolates was checked for plasmid DNA. We detected circular DNA plasmids in only two isolates of those examined. This result apparently indicates no relationship between the presence of plasmids and toxin productivity. Intracellular mycoviruses have been shown to be associated with genetic instability of the virulence in several phytopathogenic fungi, e.g., hypovirulence of *Endothia parasitica*, the causal agent of chestnut blight (Anagnostakis, 1982). Most characterized mycoviruses contain segmented double-stranded RNA (dsRNA) genomes (Gabrial, 1980). We have also studied the relationship between the occurrence of dsRNAs in mycelia and AK-toxin productivity using several isolates of AK-toxin producers and toxin-deficient mutants (Hayashi et al., 1988). About 50% of isolates showed detectable levels of dsRNAs in mycelia. However, there was no significant correlation between the presence of dsRNAs and the toxin productivity.

Gene Transcripts Different in an AK-Toxin Producer and Its Deficient Mutant

We employed an AK-toxin producer (strain 15A) and its toxin-deficient mutant (strain 15B) to examine differences in transcribed RNA species. Strain 15B is a spontaneous mutant isolated from a stock culture of strain 15A. These two

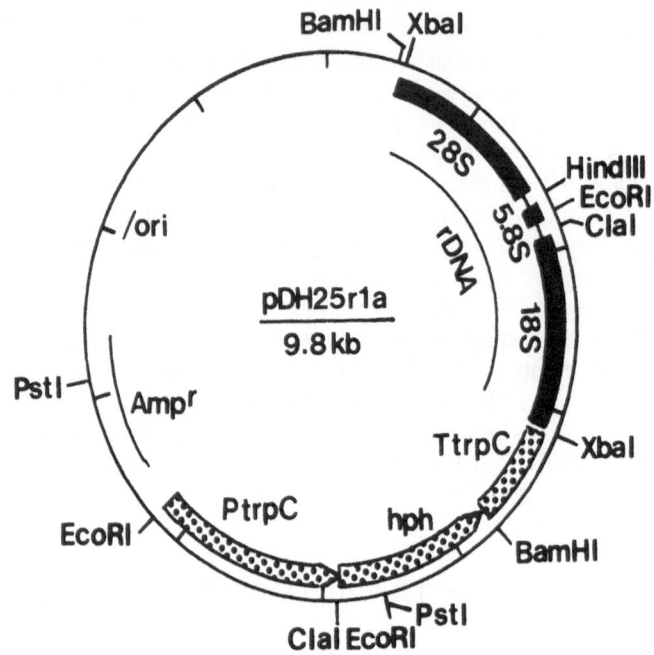

Figure 2. A vector for *A. alternata* transformation with homologous recombination mediated by a rDNA sequence. The vector pDH25r1a was constructed by inserting a rDNA fragment into a unique *Xba*I site of pDH25.

isolates are morphologically indistinguishable from one another and can be distinguished only in AK-toxin productivity and the resulting pathogenicity.

A competitive hybridization procedure (Bogorad et al., 1983) showed that specific transcripts were present in toxin producer 15A, but not in its toxin-deficient avirulent mutant 15B (Tsuge et al., 1986). A poly(A)$^+$RNA fraction from the toxin producer was cloned as cDNAs using pBR322 in *Escherichia coli* and screened by competitive hybridization. The selected cDNA clones were derived from transcripts more abundant in the producer than in its toxin-deficient mutant. Dot blot hybridization analysis with the cDNA clones revealed a correlation between the amount of the specific RNAs and toxin productivity in several *A. alternata* isolates (Tsuge et al., 1986).

Genetic Transformation of *Alternaria alternata*

Genetic transformation systems are needed for the cloning and experimental manipulation of the structural and regulatory genes involved in HST synthesis. We have developed an efficient DNA transformation system for *A. alternata* based on resistance to an antibiotic, hygromycin B.

1. Efficient Transformation of A. alternata Mediated by the Repetitive rDNA Sequences

E. coli hph gene, which encodes hygromycin B phosphotransferase (Gritz et al., 1983), was chosen as a selective marker for transformation, because it has been effectively used for transformation of eukaryotes and because hygromycin B inhibited the growth of *A. alternata* at a relatively low concentration (10 µg/ml). Plasmids containing fusions of fungal promoters with the *hph* gene appeared to be applicable to phytopathogenic fungi (Fincham, 1989; Leong et al., 1989). *A. alternata* was subjected to transformation using the plasmid vector pDH25 (Cullen et al., 1987) which consists of pBR322 and the *hph* gene under the control of *Aspergillus nidulans trpC* promoter and terminator (Mullaney et al., 1985). Transformation was achieved by introducing the plasmid DNA into the protoplasts of *A. alternata* as described for *Neurospora crassa* (Vollmer et al., 1986). Transformants grew on a selective medium containing hygromycin B (100 µg/ml) during 4 days of incubation (Tsuge et al., 1990). There were two types of transformants, giving large and small colonies, on the selectivemedium. On average, 4.5 large and 600 small colonies were obtained in a transformation experiment with 1µg of the vector. The presence of vector DNA in the transformant chromosomes was verified by the Southern hybridization of genomic DNA from the large- and small-colony transformants with *hph* and pBR322 as probes. In the large-colony transformants, the vector often integrated into the recipient chromosomes in highly rearranged tandem arrays (Tsuge et al., 1990).

Increased transformation efficiency could be expected if a homologous recombination system was employed. We thus selected the rDNA gene, which

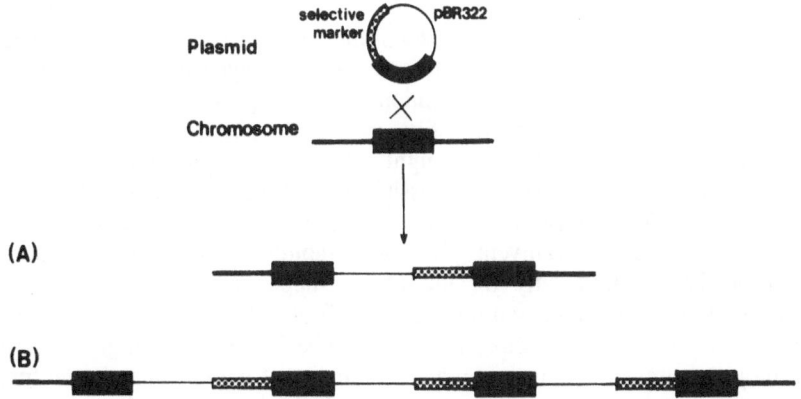

Figure 3. Schematic representation of plasmid integration into recipient chromosome by homologous recombination. A. Single crossover recombination between homologous regions on the plasmid and the chromosome. B. Integration of tandemly-arrayed multiple copies of plasmid DNA. This was frequently observed in the *Alternaria* transformants. ——— , pBR322 DNA; ▇▇ , homologous DNA; ▭▭▭ , selective marker DNA; ▬▬ , chromosomal DNA.

is highly repetitive in the genome, from our genomic library of *A. alternata* (Tsuge et al., 1989) and used it to construct new vectors (Tsuge et al., 1990). Use of these vectors gave a higher transformation efficiency than the original plasmid. The best vector pDH25rla (Fig. 2) made large-colony transformants at a frequency 20 times higher than pDH25. Probes of Southern blots with *hph* and pBR322 showed that transformation events in *A. alternata* with pDH25rla resulted from homologous recombination by a single crossover between the plasmid-borne rDNA segment and its homologue in the chromosome, where the vector DNAs were often repeated tandemly (Fig. 3). All vectors harboring rDNA fragments appeared to be integrated into the recipient chromosomes by homologous recombination. This type of homologous integration is typical of events characterized in *Saccharomyces cerevisiae* (Hinnen et al., 1978), but rather rare in filamentous fungi (Fincham, 1989; Hynes, 1986). Plasmid integration via simple homologous recombination is a prerequisite to many sophisticated genetic manipulations, including the eviction and transplacement of mutant genes (Winston et al., 1983). The system we developed here will provide a potential method for the genetic engineering of *A. alternata*.

Autonomously replicating vectors are desirable for transformation of filamentous fungi, because such vectors seem to improve transformation frequency and to simplify the isolation and cloning of genes as shown in *S. cerevisiae* (Futcher, 1988). Circular DNA plasmids were isolated from two field isolates of the Japanese pear pathotype of *A. alternata*. We are now in the process of the further characterization of the plasmids for constructing transformation vectors for *A. alternata*.

2. *Amplification of Transforming DNA in* <u>A. alternata</u> *Chromosome*

When we examined the degree of hygromycin B resistance of rDNA-vector transformants, rapidly growing sectors occasionally appeared in colonies on a medium containing a high concentration of the drug that inhibited more than 90% of the radial growth of transformants. This phenomenon suggests that amplification of the *hph* gene might have occurred in the transformant chromosomes. To confirm this possibility, transformants purified by three successive single-spore isolations were cultured on a medium containing various concentrations of hygromycin B. The rapidly growing colony sectors occasionally occurred only on the plates supplemented with the drug at concentrations that suppressed more than 90% of radial growth. Resistance of fungi isolated from such sectors was found to be 3 to 10 times higher than that of the parent transformants, and the copy number of the *hph* gene in the fungal chromosomes appeared to increase along with the increase of drug resistance. The Southern hybridization of genomic DNAs with *hph* and pBR322 probes showed that the intact vector DNA was amplified in tandemly repeated forms in the chromosomes, suggesting that *A. alternata* possesses the ability to amplify exogenous DNAs under selective conditions. This feature of *A. alternata* may allow us to effectively analyze gene functions in the fungal cells through stable expression of exogenous genes.

Gene Complementation of *Alternaria alternata* Mutants

We constructed vectors that contained the *hph* gene as a selectable marker and *A. alternata* rDNA fragments as mediators for efficient transformation of the fungus. A cosmid vector pHSTl was produced by inserting the λ phage *cos* sequence and a unique *Bam*HI site into the efficient transformation vector pDH25rla. Since the λ cos site confers the ability to accept only large (27 to 40 kb) DNA fragments on the vector, a relatively small number (approximately 3000) of transformants is expected to cover the entire *A. alternata* genome. The cosmid vector is thus suitable for the construction of *A. alternata* genomic libraries from which pathogenicity genes can be isolated by complementation of *A. alternata* mutants. A genomic library of *A. alternata* Japanese pear pathotype strain 15A was generated by ligating 27- to 40-kb genomic DNA fragments into the cosmid according to the procedure of Ish-Horowicz and Burke (1981). The complementation of AK-toxin-deficient mutants with the 15A genomic library is under way.

To validate the effectiveness of our cosmid vector system for the direct isolation of genes by complementation of appropriate recipients and because assay of the phenotype is simple and easy, we selected melanin biosynthesis genes as a model system. Three types of mutants in melanin biosynthesis were obtained after mutagenesis of strain 15A spores, and their deficiencies in the steps of the metabolic pathway were characterized (Tanabe et al., 1988) (Fig. 4). First, the mutant Brm-1 which lacks scytalone dehydratase activity (Fig. 4) was transformed by the 15A genomic cosmid library. Individual hygromycin B-resistant transformants were transferred to fresh agar media containing the antibiotics and cultured for several additional days. One transformant among 1363 tested was able to synthesize melanin as judged by the normal pigmentation of the mycelia. The transforming cosmid, designated pMBRl, was recovered from the genome by λ packaging and transduction of *E. coli* as described for *Aspergillus* by Yelton, et al., (1985). After retransformation of the mutant Brm-1 with pMBRl, 88% of hygromycin B-resistant transformants exhibited the normal melanization, indicating that an insert in the cosmid carries a gene for melanin biosynthesis. Interestingly, when the mutants Alm and Brm-2 whose deficiencies in melanin metabolism are different from the mutant

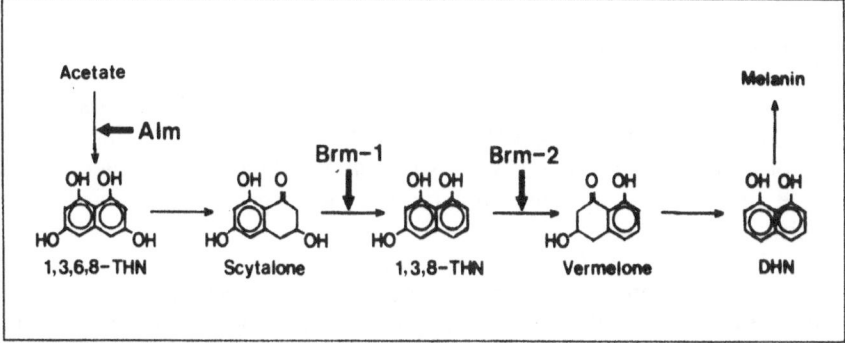

Figure 4. Deficient steps in the biosynthetic pathway of fungal 1,8-DHN melanin in melanin mutants of *A. alternata* Japanese pear pathotype. See text for details.

Brm-1 (Fig. 4) were transformed with pMBRl, 90% and 80% of drug-resistant transformants of the mutants Alm and Brm-2, respectively, also regained wild-type pigmentation of cultured mycelia. Assays for abilities of various restriction enzyme digests of pMBRl to complement melanization of each mutant showed that three genes were located as a cluster in the *A. alternata* genome. The cosmid system employed here is potentially valuable for the isolation of *A. alternata* genes, including those for the pathogenicity.

Concluding Remarks

We have applied gene manipulation techniques to the analysis of genes in the Japanese pear pathotype of *A. alternata* which causes black spot of Japanese pear by producing a host-specific AK-toxin. RFLP analysis of more than 250 field isolates of the pathogen clarified the presence of at least five types of rDNA variants in the pathogen population and provided a clue for assessing population dynamics of the pathogen. Comparison of RNA species in AK-toxin producer and non-producer mutants showed that certain RNA species are detected only in the toxin producer. However, their participation in toxin production could not be evaluated critically by analysis of transcript levels. Consequently, we have developed a transformation system of this fungus as a way to elucidate biochemical processes of toxin production by means of complementation of toxin-deficient mutants with wild-type genomic DNA. A high transformation efficiency was achieved through homologous integration of transforming DNA mediated by rDNA sequences. We constructed a genomic library of a wild-type *A. alternata* Japanese pear pathotype by using a cosmid derived from our transformation plasmid vector. As a model system, genes for fungal melanization were successfully isolated by complementation of melanin-deficient mutants with the library of the wild-type strain. The analysis of a recovered cosmid clone showed that at least three genes for melanin biosynthesis are located as a gene cluster in the *Alternaria* genome. The technology established in these studies will certainly contribute to our long-term goal, the elucidation of genetic regulation of the pathogenicity of *A. alternata* pathogens mediated by HST production.

Acknowledgements

This review is respectfully dedicated to the late Professor Syoyo Nishimura, who provided us with the opportunity to study HSTs and to learn his eminent way of thinking about HSTs as well as his serious attitude toward science. We would like to sincerely thank him for his encouragement throughout these studies.

We thank Dr. S. Ouchi, Kinki University, for a critical review of this manuscript. We are grateful to Dr. D.J. Henner, Genentech Inc., who kindly provided pDH25. Graduate students, K. Tanabe, Y. Adachi, N. Kimura, and H. Shiotani, in the Plant Pathology Laboratory, have been involved in the recent experiments reported in this paper. This work was supported in part by

Research Grants (No. 62304015 and No. 01304014) from the Ministry of Education, Science, and Culture of Japan.

References

Anagnostakis, S.A., 1982, Biological control of chestnut blight, *Science* **215**:466-471.

Bogorad, L., et al., 1983, Cloning and physical mapping of maize plastid genes, *Methods Enzymol.* **97**:524-554.

Cullen, D., et al., 1987, Transformation of *Aspergillus nidulans* with the hygromycin-resistance gene, *hph*, *Gene* **57**:21-27.

Feng, B.N., et al., 1990, Biosynthesis of host-selective toxins produced by *Alternaria alternata* pathogens. I. (8R,9S)-9,10-epoxy-8-hydroxy-9-methyl-deca-(2E,4Z,6E)-trienoic acid, as a biological precursor of AK-toxins, *Agric. Biol. Chem.* **54**:845-848.

Fincham, J.R.S., 1989, Transformation in fungi, *Microbiol. Rev.* **53**:148-170.

Futcher, A.B., 1988, The 2 μm circle plasmid of *Saccharomyces cerevisiae*, *Yeast* **4**:27-40.

Gabrial, S.A., 1980, Effect of fungal viruses on their hosts, *Annu. Rev. Phytopathol.* **18**:441-461.

Garber, R.C., et al., 1988, Organization of ribosomal RNA genes in the fungus *Cochliobolus heterostrophus*, *Curr. Genet.* **14**:573-582.

Gritz, L. and Davies, J., 1983, Plasmid-encoded hygromycin B resistance: the sequence of hygromycin B phosphotransferase gene and its expression in *Escherichia coli* and *Saccharomyces cerevisiae*, *Gene* **25**:179-188.

Hayashi, N. and Nishimura, S., 1989, Comparison of the parasitic fitness of AK- and AF-toxin producers of *Alternaria alternata*, *In* Host-Specific Toxins: Recognition and Specificity Factors in Plant Disease (Kohmoto, K. and Durbin, R.D., eds.), The Organizing Committee, Tottori University, Tottori, pp. 207-224.

Hayashi, N., et al., 1988, The presence of double-stranded RNAs in *Alternaria alternata* Japanese pear pathotype and their participation in AK-toxin productivity, *Ann. Phytopath. Soc. Japan* **54**:250-252.

Hinnen, A., Hicks, J.B. and Fink, G.R., 1978, Transformation of yeast, *Proc. Natl. Acad. Sci. USA* **75**:1929-1933.

Hynes, M.J., 1986, Transformation of filamentous fungi, *Exp. Mycol.* **10**:1-8.

Ish-Horowicz, D. and Burke, J.F., 1981, Rapid and efficient cosmid cloning, *Nucleic Acids Res.* **9**:2989-2998.

Leong, S.A. and Holden, D.W., 1989, Molecular genetic approaches to the study of fungal pathogenesis, *Annu. Rev. Phytopathol.* **27**:463-481.

Mullaney, E.J., et al., 1985, Primary structure of the *trpC* gene from *Aspergillus nidulans*, *Mol. Gen. Genet.* **199**:37-45.

Nakashima, T., et al., 1985, Isolation and structures of AK-toxin I and II, host-specific phytotoxic metabolites produced by *Alternaria alternata* Japanese pear pathotype, *Agric. Biol. Chem.* **49**:807-815.

Nakatsuka, S., et al., 1990, Biosynthetic origin of (8R,9S)-9,10-epoxy-8-hydroxy-9-methyl-deca-(2E,4Z,6E)-trienoic acid, a precursor of AK-toxins produced by *Alternaria alternata*, *Phytochemistry* **29**:1529-1531.

Nishimura, S., 1987, Recent development of host-specific toxin research in Japan and its agricultural use. *In* Molecular Determinants of Plant Diseases (Nishimura, S., et al., eds.), Japan Scientific Societies Press, Tokyo and Springer-Verlag, Berlin. pp. 11-26.

Nishimura, S. and Kohmoto, K., 1983, Host-specific toxins and chemical structures from *Alternaria* species, *Annu. Rev. Phytopathol.* **21**:87-116.

Nishimura, S. and Nakatsuka, S., 1989, Trends in host-selective toxin research in Japan, *In* Host-Specific Toxins: Recognition and Specificity Factors in Plant Disease (Kohmoto, K. and Durbin R.D., eds.), The Organizing Committee, Tottori University, Tottori, pp. 19-31.

Petes, T.D. and Botstein, D.B., 1977, Simple menderian inheritance of the reiterated ribosomal DNA of yeast, *Proc. Natl. Acad. Sci. USA* **74**:5091-5095.

Scheffer, R.P., 1983, Toxins as chemical determinants of plant disease. *In Toxins and Plant Pathogenesis* (Daly, J.M. and Deverall, B.J., eds.), Academic Press Australia, Sydney, pp. 1-40.

Scheffer, R.P. and Livingston, R.S., 1984, Host-selective toxins and their role in plant diseases, *Science* **223**:415-419.

Specht, C.A., Novotny, C.P., and Ullrich, R.C., 1983, Strain specific differences in ribosomal DNA from the fungus *Schizophyllum commune, Curr. Genet.* **8**:219-222.

Tanabe, K., Nishimura, S. and Kohmoto, K., 1988, Pathogenicity of pectic enzyme- and melanin-deficient mutants of *Alternaria alternata* Japanese pear pathotype, *Ann. Phytopathol. Soc. Japan* **54**:364 (Abstr.).

Tanabe, K., Tsuge, T., and Nishimura, S., 1989, Potential application of DNA restriction fragment length polymorphisms to the ecological studies of *Alternaria alternata* Japanese pear pathotype, *Ann. Phytopathol. Soc. Japan* **55**:361-365.

Tanaka, S., 1933, Studies on black spot disease of the Japanese pear (*Pyrus serotina* Rehd.), *Mem. Coll. Agric. Kyoto Imp. Univ.* **28**:1-31.

Tsuge, T., Hayashi, N., and Nishimura, S., 1986, Metabolic regulation of host-specific toxin production in *Alternaria alternata* pathogens (3). Instability of pathogenicity in field isolates of *Alternaria alternata* Japanese pear pathotype, *Ann. Phytopathol. Soc. Japan* **52**:488-491.

Tsuge, T., Kobayashi, H., and Nishimura, S., 1986, Metabolic regulation of host-specific toxin production in *Alternaria alternata* pathogens (4). Molecular cloning of mRNA in AK-toxin producing isolate, *Ann. Phytopathol. Soc. Japan* **52**:690- 699.

Tsuge, T., Kobayashi, H., and Nishimura, S., 1989, Organization of ribosomal RNA genes in *Alternaria alternata* Japanese pear pathotype, a host-selective AK-toxin-producing fungus, *Curr. Genet.* **16**:267-272.

Tsuge, T., Nishimura, S., and Kobayashi, H., 1990, Efficient integrative transformation of the phytopathogenic fungus *Alternaria alternata* mediated by the repetitive rDNA sequences, *Gene* **90**:207-214.

Vollmer, S. J. and Yanofsky, C., 1986, Efficient cloning of genes of *Neurospora crassa*, *Proc. Natl. Acad. Sci. USA* **83**:4869-4873.

Winston, F., Chumley, F., and Fink, G.R., 1983, Eviction and transplacement of mutant genes in yeast, *Methods Enzymol.* **101**:211-228.

Wu, M.M.J., Cassidy, J.R., and Pukkila, P.J., 1983, Polymorphisms in DNA of *Coprinus cinereus, Curr. Genet.* **7**:385-392.

Yelton, D.E., Timberlake, W.E. and Van den Hondel, C.A.M.J.J., 1985, A cosmid for selecting genes by complementation in *Aspergillus nidulans*: Selection of the developmentally regulated yA locus, *Proc. Natl. Acad. Sci. USA* **82**:834-838.

Yoder, O.C., 1980, Toxins in pathogenesis, *Annu. Rev. Phytopathol.* **18**:103-129.

Summary of Discussion of Tsuge's Paper

Durbin opened the discussion by stressing the importance of isolation and characterization of intermediates of toxin biosynthesis in the analysis of toxin-plant cell interaction and asked whether *Tsuge* found AK-toxin or its intermediates in resting spores. *Tsuge* answered that AK-toxin and its intermediates like epoxy-decatrienoic acid are most likely synthesized during conidial germination because *Nishimura's* group did not detect these compounds in dormant spores. *Bennetzen* commented that homologous recombination is common in *Saccharomyces cerevisiae*, but rare in fungi and other eukaryotes and asked if *Tsuge* knew any other example of enhanced transformation frequency by using strategies he has previously used in *Alternaria alternata*. *Tsuge* answered that vectors linked with rDNA fragments have been used in *A. nidulans* and other fungi, but transformation efficiency did not significantly increase as compared with that observed in *A. alternata*. *Chumley* then asked what the transformation frequency would be if *Tsuge* introduced a single copy of a DNA fragment like a gene for melanin biosynthesis instead of a highly repeated rDNA fragment. *Tsuge* responded that vectors linked with the DNA fragment for melanin synthesis also enhanced transformation frequency, but not to the same degree as with the rDNA fragment and inferred that the high transformation frequency with the rDNA fragment is due to the highly repeated sequence in the DNA. *Chumley* then asked whether the albino mutant *Tsuge* used is deficient in a specific enzyme for melanin biosynthesis or is due to a mutation in the regulatory gene. *Tsuge* believed that it is a mutant for melanin biosynthesis because it produced melanin in the medium containing scytalone. *Tsuyumu* asked about the possibility that more than one copy of chromosomal DNA clones had been incorporated into the transformants. *Tsuge* answered that only a single clone has so far been isolated from a transformant, but concomitant insertion of two or more copies may be possible. They had not checked the mode of incorporation of cosmid clones into the recipient chromosomal DNA, hence were not sure if they had indeed been inserted into the rDNA region. *Macko* observed that it is very important to look into every possible cofactor involved in the synthesis of toxins in a cell free system.

Chapter 11
Strategies for Characterizing and Cloning Host Specificity Genes in *Magnaporthe grisea*, the Rice Blast Fungus

Forrest G. Chumley and Barbara Valent

Although field isolates of the heterothallic Ascomycete, *Magnaporthe grisea* Barr (anamorph, *Pyricularia oryzae* Cav. or *P. grisea*), include pathogens of many grasses, individual isolates have a limited host range, parasitizing one or a few grass species (Mackill et al.,1986). Strains of the fungus that parasitize rice (*Oryza sativa*) are subdivided into races, depending on the rice cultivars they can successfully infect. The rice blast fungus shows a high degree of variability in the field; new races frequently appear with the ability to attack previously resistant rice cultivars (Ou, 1985).

Many genes in the pathogen must act together to execute the metabolic and developmental pathways that comprise the disease cycle, the ordered series of events leading from spore germination through infection, lesion development, and conidiation. Genes that control the development and function of infection structures are examples of general pathogenicity determinants, those required for the pathogen to infect any host. Genes that determine cultivar specificity, however, appear to provide important signals for a complex recognition system in the host that can trigger an effective disease resistance response (Flor, 1971; Day, 1974; Crute, 1986).

Cultivars of rice that differ from one another by the presence or absence of dominant resistance genes have been developed (Yamada et al., 1976). The resistance of any one of these cultivars is effective only against certain races of the pathogen. The genetic basis for differences in cultivar specificity between races of the rice blast pathogen has been previously unknown due to the infertility of field isolates of *M. grisea* that infect rice. Fertile *M. grisea* rice pathogens have now been developed. We have identified pathogen genes that control rice cultivar specificity, as well as genes that control pathogenicity toward a second host, weeping lovegrass (*Eragrostis curvula*). In this analysis, a gene was termed an "avirulence gene" if alleles determine virulence or avirulence in an all-or-nothing, cultivar specific fashion. Genes with no cultivar specific effects were termed "pathogenicity genes." Because obstacles to genetic analysis of rice pathogens have been overcome, the *M. grisea* system has properties that recommend it for undertaking a detailed analysis of the molecular basis for host specificity (Valent, 1990).

Avirulence Genes Identified in a Backcrossing Scheme

The interfertility of *M. grisea* strains with different host species specificities has permitted genetic analysis of those differences. A backcrossing regime was initiated by crossing the weeping lovegrass (*Eragrostis curvula*) pathogen, 4091-5-8, a highly fertile hermaphrodite, and the rice pathogen, O-135, a female sterile field isolate that also infects weeping lovegrass (Valent et al., 1990). All progeny from this cross infected weeping lovegrass, as expected, because both parents are weeping lovegrass pathogens. However, only 6 of 59 progeny were pathogens of rice, indicating that the two parental strains differ by several genes that control the ability to infect rice.

In the backcrossing regime, the rice pathogen was the recurrent parent. Female parents for the backcross generations were progeny strains chosen for having fertility comparable to that of 4091-5-8. The ratios of pathogenic to nonpathogenic (and virulent to avirulent) progeny through the backcross generations suggested that O-135 and 4091-5-8 differ in two types of genes that control the ability to infect rice. First, polygenically inherited factors determine the extent of lesion development on rice. Second, single avirulence genes govern, in an all-or-nothing fashion, virulence toward specific cultivars of rice. Several crosses confirmed the segregation of three unlinked genes, *Avr1-CO39*, *Avr1-M201* and *Avr1-YAMO* that determine avirulence on rice cultivars CO39, M201, and Yashiro-mochi, respectively. *Avr1-M201* may be linked to the mating type locus, *Mat1*.

The avirulence alleles of the cultivar specificity genes identified in these crosses appear to have been inherited from the parent that is a nonpathogen of rice, because the rice pathogen parent is virulent on all three cultivars. We identified avirulence genes for each cultivar used in these experiments, suggesting a high potential for identification of other avirulence genes derived from the nonpathogen of rice. Yaegashi, et al., (1981) reported similar results suggesting that a finger millet pathogen, WGG-FA40, carries an avirulence gene corresponding to the rice blast resistance gene, *Pi-a*.

Avirulence Genes Identified in Other Crosses

A separate series of crosses also identified single genes that control cultivar specificity (Valent, Farrall and Chumley, in preparation). Both parents in Cross 4360 (Table 1) are virulent on rice cultivars CO39, M201, and Sariceltik, and all progeny are virulent on these three cultivars. The parents differ in specificity toward rice cultivars Yashiro-mochi and Maratelli. Data from five complete tetrads (a representative tetrad is shown in Table 1) and from 30 random ascospore cultures suggest that two unlinked avirulence genes segregated among the progeny of this cross: *Avr2-YAMO* (alleles determine virulence or avirulence on cultivar Yashiro-mochi) and *Avr1-MARA* (alleles determine virulence or avirulence on cultivar Maratelli). The segregation of both genes has been confirmed in at least two subsequent crosses. These genes are not linked to the mating type locus. The parents of cross 4360 differ in specificity toward five additional cultivars; further analysis of the progeny will probably reveal the segregation of additionial avirulence genes.

Table 1. Lesion types produced by parents of cross 4360 and progeny from one tetrad.

| Strain | Mating type | Lesion type on rice cultivar | | | Lesion type on weeping lovegrass |
		Sariceltik	Yashiro-mochi	Maratelli	
Parents:					
4244-7-8	1	5[a]	0[a]	0	5
6043	1	5	5	5	0
Progeny from tetrad 1:					
4360-1-1	2	5	0	5	0
4360-1-4	2	5	0	5	0
4360-1-2	2	5	5	0	5
4360-1-8	2	5	5	0	5
4360-1-3	1	5	5	5	0
4360-1-7	1	5	5	5	0
4360-1-5	1	5	0	0	5
4360-1-6	1	5	0	0	5

[a]A scale of lesion types has been defined according to the size of lesions formed, ranging from Type 0, no visible symptoms, to Type 5, the largest spreading lesions characteristic of the host variety (Valent et al. 1986; Valent et al. 1990).

Genetic crosses were performed to determine if *Avr2-YAMO* is allelic to *Avr1-YAMO*, described in the previous section. Data from three different crosses indicated that the two genes are not linked to one another.

Genes for Pathogenicity to Weeping Lovegrass

The parents of cross 4360 differ in the ability to infect weeping lovegrass (see Table 1). Single gene segregation of pathogenicity toward weeping lovegrass was observed among the random ascospore progeny and in the tetrads (Table 1). Subsequent crosses confirmed simple segregation of the gene now named *Pwl2*. Preliminary data suggest that *Pwl2* is not linked to the weeping lovegrass pathogenicity determinant, *Pwl1*, previously identified in a cross between a weeping lovegrass pathogen (Valent et al. 1986; Valent et al., 1987).

The genes *Pwl1* and *Pwl2* must be considered "pathogenicity genes," with no cultivar specific effects, because cultivars of weeping lovegrass are not available. However, it remains an intriguing possibility that these genes may function in a manner analogous to avirulence genes.

Unstable Genes in *M. grisea*

The host specificity genes we have identified are described in Table 2. Two of these genes, *Pwl2* and *Avr2-YAMO*, appear to be unstable. That is, spontaneous pathogenic or virulent mutants derived from strains carrying the nonpathogenic or avirulent alleles of these genes appear frequently in standard assays. Two other *M. grisea* genes appear to be unstable in several strains: SMO^+, a gene that controls spore morphology (Hamer et al., 1989), and BUF^+, a gene that encodes an enzyme involved in biosynthesis of the gray pigment, melanin (Chumley, Valent, in press). However, many other genes appear to mutate at normal frequencies, including ALB^+ and RSY^+, which encode melanin biosynthetic enzymes. Cloning stable and unstable genes will yield clues to the molecular basis for genetic instability in *M. grisea*. The mechanism by which new races of a pathogen arise is an intriguing problem. Although the extent of instability in race determinants has been debated (Latterell, 1975; Ou, 1985), it is clear that new races rapidly appear in the field when new blast resistant rice cultivars are introduced.

Cloning Genes for Host Specificity

We are pursuing three independent approaches for cloning *M. grisea* genes that govern cultivar or host species specificity. The first approach is an effort to clone avirulence genes by complementation of function. A gene library has been prepared in the cosmid vector pMOcosX (M.J. Orbach, unpublished results) using DNA from strain 4392-1-6, which carries *Avr1-CO39*, *Avr1-M201*, *Avr1-YAMO*, *Avr2-YAMO*, and *Avr2-MARA*. The library has been stored as 4400 individual clones. We are screening individual cosmid clones for the presence of avirulence genes by introducing them one-by-one into recipient strain CP983 (which is virulent on all five cultivars) and screening transformants for loss of virulence on the five rice varieties. We have assumed the alleles for avirulence will be dominant; we expect to isolate one of the avirulence genes per 600 to 1000 clones screened.

The second and third approaches rely on cloning avirulence and pathogenicity genes by chromosome walking from physical genetic markers linked to the gene of interest. The second approach utilizes classical RFLP markers; the third approach utilizes MGR sequence polymorphisms (Hamer et al. 1989). The parents for the RFLP mapping population (74 random progeny of the cross in Table 1) are strains 4224-7-8 [*Avr2-YAMO*, *Avr1-MINE* (cultivar Minehikari), *Avr1-TSUY* (cultivar Tsuyuake), *Avr1-MARA*] and 6043 (*Pwl2*; Leung et al., 1988). Cosmid clones are being used as probes for detecting RFLPs; about 30% of the cosmids yield useful markers with at least one of the eight restriction enzymes tested. With about 30 RFLP markers mapped, linkage has been detected between *Avr2-YAMO* and flanking RFLPs that lie 5.9 cM and 13.4 cM distant from the gene. (J. Sweigard, A. Walter, B. Valent, F. Chumley, unpublished results). MGR sequence polymorphisms have served as reliable genetic markers in the backcrossing scheme described above (Valent et al., 1990). One such marker maps approximately 7.6 cM from *Avr1-CO39*;

Table 2. Genes for host specificity identified in *M. grisea*.

Gene	Host	Source[a]	Stability[b]
Avr1-C039	C039	Weeping lovegrass pathogen	Stable
Avr1-M201	M201	Weeping lovegrass pathogen	Stable
Avr1-YAMO	Yashiro-mochi	Weeping lovegrass pathogen	Stable
Avr2-YAMO	Yashiro-mochi	Rice pathogen	Unstable
Avr1-MARA	Maratelli	Unknown	Unknown
Pwl1	Weeping lovegrass	Goosegrass pathogen	Stable
Pwl2	Weeping lovegrass	Unknown	Unstable

[a]The "source" listed for each gene is based on the infectivity of the parents in the cross in which the gene was identified. The source indicated is the parent strain that carried the allele for nonpathogenicity or avirulence.

[b]A host specificity determinant is listed as "unstable" if virulent or pathogenic mutants appear at a high frequency, as described in the text.

single copy DNA adjacent to the MGR sequence has been cloned to serve as the starting point for chromsome walk to clone the *avr* gene (M.J. Orbach, unpublished results).

Discussion

Host specificity genes of *M. grisea* that have been confirmed by segregation in three or more crosses are listed in Table 2. Leung et al., (1988) report evidence for additional genes. Nonpathogens of rice contain genes that function as cultivar specific avirulence genes when they are introduced into strains that infect rice. Similar observations have been published concerning bacterial plant pathogens (Whalen et al., 1988; Kobayashi et al., 1989). The ease with which single genes for cultivar specificity have been identified in the crosses reported here suggests that these genes are numerous.

Whether or not the cultivar specificity genes we have defined may be classical avirulence genes remains to be determined. We can provide no information on dominance relationships between alleles of these genes because *M. grisea* does not form stable vegetative diploids or heterokaryotic conidia (Crawford et al., 1986). Further genetic analysis of both the host and the pathogen will be required to define any functional correspondence that may exist between a particular pathogen gene and a particular resistance gene in rice. The rice cultivar Yashiro-mochi, one of the ten Japanese differential cultivars (Yamada et al., 1976), is reported to have the blast resistance gene, *Pi-ta*.

Genetic crosses of Yashiro-mochi will determine if the pathogen genes *Avr1-YAMO* and *Avr2-YAMO* interact with the blast resistance gene, *Pi-ta*, with previously unidentified resistance genes, or with any identifiable gene for resistance. The cultivars CO39, M201, and Maratelli have not previously been reported to carry blast resistance genes. Genetic crosses involving these cultivars will determine if they contain unidentified blast resistance genes.

Our efforts now are focused on cloning avirulence genes as a first step toward understanding the molecular mechanisms that determine cultivar specificity. With cloned avirulence genes in hand, we plan to address the following questions. What gene products are encoded by avirulence genes? How might these gene products interact with the products of host resistance genes? How is the expression of avirulence genes controlled? What molecular genetic events accompany the appearance of new races in the field? What is the difference between stable and unstable avirulence genes? How do rice pathogen avirulence genes differ from those present in nonpathogens of rice? The goal of understanding the molecular basis for race-cultivar specificity remains a major challenge in biology.

References

Crute, I.R. and Fraser, R.S.S., 1986, The genetic basis of relationships between microbial parasites and their hosts. *In* Mechanisms of Resistance to Plant Diseases. Martinius Nijhoff and W. Junk, Dordrecht, pp. 80-142.

Crawford, M.S., Chumley F.G., Weaver C.G., and Valent, B., 1986, Characterization of the heterokaryotic and vegetative diploid phases of *Magnaporthe grisea*, *Genetics* 114:1111-1129.

Chumley, F.G., and Valent, B., 1990, Genetic analysis of melanin deficient nonpathogenic mutants of *Magnaporthe grisea*, *Mol. Plant-Microbe Interac.* (in press).

Day, P.R., 1974, Genetics of Host-Parasite Interaction. W.H. Freeman and Co., San Francisco.

Flor, H.H., 1971, Current status of the gene-for-gene concept, *Annu. Rev. Phytopathol.* 9:275-296.

Hamer, J.E., Farrall, L., Orbach, M.J., Valent, B., and Chumley, F.G., 1989, Host-species specific conservation of a family of repeated DNA sequences in the genome of a fungal plant pathogen, *Proc. Natl. Acad. Sci. USA* 86: 9981-9985.

Hamer, J.E., Valent, B., Chumley, F.G., 1989, Mutations at the SMO genetic locus affect the shape of diverse cell types in the rice blast fungus, *Genetics* 122:351-361.

Kobayashi, D.Y., Tamaki, S.J., Keen, N.T., 1989, Cloned avirulence genes from the tomato pathogen *Pseudomonas syringae* pv. *tomato* confer cultivar specificity on soybean. *Proc. Natl. Acad. Sci. USA* 86:157-161.

Latterell, F.M., 1975, Phenotypic stability of pathogenic races of *Pyricularia oryzae*, and its implication for breeding blast resistant rice varieties. *In* Proceedings of the Seminar on Horizontal Resistance to Blast Disease of Rice. Colombia Series CE-No. 9 Centro Internacional de Agricultura Tropical, Cali, pp. 199-234.

Leung, H., Borromeo, E.S., Bernardo, M.A., Notteghem, J.L., 1988, Genetic analysis of virulence in the rice blast fungus *Magnaporthe grisea*, *Phytopathology* 78:1227-1233.

Mackill, A.O., Bonman, J.M., 1986, New hosts of *Pyricularia oryzae*, *Plant Disease* 70:125-127.

Ou, S.H., 1985, *In* Rice Diseases. Commonwealth Agricultural Bureaux, Slough, UK. pp. 109-201.

Valent, B., 1990, Rice blast as a model System for Plant Pathology, *Phytopathology* 80:33-36.

Valent, B., Chumley. F.G., 1987, Genetic Analysis of Host Species Specificity in *Magnaporthe grisea*. UCLA Symp. Mol. Cell. Biol. (New Series) 48:83-93.

Valent, B., Crawford, M.S., Weaver, C.G., Chumley, F.G., 1986, Genetic studies of pathogenicity and fertility of *Magnaporthe grisea*, *Iowa State J. Res.* 60:569-594.

Valent, B., Farrall, L., Chumley, F.G., 1990, *Magnaporthe grisea* genes for pathogenicity and virulence identified through a series of backcrosses, *Genetics* (In press).

Whalen, M.C., Stall, R.E., Staskawicz, B., 1988, Characterization of a gene from a tomato pathogen determining hypersensitive resistance in non-host species and genetic analysis of this resistance in bean, *Proc. Natl. Acad. Sci. USA* 85:6743-6747.

Yaegashi, H., Asaga, K., 1981, Further studies on the inheritance of pathogenicity in crosses of *Pyricularia oryzae* with *Pyricularia* sp. from finger millet, *Ann. Phytopathol. Soc. Jpn.* 47:677-679.

Yamada, M., Kiyosawa, S., Yamaguchi, T., Hirano, T., Kobayashi, T., Kushibuchi, K., Watanabe, S., 1976, Proposal of a new method for differentiating races of *Pyricularia oryzae* Cavara in Japan, *Ann. Phytopathol. Soc. Jpn.* 42:216-219.

Summary of Discussion of Chumley's Paper

Mills initiated the questioning by asking if more could be said about the B chromosomes. *Chumley* responded by stating that rice pathogens contain anywhere from one to three or more B chromosomes depending on whether they are all resolved. We are calling them B chromosomes because they do not seem to be transmitted normally in genetic crosses and they act as if they are acentric. They appear to be somewhat typical in fungal cells where they have been looked for. They do not seem to be present in *P. grisea* isolates that have been studied. When probes are made of individually isolated B chromosomes, they appear to hypridize only to isolated B chromosomes. *Hashiba* asked whether transposons can be used in mutational analysis. *Chumley* said that we would like to be able to do insertional mutagenesis, but the element has no selective phenotype and it is present in high copy number and therefore may not be very helpful in that regard. It may be valuable for use as a marker for chromosome segments. *Bushnell* stated that in the dikaryon of cereal rust fungi there is more indication that avirulence may be semidominant and not dominant. In complementation experiments, it becomes unclear how the addition of one avirulence gene can shift the phenotype. *Chumley* responded that there are lots of things that can sink the complementation cloning approach he described but that they want to proceed for at least the next year. *Van Alfen* asked whether they have looked for any reverse transcriptase activity or movement of these elements. *Chumley* responded that they have not seen any evidence of movement. They intend to clone the ends, the 3' end in particular, to see if there are any structures that would be typical of these line-1 type sequences, using reverse PCR and primers from within the region. Protein extracts are also being made to see if there is reverse transcriptase activity. We are anxious to see if virus-like particles in this fungus contain sequences that are homologous to these elements. It is not known whether line-1 elements move through a viral intermediate so we might have the opportunity to make a contribution to understanding the biology of line-1 elements in this fungus.

Chapter 12
Role of Host-Specific Toxins in the Pathogenesis of *Alternaria alternata*

Hiroshi Otani, Keisuke Kohmoto, Motoichiro Kodama, and Syoyo Nishimura[†]

Host recognition and its specificity in host-parasite interactions are the most attractive subjects in the field of physiological plant pathology. Recently, several models have been proposed as to the mechanisms responsible for determining disease specificity. One of these models comes from studies of host-specific or host-selective toxins (HSTs) produced by fungal pathogens. A mechanism that determines specificity in the diseases involving HST is comprised of three basic processes (Nishimura et al., 1983; Kohmoto et al., 1987, 1989): 1) Spores of a fungal parasite release HST, a host recognition factor, on germination, 2) the released signal factor selectively binds to receptor sites in the host cells, and 3) the accessible state or susceptibility of host cells to possible hyphal invasion is disposed by the signal transduction.

There are now at least fourteen fungal plant pathogens that produce HST (Scheffer, 1983). Of these HST producers, seven examples are from *Alternaria* pathogens which are considered to be distinct pathotypes of the collective species, *A. alternata* (Fr.) Keissler (Nishimura et al., 1983). In this paper, we will review our recent work on the initial action sites of *Alternaria* HSTs and their subsequent essential events which lead to induction of susceptibility, and discuss their role in pathogenesis.

Primary Action Sites of *Alternaria* HSTs

On the basis of results obtained from biochemical, physiological, and ultrastructural studies on the primary action site, *Alternaria* HSTs are classified into three groups (Table 1). ACT-, AF- and AK-toxins produced by tangerine, strawberry, and Japanese pear pathotypes of *A. alternata*, respectively, exert primary effects on the plasma membranes of susceptible cells. A rapid increase in electrolyte loss from tissues and the occurrence of invagination of plasma membranes are common characteristics of this group (Kohmoto et al., 1987). In addition, AF

[†]Deceased May 27, 1989.

Table 1. *A. alternata* HSTs known to date (1990).

Disease	*A. alternata* pathotypes	HST	Target site
Alternaria blotch of apple	Apple	AM-toxin	Chloroplast and plasma membrane
Alternaria stem canker of tomato	Tomato	AL-toxin	Mitochondrion
Black spot of Japanese pear	Japanese pear	AK-toxin	Plasma membrane
Black spot of strawberry	Strawberry	AF-toxin	Plasma membrane
Brown spot of rough lemon	Rough lemon	ACR-toxin	Mitochondrion
Brown spot of tangerine	Tangerine	ACT-toxin	Plasma membrane
Brown spot of tobacco	Tobacco	AT-toxin	Mitochondrion

and AK-toxins induce a rapid depolarization of the respiration-dependent membrane potential sustained by the H^+-pump (Namiki et al., 1986; Otani et al., 1989a).

Recently, Otani et al. (1989b) isolated plasma membranes from young fruits of Japanese pear and determined the effect of AK-toxin on ATPase activity of the isolated plasma membranes. However, no direct effect of the toxin was detected on the activity of membrane H^+-ATPase. An experiment was designed to find the receptor sites for AK-toxin in the plasma membranes of susceptible cells (Otani et al., 1989b). The plasma membrane fractions isolated from susceptible and resistant pear fruits were mixed with AK-toxin I solution (0.02 μM), which is approximately four times greater than the minimum concentration required to cause necrosis on susceptible pear leaves. The mixture was spotted on susceptible pear leaves, and necrosis formation on the leaves was determined after incubation for 48-h. The toxin-induced necrosis was reduced about 50% as compared with the control in the solution mixed with susceptible membrane fractions at concentrations of 0.9 to 3.6 mg protein/ml (Table 2). On the other hand, some reduction of necrosis was observed when AK-toxin was mixed with the membrane fraction of resistant cultivars, unless a massive amount of membrane (3.6 mg protein/ml) was used. When the membrane fractions were pretreated with SH-reagents that protect plasma membrane from AK-toxin action (Otani et al., 1985), the reduction of necrosis disappeared only in susceptible membrane fractions at 0.9 to 2.7 mg protein/ml. These results indicate that AK-toxin may bind to specific substances containing sulfhydryl radicals that exist in susceptible plasma membranes, although AK-toxin also binds non-specifically to susceptible and resistant plasma membranes. The isolation and characterization of the specific binding substances are under way.

The second group is comprised of HSTs which primarily affect mitochondrial function of susceptible cells. ACR-toxin (ACRL-toxin) produced by *A. alternata* rough lemon pathotype induces ultrastructural changes in mitochondria of the susceptible rough lemon (Kohmoto et al., 1984). A mitochondrial site of the toxin action is confirmed by the following experiments with mitochondria isolated from rough lemon leaves: 1) the toxin causes uncoupling of oxidative phosphorylation and changes in membrane potential, and 2) the toxin inhibits malate oxidation, apparently because of the lack of NAD^+ in the matrix (Akimitsu et al., 1989). AL-toxin (AAL-toxin) produced by *A. alternata* tomato pathotype and AT-toxin produced by *A. alternata* tobacco pathotype also appear to affect the mitochondria of susceptible cells. The first ultrastructural changes in host cells detected by treatments with AL and AT-toxins were on mitochondria (Park et al., 1981; Kodama et al., 1985). Gilchrist (1983) indicated, however, that AL-toxin inhibits aspartate carbamoyltransferase of susceptible tomatoes. The precise action sites of AL and AT-toxins still remain unclear.

The representative of the third group is AM-toxin, produced by *A. alternata* apple pathotype. Two primary sites have been suggested for AM toxin action by electron microscopic and physiological studies (Kohmoto et al., 1982). One site is in the chloroplast, where the toxin induces detachment and vesiculation of grana lamellae and a decrease in photosynthetic CO_2 fixation. The other is in the plasma membrane, where the toxin causes invagination of membranes and electrolyte loss. Based on varietal differences in susceptibility to the pathogen and sensitivity to the toxin, apple cultivars can be divided into three groups: susceptible, moderately resistant, and resistant (Kohmoto et al., 1977). Recently, Shimomura et al. (1989) extended these studies to include the

Table 2. Toxicity of AK-toxin mixed with plasma membrane fractions from susceptible and resistant pear fruits.

Plasma membrane fraction (mg protein/ml)	Necrotic area[a] (mm^2)				
	Susceptible cv.			Resistant cv.	
	Nijisseiki	Osa-Nijisseiki	Shinsui	Kosui	Hosui
3.6	7.4(40)[b]	13.0(44)	8.5(40)	8.7(46)	11.0(59)
2.7	4.5(24)	13.0(44)	6.8(32)	19.2(101)	13.4(72)
1.8	8.0(43)	18.6(63)	11.7(55)	19.2(101)	20.3(109)
0.9	10.0(54)	18.9(64)	8.9(42)	17.7(93)	16.4(88)
Control	18.6(100)	29.5(100)	21.3(100)	19.0(100)	18.6(100)

[a]Susceptible pear leaves were treated with mixture of AK-toxin I (0.02 µM) and plasma membrane fractions, and necrotic area appeared on the leaves was measured after 48-h.
[b]Per cent of control.

toxin action on the choloroplast and plasma membrane. AM-toxin I at 10^{-8} M inhibited CO_2 fixation and caused electrolyte loss as well as necrosis in susceptible leaves. Toxin-induced electrolyte loss and necrosis in moderately resistant leaves were observed at $10^{-5}M$, while CO_2 fixation was significantly inhibited at $10^{-6}M$. Strangely enough, $10^{-5}M$ toxin inhibited CO_2 fixation in resistant apple and some non-host leaves without causing electrolyte loss and necrosis. On the other hand, susceptible apple cultivars exhibited a marked tissue specificity in relation to the toxin action (Maeno et al., 1984). No reactions were detected in susceptible non-green tissues such as petals, while green and non-green tissues of moderately resistant apple cultivars had almost the same sensitivity to the toxin. More recently, Shimomura et al. (1990) examined the effect of AM-toxin on green and white calli induced from petal and leaf. The result (Table 3) showed that white calli of susceptible cultivars were completely insensitive to the toxin. The complicated mode of action of AM-toxin is still under investigation.

Early Events in HST Action and Induction of Susceptibility

Recently, the nature of some components involved in the sequence of HST-induced events has been partially characterized by examining counteractive effects of various treatments on toxin action. Most of these results were reported in our recent reviews (Kohmoto et al., 1987, 1989). Thus, the subjects discussed here are mainly focused on the events associated with induction of susceptibility by three *Alternaria* HSTs which have different action sites.

Table 3. Responses of green and white calli induced from apple leaf and petal to AM-toxin.

Apple cv.	Source of callus (Response to toxin)[a]	Color of callus	Response to toxin [a]
Red Gold (Susceptible)	Leaf (+)	Green White	+ -
Indo (Susceptible)	Petal (-)	White	-
Jonathan (Moderately resistant)	Leaf (+)	Green White	+ +
	Petal (+)	Green White	+ +

[a]+; Sensitive to toxin, -; Insensitive to toxin.

1. AK-Toxin

The SH-alkylating reagents bromoacetic acid, iodoacetamide, and iodomethane exerted a markedly protective effect on both AK-toxin-induced electrolyte loss and veinal necrosis in susceptible pear leaves, provided the leaves were treated before toxin exposure (Otani et al., 1985). Susceptible pear leaves were treated with SH-reagents and inoculated with spores of the pathogen. After 24-h of incubation, fungal behavior on the leaves was observed. The chemicals did not affect germination and appressorial formation of the fungus, but significantly reduced the fungal infection to tissues and the number of lesions.

Copper- and iron-chelating agents, such as salicylaldoxime and sodium sulfide, had a countereffect on AK-toxin-induced necrosis, but not suppressed electrolyte loss (Otani et al., 1985). This effect was detectable within 10-h after toxin exposure. Unlike the case of SH-reagents, treatment with these chemicals did not affect the fungal infection on susceptible leaves.

2. ACR-Toxin

Although the primary target of ACR-toxin is the mitochondrion, it also induces a rapid increase in electrolyte loss from leaf tissues and eventually causes water congestion and veinal necrosis on susceptible leaves within 24-h after exposure to the toxin. Both electrolyte loss and necrosis are markedly suppressed in light (Akimitsu et al., 1988). On the other hand, the toxin action on isolated mitochondria was not influenced by light. Effect of light on infection behavior was investigated (Kohmoto et al., 1988). In further investigations (The Japanese Quarantine Act prohibits the use of pathogenic spores) we mixed saprophytic *A. alternata* spores with ACR-toxin solution. The spore suspension was sprayed on susceptible leaves and incubated in the dark and light. In the dark, necrosis became visible after 24-h incubation; by 48-h it extended to the whole leaf, but was completely suppressed in the light. The rate of infection hypha formation in the light was also suppressed.

3. AM-Toxin

SH-reagents counteractive to AK-toxin action, when applied before AM-toxin exposure, gave a remarkable protection against the toxin-induced electrolyte loss, and reduced necrosis on susceptible apple leaves (Shimomura et al., 1989). On the contrary, the toxin-induced inhibition of photosynthetic CO_2 fixation, an early event in the toxin action on chloroplast, was not affected by these chemicals. When susceptible leaves were treated with SH-reagents and inoculated with pathogenic spores, the pathogen failed to invade the tissues, indicating that the toxin-induced dysfunction of plasma membranes, rather than chloroplasts, is necessary for the induction of susceptibility.

AM-toxin action was affected by light (Tabira et al., 1989). Continuous irradiation of light to susceptible leaves after toxin exposure inhibited toxin-induced necrosis. When light irradiation was interrupted with darkness for specified periods of time immediately after toxin exposure, a period of darkness longer than 5-h was required for necrosis development. However, the time required for necrosis was shortened to about 3-h if light-cut-off started later than 2-h after toxin exposure. The action spectrum for the photo-protection was estimated to be 570 to 680 nm, the most effective wavelength being near 602 nm. The effective light was very specific to the necrosis induced by AM-toxin, and was not affected by treatment with photosynthetic inhibitors. Moreover, light did not affect toxin-induced electrolyte loss and reduction of photosynthetic CO_2 fixation. When the leaves were inoculated with pathogenic spores or nonpathogenic spores plus AM-toxin, light had no protective effect on their invasion.

The effects of various treatments on actions of three HSTs are summarized in Table 4. Interestingly, the protection of electolyte loss induced directly or indirectly by HSTs led to the suppression of fungal colonization. However, even if the process of host cell death after plasma membrane dysfunction was blocked, fungal infection was easily established. These results indicate that toxin-induced plasma membrane disorder is a key event in early pathogenesis.

Host Resistance and Its Suppression by HST

As stated in the preceding section, HST is required not for killing host cells prior to fungal infection but for causing dysfunction to the plasma membrane system. A question arose: why did the membrane dysfunction by HST permit the pathogen to easily establish itself in host tissues? We hypothesized that the key role of HST in pathogenesis is to suppress the general resistance mechanisms in the host (Nishimura et al., 1983). Several lines of evidence from studies with AK-toxin supported this hypothesis (Hayami et al., 1982; Tanabe et al., 1986; Otani et al., 1988).

Pre-inoculation with spores of various fungi nonpathogenic to Japanese pear reduced the subsequent infection by *A. alternata* Japanese pear pathotype. Such an inhibition of the infection was also induced by pre-treatment with spore germination fluids of these nonpathogens. Spore germination fluid of the pathogen gave the same protection when AK-toxin had previously been removed. These results suggest that pear leaves possess a latent resistance mechanism to fungal invasion, which could be induced by the factor released from the germinating spores of fungi regardless of their pathogenicity. A

Table 4. Comparison of effects of various treatments on actions of three *Alternaria* HSTs.

HST	Treatment	Toxin action to:[a]				Fungal coloni- zation[a]
		Electrolyte loss	Mitochondrial oxidation	CO_2 fixation	Cell viability	
ACR	Light	+	-		+	+
AK	SH-reagent	+			+	+
	Copper and iron chelating agent	-			+	-
AM	SH-reagent	+		-	+	+
	Light	-		-	+	-

a:+; Protected, -; Not protected.

resistance-inducing factor was isolated from spore germination fluid of *A. alternata* by column chromatography with Con A-Sepharose 4B, Sephadex G-25 and Sephacryl S-300, and was termed "inducer." The inducer was sensitive to β-glucosidase and α-mannosidase, but was insensitive to β-galactosidase, α-glucosidase and proteinase K, indicating that a moiety containing β-glucose and α-mannose is responsible for inducer activity. Chemical analysis showed that the inducer is a polysaccharide with molecular weight of approximately 40 kDa, and consists primarily of mannose with a limited amount of galactose, glucose, rhamnose and xylose. The inducer had activity on pear leaves at a concentration of 1 μg/ml, and required an interval of 4 to 6-h for the induction of resistance to be expressed. To test whether the inducer is host-specific, apple, strawberry and tobacco leaves were treated with the inducer, and then inoculated with the respective *A. alternata* pathogen. On these plants, the fungal infection was markedly reduced by inducer treatment, indicating that inducer non-specifically triggers the host resistance mechanisms.

Although the fungi belonging to *A. alternata* have a general potential for aggressiveness, they are usually able to invade only susceptible tissues. However, if host resistance was not induced in the tissues, *A. alternata* may easily invade the tissues by its aggressiveness. Therefore, the relationship between the suppression of resistance induction and the degree of infection was investigated. Susceptible and resistant pear leaves were heat-treated in water at 50°C for 20 seconds, and inoculated with spores of virulent and avirulent *A. alternata*. These isolates, regardless of their pathogenicity, colonized the heated

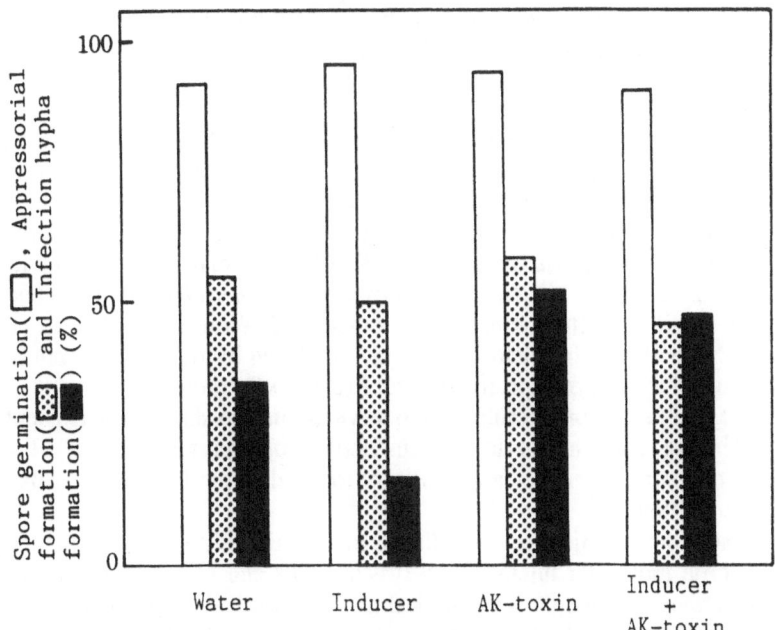

Figure 1. Effect of AK-toxin on induction of resistance in susceptible pear leaves by inducer. Pear leaves were treated with inducer plus AK-toxin, AK-toxin, inducer, and water for 6-h at 28°C, and inoculated with spores of *A. alternata* Japanese pear pathotype. Fungal behavior was observed 24-h after inoculation.

susceptible and resistant leaves. When pear leaves were heated and then treated with the inducer, no effect of the inducer was detected. On the contrary, the heat treatment was ineffective when treatment was made more than 4-h after the exposure to the inducer. These results indicate that the induction of resistance in pear tissues by the inducer requires at least 4 hr, and the induction of resistance is suppressed by heat treatment, enabling *A. alternata* to invade the tissues.

AK-toxin is released from germinating spores of virulent spores, but not from avirulent spores. When avirulent spores are inoculated together with AK-toxin, the spores could invade the susceptible tissues as if they are virulent. Thus the AK-toxin from germinating spores is necessary for invasion of *A. alternata*. To examine whether AK-toxin suppresses the induction of resistance by the inducer, susceptible pear leaves were treated with solutions of AK-toxin plus inducer, AK-toxin, inducer, and water. After incubation for 6-h, the leaves were inoculated with virulent spores and infection behavior in the leaf tissues was observed. Although the formation of infection hypha was reduced on the leaves treated with the inducer, the activity of inducer was suppressed completely when it was treated together with AK-toxin (Fig. 1). Thus, the role of AK-toxin in pathogenesis is to suppress the induction of resistance in the host plants.

Concluding Remarks

Many attempts have been made to find the primary target site of HSTs, because the knowledge of the mechanisms underlying the initial physiological event in toxin-treated host cells is essential for understanding the mutual recognition between hosts and pathogens. Plasma membranes, mitochondria, and chloroplasts of *Alternaria* HSTs were suggested as possible target organelles for the primary toxin action. However, we cannot draw this conclusion until the specific sites of toxin action in host cells are isolated and characterized chemically. Recently, Dewey et al. (1988) reported that a 13 kDa maize mitochondrial protein is associated with sensitivity to HMT-toxin (T-toxin), an HST produced by *Biporaris maydis* race T. When the maize gene coding for the 13 kDa protein was cloned and expressed in *Escherichia coli*, the bacteria became sensitive to the toxin. On the other hand, Wolpert and Macko (1989) reported that HV-toxin (victorin), an HST produced by *B. victoriae*, binds in a covalent and a genotype-specific manner to a 100 kDa protein from oat leaves, a possible toxin receptor. Results with our AK-toxin also showed the existence of toxin-binding components in the susceptible plasma membranes. Further studies of toxin action sites are necessary for evaluating the molecular basis of host recognition.

The ultimate objective of studies of toxin action is not only to characterize the specific molecular interaction between toxins and host targets, but also to elucidate the sequence of physiological and biochemical events involved in the early pathogenesis. In the present paper, we showed that *Alternaria* HSTs which have different action sites caused directly or indirectly plasma membrane dysfunctions in the host cells and consequently induced a susceptible state in these cells without necrosis. Furthermore, we pointed out that the key role of HST in pathogenesis is to suppress the induction of host resistance.

Based on these results, we can now depict a scheme for the mechanism of *A. alternata* infection. The pathogen releases both HST and inducer at the spore germination stage. HST binds immediately to the HST-specific site in susceptible cells. A signal of HST recognition at the site is transmitted directly or indirectly to plasma membranes. Because of the plasma membrane dysfunction, a channel to suppress the induction of resistance is switched on. This process may occur prior to induction of host resistance by inducer, because the induction requires at least 4-h. When no resistance is induced, the pathogen can invade the tissues easily. In resistant genotypes, host resistance is induced in the cells by inducer because of their insensitivity to HST. At present, however, the scheme is still hypothetical in many respects, and requires further evaluation.

Acknowledgements

This work was supported, in part, by grants from the Ministry of Education, Science and Culture, Japan (Nos. 63440010 and 01304014).

References

Akimitsu, K., et al., 1988, Inhibition by light of actions of ACR-toxin and HMT-toxin in susceptible leaf tissues, *Ann. Phytopathol. Soc. Japan* 54:362-363 (Abstr.).

Akimitsu, K., et al., 1989, Host-specific effects of toxin from the rough lemon pathotype of *Alternaria alternata* on mitochondria, *Plant Physiol.* 89:925-931.

Dewey, R., et al., 1988, A 13KD maize mitochondrial protein in *Escherichia coli* confers sensitivity to *Biporaris maydis* toxin, *Science* 239:293-295.

Gilchrist, D.G., 1983, Molecular modes of action, *In* Toxins and Plant Pathogenesis (Daly, J.M. and Deverall, B.J., eds.), Academic Press, Sydney, pp. 81-136.

Hayami, C., et al., 1982, Induced resistance in pear leaves by spore germination fluids of nonpathogens to *Alternaria alternata*, Japanese pear pathotype, and suppression of the induction by AK-toxin, *J. Fac. Agric. Tottori Univ.* 17:9-18.

Kodama, M., et al., 1985, Ultrastructural changes in host cells treated with AT-toxin, a host-specific toxin from the tobacco pathotype of *Alternaria alternata*, *Ann. Phytopathol. Soc. Japan* 51:379 (Abstr.).

Kohmoto, K., Taniguchi, T. and Nishimura, S., 1977, Correlation between the susceptibility of apple cultivars to *Alternaria mali* and their sensitivity to AM-toxin I, *Ann. Phytopathol. Soc. Japan* 43:65-68.

Kohmoto, K., Nishimura, S., and Otani, H., 1982, Action sites for AM-toxins produced by the apple pathotype of *Alternaria alternata*, *In* Plant Infection: The Physiological and Biochemical Basis (Asada, Y., Bushnell, W.R., Ouchi, S., and Vance, C.P., eds., Japan Sci. Soc. Press, Tokyo and Springer-Verlag, Berlin, pp. 81-136.

Kohmoto, K., et al., 1984, Ultrastructural changes in host leaf cells caused by host-selective toxin of *Alternaria alternata* from rough lemon, *Can. J. Bot.* 62:2485-2492.

Kohmoto, K., Otani, H. and Nishimura, S., 1987, Primary action sites for host-specific toxins produced by *Alternaria* species, *In* Molecular Determinants of Plant

Diseases (Nishimura, S., Vance, C.P. and Doke, N., eds.), Japan Sci. Soc. Press, Tokyo and Springer-Verlag, Berlin, pp. 127-143.

Kohmoto, K., et al., 1988, Relationship between disorder of plasma membranes and induction of accessibility to fungal invasion, *Ann. Phytopathol. Soc. Japan* **54**:363 (Abstr.).

Kohmoto, K., et al., 1989, Host recognition: Can accessibility to fungal invasion be induced by host-specific toxins without necessitating necrotic cell death? *In* Phytotoxins and Plant Pathogenesis, NATO ASI Series, vol. H27 (Graniti, A., Durbin, R. D. and Ballio, A., eds.), Springer-Verlag, Berlin, pp. 249-265.

Maeno, S., et al., 1984, Different sensitivities among apple and pear cultivars to AM-toxin produced by *Alternaria alternata* apple pathotype, *J. Fac. Agric. Tottori Univ.* **19**:8-19.

Namiki, F., et al., 1986, Studies on host-specific AF-toxins produced by *Alternaria alternata* strawberry pathotype causing Alternaria black spot of strawberry (5). Effect of toxin on membrane potential of susceptible plants as assessed by electrophysiological method, *Ann. Phytopathol. Soc. Japan* **52**:610-619.

Nishimura, S. and Kohmoto, K., 1983, Roles of toxins in pathogenesis, *In* Toxins and Plant Pathogenesis (Daly, J.M. and Deverall, B.J., eds.), Academic Press, Sydney, pp.137-157.

Otani, H., et al., 1985, Two different phases in host cell damages induced by AK-toxin of *Alternaria alternata* Japanese pear pathotype, *J. Fac. Agric. Tottori Univ.* **20**:8-17.

Otani, H., et al., 1988, Suppression of resistance by toxins, Abstracts in 5th ICPP held in Kyoto, p. 220.

Otani, H., et al., 1989a, Effect of AK-toxin produced by *Alternaria alternata* Japanese pear pathotype on membrane potential of pear cells, *Ann. Phytopathol. Soc. Japan* **55**:466-468.

Otani, H., Kohmoto, K. and Nishimura, S., 1989b, Action sites for AK-toxin produced by the Japanese pear pathotype of *Alternaria alternata*, *In* Host-Specific Toxins: Recognition and Specificity Factors in Plant Disease (Kohmoto, K. and Durbin, R. D., eds.), Tottori Univ. Press, Tottori, pp. 107-120.

Park, P., et al., 1981, Comparative effects of host-specific toxins from four pathotypes of *Alternaria alternata* on the ultrastructure of host cells, *Ann. Phytopathol. Soc. Japan* **47**:488-500.

Scheffer, R.P., 1983, Toxins as chemical determinants of plant disease, *In* Toxins and Plant Pathogenesis (Daly, J.M. and Deverall, B.J., eds.), Academic Press, Sydney, pp. 1-40.

Shimomura, N., et al., 1989, Two action sites of AM-toxin and their pathological significance, *Ann. Phytopathol. Soc. Japan* **55**: 482 (Abstr.).

Shimomura, N., et al., 1990, Different responses of susceptible and moderately resistant apple cultivars to AM-toxin, Abstracts in 1990 Annual Meeting of Phytopathol. Soc. Japan held in Matsuyama, p. 36.

Tabira, H., et al., 1989, Light-induced insensitivity of apple and Japanese pear leaves to AM-toxin from *Alternaria alternata* apple pathotype, *Ann. Phytopathol. Soc. Japan* **55**:567-578.

Tanabe, K., et al., 1986, Purification and biological activity of inducer for host resistance produced by *Alternaria alternata*, *Ann. Phytopathol. Soc. Japan* **52**:527 (Abstr.).

Wolpert, T.J. and Macko, V., 1989, Specific binding of victorin to a 100-kDa protein from oats, *Proc. Natl. Acad. Sci. USA* **86**:4092-4096.

Summary of Discussion of Kohmoto's Paper

Oku initiated the discussion by asking whether inducer induces phytoalexin or host responses which are responsible for inhibiting infection by the fungus. *Kohmoto* replied that in their system, especially the leaves, no phytoalexin was detected. Instead, infection inhibiting factors were detected that do not have any antibiotic activity but only regulate infection peg formation. Recently two or three phenolic substances were isolated, one with very similar properties to *Kohmoto's* inhibitor. *Yamada* stated that in the pea system pretreatment produces an effect in about 3 hours. He then asked how long these HSTs are effective in suppressing or inhibiting the resistance response. *Kohmoto* responded that was not possible to study this because HST at later times kills the host cell and that that was a special case. It usually kills the cells under natural conditions. *Yamada* then asked whether the concentration of toxin was similar to that under natural conditions when the host was treated. *Kohmoto* replied that he thought that to be the case. He said that because the toxin is highly potent, it is usually used at nanomolar concentrations. He added that professor *Nishimura* had already determined the production of toxin by spores. Actual concentration at the penetrating cite is still undetermined. *Alexander* then asked whether any one has investigated whether or not treatment of plant leaves by spore germination fluid induces plant genes. *Kohmoto* responded that this has been looked at in other systems but not in theirs. *Chumley* wondered about the range of specificities of resistance inducing factors, and asked whether the various *A. alternaria* isolates produce resistance-inducing factors and do they show limited host specificities. *Kohmoto* replied that the specificity is very broad, and that the elicitor could also induce pisatin in pea leaves. However, it is not known whether each isolate produces the same or different substances. *Nester* asked if it was possible to measure actual binding of the toxin to the various organelles such as chloroplasts or mitochondria. *Kohmoto* responded that they have not done that but would like to. *Vance* referred to another toxin, the AK toxin, and asked whether the two-fold higher binding by susceptible cells as compared with the resistant ones was significant. He further asked if *Kohmoto* was surprised to see binding by the resistant cells and commented that he would have expected a larger difference in binding between the susceptible and resistant cells. *Kohmoto* said the binding was just total binding and not binding specific for HST, because it is not binding displaced by another ligand. Before the critical estimates of binding between toxin and membrane are made, he indicated that they have to measure the specific binding.

Chapter 13
Suppressor Production as a Key Factor for Fungal Pathogenesis

Tomonori Shiraishi, Tetsuji Yamada, Hachiro Oku, and Hirofumi Yoshioka

Plants are endowed with diverse mechanisms that protect them from pathogenic microorganisms. Active defense, including formation of many chemical and physical barriers (e.g., phytoalexin, infection inhibitor, pathogenesis related proteins, lignin, callose, etc.), is considered the main part of the resistance mechanism, because negating of such defense reactions by prior treatment with several metabolic inhibitors or pre-inoculation with compatible fungi allows pathogenic fungi to invade non-host plants. Barriers induced in plant tissues after fungal invasion seem to block penetration, growth, and reproduction of the pathogen. Resistance-inducing substances called inducers or elicitors are released from spores into spore-germination fluid of both pathogenic and nonpathogenic fungi (Hayami et al., 1982; Shiraishi et al., 1978b). As far as the authors know, there is no pathogen that does not produce elicitors. Elicitors from pathogenic fungi are also able to induce resistance in their host.

These facts prompted us to theorize that host-fungus specificity, namely species specificity, cannot be explained solely by elicitor production, but rather is determined by the substances which suppress the host-resistance expression. Since the latter half of the 1970s, suppressing substances (suppressor) have been found in spore-germination fluids and culture filtrates from several fungi (Table 1; toxins and suppressors from infected plants are excluded from this table).

Unlike host-specific toxins, suppressors cause no visible damage in tissues or isolated protoplasts, but suppress the accumulation of phytoalexins (Kessmann et al., 1986; Oku et al., 1977; Shiraishi et al., 1978a; Ziegler et al., 1982), block hypersensitive response (Doke, 1975; Doke et al., 1980; Doke et al., 1979; Storti et al., 1988), and allow even the avirulent fungi to invade the treated tissues (Oku et al., 1980, 1987; Shiraishi et al., 1978a,b). However, the action of suppressor in accessibility induction (Ouchi et al., 1974; Ouchi et al., 1981) is similar to that of host specific toxins (Comstock et al., 1973; Hayami et al., 1982; Otani et al., 1975; Yamamoto et al. 1984; Yoder et al., 1969), and/or the toxin of *Pyricularia oryzae* (Arase et al., 1988) or Ophiobolin A (Xiao et al., 1989). Therefore, one of the most important parasitic adaptations of fungi seems to be their abilities to negate active defense and condition plant cells to be accessible to infection.

In this treatise, we will review the role of the suppressor, especially, in the infection of pea plant by *Mycosphaerella pinodes*.

Production of Suppressor and Elicitor During the Infection Process of *Mycosphaerella pinodes*

Since *Mycosphaerella pinodes* usually infests and infects with pycnospores, the initial recognition in pea plant is most likely mediated by factors in spore germination fluid. The germination fluid from a virulent strain of suspended spores (IF030342, ATCC42741, OMP-1) was collected periodically, filtered to exclude spores, and separated into high (>20,000 Da) and low-molecular-weight fractions (<20,000 Da) by the method described previously (Hiramatsu et al., 1986). The polysaccharide elicitor (ca. 70,000 Da; Thanutong et al., 1982) and glycopeptide suppressor (<5000 Da) were included in the high- and low-molecular-weight fractions, respectively. The activities of these components were measured by the quantitative analysis of accumulated pisatin, a major phytoalexin of pea (Masuda et al., 1983). The elicitor activity was found in spore-germination fluid 3-h after suspending the spores and then increased gradually up to 12-h. The change of elicitor production by OMP-1 was similar to that in *M. ligulicola* (cause of chrysanthemum ray blight, strain OML) and hypovirulent mutant of *M. pinodes* (strain OMP-X76). In contrast, suppressor activity was detected immediately after OMP-1 spores were suspended in water, and gradually increased until 9-h. However, the suppressor activity of OMP-X76 was lower than that of OMP-1, and OML produced a small quantity until 24-h. The suppressor activity was also detected in the drop diffusate recovered from OMP-1 spore suspension placed on the living pea leaves. OML infection was established on pea leaves treated with OMP-1 spore suspension fluid recovered at any interval, but such effect was not observed with OML spore-suspension fluid. These facts suggest that the determination of specificity may be explained by the production of a suppressor, but not by the elicitor. Since a part of the suppressor was secreted immediately after pycnospores contacted the pea-leaf surface, it may be that this conditions pea cells to accept the fungus well before penetration.

Accessibility Induction by the Suppressor from *M. pinodes*

The concentrated, low-molecular-weight fraction obtained by Sephadex G-15 gel filtration was separated by TLC (silica gel GF254, Merck) utilizing a solvent

Table 1. Suppressor from phytopathogenic fungi.

Fungus	Nature	Tested plant	Reference
Phytophthora infestans	Glucan,Phosphoglucan	Potato	Doke, 1975
P. infestans	?	Tomato	Storti et al. 1988
P.megasperma glycinea	Mannan-glycoprotein	Soybean	Ziegler & Pontzen 1982
Mycosphaerella pinodes	Glycopeptide	Pea	Oku et al. 1977
M. ligulicola	Glycopeptide?	Chrysanthemum	Oku et al. 1987
M. melonis	Glycopeptide?	Cucumber	Oku et al. 1987
Ascochyta rabiei	Glycoprotein	Chick pea	Kessmann & Barz 1986

system of ethanol:acetic acid:water (4:1:1). Two ninhydrin-positive and active fractions were obtained, and the Rf values were 0.04 (F2) and 0.3 (F5), respectively. The activity of F5 was reduced by pronase treatment. The preliminary treatment of pea leaves with F2 (40 µg/ml, BSA equiv.) or F5 (50 µg/ml, BSA equiv.) markedly increased the infection frequency of a non-pathogen, *Stemphylium sarcinaeforme* (pathogen causing *Stemphylium* leaf spot of red clover) (Shiraishi et al., 1978a,b). The other nonpathogens such as *Mycosphaerella ligulicola, M. melonis,* and *Alternaria alternata* were also able to establish infection on F5-treated pea leaves. Thus, the suppressor may condition pea plants to accept even nonpathogenic fungi.

The specificity of F5 in biological activity coincided with the host range of *M. pinodes* (Oku et al., 1980, 1987). Conidiospores of *A. alternata* were able to establish infection on 5 out of 12 species of F5-treated leguminous plants (*Trifolium pratense, Pisum sativum, Medicago sativa, Milletia japonica,* and *Lespedeza bicolor*) which were invaded in varying degrees by *M. pinodes*. The same results were obtained with the crude, low-molecular-weight fraction and/or germination fluid of *M. pinodes*. Therefore, the suppressor is most likely the determinant of *M. pinodes* specificity as it induces accessibility on only the host plant of this fungus.

Suppression of Chemical Barrier Formation in Pea Tissues Induced by the Elicitor

At least two chemical barriers, a yet-unidentified infection inhibitor and pisatin, were induced in pea tissues treated with high-molecular-weight polysaccharide elicitors in spore-germination fluid from *M. pinodes, M. ligulicola,* or *M. melonis.*

1. Infection Inhibitor

Infection by *M. pinodes* was markedly inhibited on uninjured pea leaves by treating with OMP-1 elicitor (100 µg/ml, glucose equiv.) 1-h before inoculation. The same type of local resistance to *M. pinodes* was also induced by elicitors from other species of *Mycosphaerella* (Table 2) (Oku et al., 1987; Yamamoto et al., 1986). Similar resistance was induced on chrysanthemum petal and cucumber cotyledons treated with elicitors. The inhibition zone is localized in tissues beneath the applied droplets. Pisatin was not detected in the solution recovered from elicitor-treated pea leaves, but a kind of infection inhibitor (molecular weight 329) was extracted with ethylacetate. The infection inhibitor from pea leaves hardly affected germination, germ-tube elongation, or appressorial formation of *M. pinodes* spores on heat-killed pea epidermis or cellophane sheet; but it prevented penetration. Thus, a high-molecular-weight elicitor added to intact tissues apparently induces a local defense reaction very rapidly (within 1-h). However, in the presence of the suppressor, no infection inhibitor was detected, and *M. pinodes* established infection to the same extent as the control. The result is similar to the case reported of the interaction between the inducer (elicitor) and host-specific toxins (AK- and AF-toxin) of

Alternaria alternata (Hayami et al., 1982; Yamamoto et al., 1984). However, *Erysiphe pisi*, pea powdery mildew fungus, was insensitive to this type of pea resistance (Oku et al., 1986).

2. Pisatin

Pisatin accumulation was induced in pea tissues nonspecifically by injury plus elicitors from the three species of *Mycosphaerella* (Table 2). The accumulation is due to the activation of the synthetic pathway, but not to inactivation of degradation. In pea epicotyls, pisatin accumulation, the activation of phenylalanine ammonia-lyase (PAL), and accumulation of PAL- and chalcone synthase (CHS)-mRNA were detected 6-9,3, and/or 1-h after the elicitor treatment, respectively. These processes, however, were delayed at least 3-h in the presence of OMP-1 suppressor with elicitor (Yamada et al., 1989), indicating that the suppressor temporarily interrupts the process of pisatin production. In other words, once recognized, the action of elicitor becomes irreversible in pea tissues because the eliciting activity is gradually recovered in the presence of the suppressor.

Lineweaver-Burk plot analysis of pisatin accumulation after treatment with various concentrations of elicitor and suppressor suggests that the action of the suppressor does not compete with the fungal elicitor (unlikely competition on a putative receptor site). The suppressor also decreased pisatin accumulation significantly even when added 12 to 15-h after the elicitor treatment when the pisatin synthetic pathway had already been activated. This result indicates two possibilities: that the suppressor acts upon multiple sites during the processes involved in elicitor recognition through pisatin synthesis; and/or the suppressor acts on more basic functions of pea cells.

Additionaly, the crude suppressor also blocked the accumulation of phytoalexins of pea and red clover, but failed to inhibit the accumulation of glyceollin of soybean and phaseollin of kidney bean (Fig. 1). These results also show the specific activity of the OMP-1-suppressor.

Table 2. Induction of pisatin accumulation and local resistance to *Mycosphaerella pinodes* in pea leaves by elicitors from three species of *Mycosphaerella*.

Treatment with elicitor from	Infection (% to control)[a] on uninjured leaves	Pisatin accumulation (% to control)[b]
M. pinodes	74	1353
M. ligulicola	50	663
M. melonis	38	1192

[a] The infection frequency was determinied 18-h after the inoculation with *M. pinodes* pycnospores on uninjured pea leaves pretreated with elicitor (100 µg/ml, glucose equiv.) for 2-h.
[b] Pisatin concentration was determined in injured pea leaves treated with elicitor (500 µg/ml, glucose equiv.) for 18-h.
All data were significantly different from the water control ($p<0.05$).

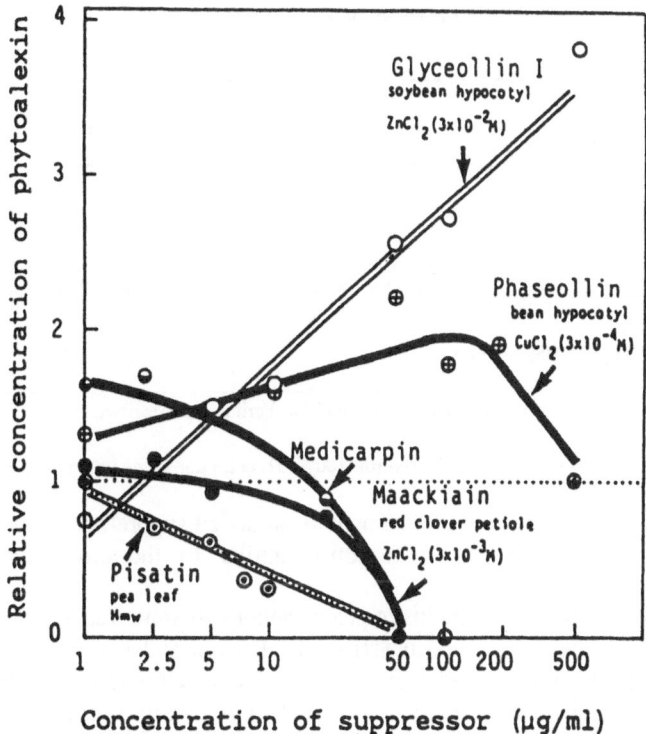

Figure 1. Effect of suppressor from *Mycosphaerella pinodes* on phytoalexin production in legumes treated with elicitor. The phytoalexin accumulation was determined 18-h after the treatment by HPLC (Masuda et al., 1983). Relative concentration of phytoalexin = conc. of phytoalexin in tissues treated with the mixture of the suppressor and elicitor/conc. of phytoalexin in tissues treated with elicitor only. Note that the induction of pisatin in pea, and medicarpin and maackiain in red clover, were suppressed in the concomitant presence of suppressor.

The Primary Action Site of OMP-1 Suppressor

It is hypothesized that pea plasma membrane initially recognizes the suppressor, which results in interfering with defense expression. Pea plasma membrane was prepared from etiolated epicotyls using an aqueous, two-polymer phase system and a stepwise sucrose gradient centrifugation (Yoshida et al., 1986). The plasma-membrane-rich fraction was recovered since its ATPase activity (Perlin and Spanswick, 1981) was sensitive to orthovanadate and dicyclo-hexylcarbodiimide, but insensitive to nitrate and azide (Table 3). The activity depended on magnesium ion but was little affected by potassium ion. The substrate specificity of the fraction was in the order of ATP>CTP>ADP>GTP. The OMP-1 suppressor markedly inhibited the ATPase activity at a concentration of 50 µg/ml (bovine serum albumin, equiv.) which suppressed pisatin accumulation in leaves and/or epicotyls induced by the elicitor. The kinetic analysis of inhibition by the suppressor indicated that the

Table 3. Effect of inhibitors and suppressor of *Mycosphaerella pinodes* on ATPase activity in pea plasma membrane.

Inhibitor (concentration)	Relative ATPase activity[a](% to control)
Na3VO4 (1 mM)	16[b]
DCCD (1 mM)	52[b]
NaNO3 (50 mM)	105
NaN3 (100 μM)	90
EDTA (10 mM)	27[b]
Suppressor (50 μg/ml, BSA equiv.)	37[b]

[a] The reaction was carried out by the method of Perlin and Spanswick (1981) with slight modification.
[b] Significant difference ($p<0.05$) to the control (100%=31.4+1.3 μM Pi/mg protein/h at 37°C).

effect was uncompetitive since the activity of added suppressor fluctuated both in Vmax- and Km-values, a response similar to the effect induced by orthovanadate.

Phosphorylation of proteins in pea plasma membrane was determined with [r-^{32}P]ATP using the method described by Briskin and Leonard (1982). The phosphorylated proteins were fractionated by electrophoresis with 4-20% gradient SDS polyacrylamide gel. The phosphorylation of ATPase seemed to be inhibited by the addition of suppressor which interfered ^{32}P-incorporation into, at least 78-, 64- and 42-kDa specific proteins that were phosphorylated in the presence of a strong inhibitor of protein kinase, K-252a (Kase et al., 1987). The proteins released radioactive phosphorus by chasing with cold ATP. It was suggested that the inhibition of ATPase activity by the suppressor resulted from blocking the formation of phosphorylated intermediate of the enzyme.

The pH change on pea leaf surface were determined periodically after adding suppressor and/or elicitor. The pH of water or elicitor solution (100/250 μg/ml, glucose equiv.) on leaflet decreased immediately after treatment, but such change was strongly inhibited in the presence of suppressor (50 μg/ml, bovine serum albumin equiv.) as it was when 1 mM orthovanadate was used. The result shows that the suppressor may block the proton pump ATPase in pea plasma membrane. The inhibitory effect of the suppressor disappeared within 2 to 3-h after the treatment. This phenomenon seems to explain the 3-h delay of PAL and CHS-gene expression in the presence of the suppressor. In addition, orthovanadate also delayed pisatin accumulation in elicitor-treated pea tissues for 6-h.

Concluding Remarks

The abilities to penetrate plant tissues, overcome host resistance, and evoke disease are crucial to all phytopathogenic fungi (Oku, 1980). In this communication, we showed that the suppressor is able to overcome host resistance. All pathogenic fungi possess means to penetrate host-plant tissues mechanically and/or chemically, through stomata, cuticle, or injured site. The penetrating pathogen encounters passive or active host defense. If the pathogen cannot escape or suppress the defense expression, it is blocked at the penetration site. As described previously (Oku et al., 1975; Shiraishi et al., 1978a,b), pisatin prevents fungal penetration at a far lower concentration than that required for the inhibition of germination and germ-tube elongation. Infection inhibitors also block penetration (Yamamoto et al., 1986). The production of these substances, which are induced by elicitors of both the pathogenic and nonpathogenic fungi of pea, was specifically suppressed by the suppressor from *M. pinodes*. In addition, the established fungus also needs the ability to overcome expressed resistance such as pisatin-degrading ability (VanEtten et al., 1989) for colony expansion because the second phase of resistance, including pisatin production, is induced by the infection.

The suppressor of *M. pinodes* cannot be included in the category of toxins because it causes no visible damage or collapse of host tissues or isolated protoplasts even at high concentrations. However, the biological function of the suppressor is close to that of host-specific toxins in that accessibility (susceptibility) is induced in spite of differences in their structures and toxicity.

The activity of the *M. pinodes* suppressor is probably not cultivar specific because the pathogen is able to invade all pea cultivars so far tested. Cultivar-race-specific suppressors were found in *Phytophthora* species. The hypersensitive response in potato induced by mycelial-wall components was suppressed by water soluble glucans from *Phytophthora infestans* races depending on the degree of compatibility (Doke et al., 1980), and an extracellular invertase in culture filtrate from *P. megasperma* f.sp. *glycinea* inhibited glyceollin accumulation as it reflected the race-cultivar specificity (Ziegler et al., 1982). It is therefore probable that OMP-1 suppressor might reveal cultivar specificity like host-specific toxins, and this hypothesis could be tested if a Mycosphaerella blight-resistant cultivar becomes available.

Our main question is how the suppressor controls the expression of host resistance. As described above, the suppressor inhibited or delayed at least two steps of pea-resistance expression induced by the elicitor. In the latter case, the delay in the resistance response seems to result from the 3-h suppression of gene expression of PAL and CHS participating in pisatin biosynthesis. However, it is believed that the suppressor plays another role in pisatin production, because the suppressor allows avirulent pathogens to invade pea tissues which might also produce other chemical and physical barriers (Mauch et al., 1988) in the early stages of interaction. Using this information as a guide, two possible modes of action were considered. The first was that the suppressor has plural sites for action in the process of resistance expression, and the second is that the suppressor acts on more fundamental functions of pea cells. The data described in the preceding section support the latter possibility. Namely, the effect of suppressor seems to inhibit the ATPase, "master enzyme" (Serrano, 1989), in pea plasma membrane responsible for many important functions, perhaps including resistance reaction. In regard to other systems, the glucan (or phosphoglucan) suppressor from the late blight fungus of potato suppressed NADPH-dependent superoxide generation in microsomal fraction of potato (Doke, 1985). Several host specific toxins were reported to inhibit electrogenic proton extrusion or mitochondrial function in susceptible cultivars (Akimitsu et al., 1989; Bednarski et al., 1977; Federico et al., 1980; Holden et al., 1989).

Together with these reports and our data, we would like to propose the concept that, "in the early infection court, the pathogenic fungi secrete the substances which suppress the basic function, for example, ion transport or energy production, supported by membrane systems of host cells, to disturb host homeostasis and avoid the host defense expression, ultimately enabling the pathogen establishing in the host tissues."

Acknowledgements

We are indebted to Profs. S. Ouchi, Kinki University, and H. Kunoh, Mie University and Dr. Y. Ichinose for their valuable discussion and Mr. Y. Yamamoto, Y. Masuda, and Y. Todoh for their technical assistance. This work was supported by Grants-in-Aid for Scientifica Research from the Ministry of Education, Science and Culture of Japan.

References

Akimitsu, K., et al., 1989, Host specific effects of toxin from the rough lemon pathotype of *Alternaria alternata* on mitochondria, *Plant Physiol.* **89**:925-931.

Arase, S., Tanaka, E., and Nishimura, S., 1988, Production of susceptibility-inducing factors in spore germination fluids of *Pyricularia oryzae*, *In* Host Specific Toxins (Kohmoto, K and Durbin, R.D. eds.), Tottori Univ. Press, Tottori, pp. 59-73.

Bednarski, M.A., and Scheffer, R. P., 1977, Effect of toxin from *Helminthosporium maydis* T on respiration and associated activities in maize tissue, *Physiol. Plant Pathol.* **11**:129-141.

Briskin, D.P., and Leonard, R.T., 1982, Partial characterization of phosphorylated intermediate associated with the plasma membrane ATPase of corn roots, *Proc. Natl. Acad. Sci. USA* **79**:6922-6926.

Comstock, J.C. and Scheffer, R.P., 1973, Role of host-selective toxin in colonization of corn leaves by *Helminthosporium carbonum*, *Phytopathology* **63**:24-29.

Doke, N., 1975, Prevention of the hypersensitive reaction of potato cells to infection with an incompatible race of *Phytophthora infestans* by constituents of the zoospores, *Physiol. Plant Pathol.* **7**:1-7.

Doke, N., 1985, NADPH-dependent O_2^- generation in membrane fractions isolated from wounded potato tubers inoculated with *Phytophthora infestans*, *Physiol. Plant Pathol.* **27**:311-322.

Doke, N., Garas, N.A., and Kuc, J., 1979, Partial characterization and aspects of the mode of action of the hypersensitivity-inhibiting factor (HIF) isolated from *Phytophthora infestans*, *Physiol. Plant Pathol.* **15**:127-140.

Doke, N., and Tomiyama, K., 1980, Suppression of the hypersensitive response of potato tuber protoplasts to hyphal wall components by water soluble glucans isolated from *Phytophthora infestans*, *Physiol. Plant Pathol.* **16**:177-186.

Federico, R., et al., 1980, Inhibition of fusicoccin-induced electrogenic proton extrusion in susceptible maize by *Helminthosporium maydis* race T toxin, *Plant Sci. Letts.* **17**:129-134.

Hayami, C., et al., 1982, Induced resistance in pear leaves by spore germination fluids of nonpathogens to *Alternaria alternata*, Japanese pear pathotype and suppression of induction by AK-toxin, *J. Fac. Agric. Tottori Univ.* **17**:9-18.

Hiramatsu, M., et al., 1986, Regulation of pisatin biosynthesis in pea leaves by elicitor and suppressor produced by *Mycosphaerella pinodes*, *Ann. Phytopathol. Soc. Japan* 52:53-58.

Holden, M.J., and Sze, H., 1989, Effect of *Helminthosporium maydis* race T toxin on electron transport in susceptible corn mitochondria and prevention of toxin actions by dicyclohexylcarbodiimide, *Plant Physiol.* 91:1296-1302.

Kase, H., et al., 1987, K-252 compounds, novel and potent inhibitors of protein kinase C and cyclic nucleotide-dependent protein kinases, *Biochem. Biophys. Res. Commun.* 142:436-440.

Kessmann, H., and Barz, W., 1986, Elicitation and suppression of phytoalexin and isoflavone accumulation in cotyledons of *Cicer arietinum* L. as caused by wounding and by polymeric components from the fungus *Ascochyta rabiei*, *J. Phytopathol.* 117:321-335.

Masuda, Y., et al., 1983, A rapid and accurate analysis of isoflavonoid phytoalexins by high performance liquid chromatography, *Ann. Phytopathol. Soc. Japan* 49:558-560.

Mauch, F., Hadwiger, L.A., and Boller, T., 1988, Antifungal hydrolases in pea tissue, I., *Plant Physiol.* 87:325-333.

Oku, H., 1980, Determinant for pathogenicity without apparent phytotoxicity in plant diseases, *Proc. Japan Acad.* 56(Ser.B):367-371.

Oku, H., Shiraishi, T., and Ouchi, S., 1975, The role of phytoalexin as the inhibitor of infection establishment in plant disease, *Naturwissenschaften* 62:486.

Oku, H., Shiraishi, T., and Ouchi, S., 1977, Suppression of induction of phytoalexin, pisatin by low-molecular-weight substances from spore germination fluid of pea pathogen, *Mycosphaerella pinodes*, *Naturwissenschaften* 64:643.

Oku, H., Shiraishi, T., and Ouchi, S., 1986, Specificity of local resistance induced in pea leaves by elicitor isolated from *Mycosphaerella pinodes*, *Ann. Phytopathol. Soc. Japan* 52:347-348.

Oku, H., Shiraishi, T., and Ouchi, S., 1987, Role of specific suppressors in pathogenesis of *Mycosphaerella* species, *In* Molecular Determinants in Plant Diseases (Nishimura, S., et al., eds.), Japan Sci. Soc. Press, Tokyo/Springer-Verlag, Berlin, pp. 145-156.

Oku, H., et al., 1980, A new determinant of pathogenicity in plant disease, *Naturwissenschaften* 67:310.

Otani, H., et al., 1975, Nature of specific susceptibility to *Alternaria kikuchiana* in Nijisseiki cultivar among Japanese pears (V), *Ann. Phytopathol. Soc. Japan* 41:467-476.

Ouchi, S., et al., 1974, Induction of accessibility and resistance in leaves of barley by some races of *Erysiphe graminis*, *Phytopathol. Z.* 79:24-34.

Ouchi, S., and Oku, H., 1981, Susceptibility as a process induced by pathogens, *In* Plant Disease Control (Staples, R.C. and Toenniessen, G.H. eds.), Wiley, New York, pp. 33-44.

Perlin, D.S., and Spanswick, R.M., 1981, Characterization of ATPase activity associated with corn leaf plasma membrane, *Plant Physiol.* 68:521-526.

Serrano, R., 1989, Structure and function of plasma membrane ATPase, *Annu. Rev. Plant Physiol. Plant Mol. Biol.* 40:61-94.

Shiraishi, T., et al., 1978a, Inhibitory effect of pisatin on infection process of *Mycosphaerella pinodes* on pea, *Ann. Phytopathol. Soc. Japan* 44:641-645.

Shiraishi, T., et al., 1978b, Elicitor and suppressor of pisatin induction in spore germination fluid of pea pathogen, *Mycosphaerella pinodes*, Ann. Phytopathol. Soc. Japan **44**:659-665.

Storti, E., et al., 1988, A potential defense mechanism of tomato against the late blight disease is suppressed by germinating sporangia-derived substances from *Phytophthora infestans*, J. Phytopathol. **121**:275-282.

Thanutong, P., et al., 1982, Isolation and partial characterization of an elicitor of pisatin production from spore germination fluid of pea pathogen, *Mycosphaerella pinodes*, Sci. Rep. Fac. Agric. Okayama Univ. **59**:1-9.

VanEtten, H.D., Matthews, D.E., and Matthews, P.S., 1989, Phytoalexin detoxification: importance for pathogenicity and practical implications, *Annu. Rev. Phytopathol.* **27**:143-164.

Xiao, J.Z., Nishimura, S., and Tsuda, S., 1989, The role of ophiobolins in pathogenesis of the producer, *Cochliobolus miyabeanus*, Ann. Phytopathol. Soc. Japan **55**:477 (Abst.).

Yamada, T., et al., 1989, Suppression of pisatin, phenylalanine ammonialyase mRNA, and chalcone synthase mRNA accumulation by a putative pathogenicity factor from the fungus *Mycosphaerella pinodes*, Mol. Plant-Microbe Interac. **2**:256-261.

Yamamoto, M., et al., 1984, Studies on host-specific AF-toxins produced by *Alternaria alternata* strawberry pathotype causing Alternaria black spot of strawberry (2) Role of toxins in pathogenesis, Ann. Phytopathol. Soc. Japan **50**:610-619.

Yamamoto, Y., et al., 1986, Non-specific induction of pisatin and local resistance in pea leaves by elicitors from *Mycosphaerella pinodes*, *M. melonis* and *M. ligulicola* and effect of suppressor from *M. pinodes*, J. Phytopathol. **117**:136-143.

Yoder, O.C., and Scheffer, R.P., 1969, Role of toxin in early interactions of *Helminthosporium victoriae* with susceptible and resistant oat tissue, Phytopathology **59**:1954-1959.

Yoshida, S., et al., 1986, Properties of plasma membrane isolated from chilling-sensitive etiolated seedlings of *Vigna radiata* L., Plant Physiol. **80**:152-160.

Ziegler, E., and Pontzen, R., 1982, Specific inhibition of glucan-elicited glyceollin accumulation in soybeans by an extracellular mannan-glycoprotein of *Phytophthora megasperma* f.sp. *glycinea*, Physiol. Plant Pathol. **20**:321-331.

Summary of Discussion of Shiraishi's Paper

Keen initiated the questioning regarding the suppressors by asking *Shiraishi* to share his thoughts on 1) the chemical structure of the suppressor and 2) the genetic tests of the role of the peptide. *Shiraishi* responded by saying that the glycopeptide had a molecular weight of about 3500 and the peptide moiety had about 30 amino acids containing asparagine and glycine and a sugar chain that is thought to contain a mannose. A genetical approach has not been used primarily because they had no cultures that are totally avirulent and no pea cultivars that are totally resistant. *Durbin* asked that since *Shiraishi* compares his suppressor to the classical host-specific toxin (HST) why was it not called an HST? *Shiraishi* responded by indicating that in pathogenicity or parasitic adaptation the pathogen has the ability to penetrate, to overcome host resistance and to evoke disease. The suppressor plays an important role especially in the second and third aspects. *Oku* commented that *M. pinoides* kills the pea plant and makes a brown spot, but as far as the suppressor is concerned there are no visible symptoms or injury observed by treatment. *Durbin* suggested that from a conceptual point of view they may not be so different. One may be suppressing an aspect of secondary metabolism and in the other case it is perhaps more directly related to primary interaction. *Ouchi* stated that the HST reaction must be host specific and this is not, as it affects several species even though the basic mechanism may be similar. In these analyses, different cultivars were not used. Therefore, it is species specific. *Yamada* stated that plants can recover from the action of the suppressor and that is what was shown. HSTs are detrimental to the cell but the suppressor is not. This aspect of the mode of action appears to be different. *Durbin* observed that one could explain it by the presence of receptors with different kinetic properties such as binding constants.

Hammerschmidt followed with two questions dealing with the infection inhibitor isolated from pea leaves pretreated with the 329 compound. He asked whether *Shiraishi* had an idea of what kind of compound that was. *Shiraishi* said it was a lipophilic compound. This compound has no effect on germination or apprasorium formation but one sees no penetration. Had he looked at the apprasoria at the ultrastructural level after treatment with this compound to see if they are fully matured in their development as compared with the water controls. *Shiraishi* said he had not. *Yoshikawa* asked what range of plant ATPase activities was inhibited by the suppressor. *Shiraishi* responded by stating that the suppressor inhibits kidney bean, soybean, and cowpea plasma membrane ATPase activity, indicating that the preliminary data show that in the in vivo system, specificity is observed but not with the in vitro system. *Macko* asked if pisatin accumulation is induced or the other signs of the defense reaction are induced just by injury or by inorganic salts, and does the suppressor suppress these? *Shiraishi* said that pisatin accumulation is delayed by the suppressor treatment, UV and copper ions. *Chumley* asked whether *Shiraishi* thought the levels of suppressor used in these experiments represent levels that might be expected? *Shiraishi* said yes, but higher concentrations are required for pisatin suppression. *Keen* said that it appears that in soybean suppressor acts as elicitor, would *Shiraishi* care to comment on that? *Shiraishi* said yes, it appears so. *Vance* asked how *Shiraishi* applied the suppressor to the tissue, did he do that in sort of a drop diffusate test or did he apply it to excised leaves? *Shiraishi* responded by stating that they tried both; when a non-pathogen was used, the suppressor was put into spore germination droplets.

Plant's Response

Part's Responses

Chapter 14
Molecular Aspects of Elicitation of Host Defense Reactions

Masaaki Yoshikawa and Youji Takeuchi

Disease resistance in many plant-fungal pathogen interactions has been attributed to inducible production of low molecular weight antibiotic compounds, phytoalexins (Keen, 1981). The induction of phytoalexins in infected plants is presumed to be mediated by an initial recognition process between plants and pathogens which involves the detection of certain unique molecules of pathogen origin, termed elicitors, by recognitional receptor-like molecules in plants, thereby setting off a cascade of biochemical events leading ultimately to phytoalexin accumulation (Yoshikawa, 1983; Yoshikawa et al., 1982). Detailed mechanisms involved in each biochemical process leading to the phytoalexin production, however, are poorly understood. The subject described here is focused mainly on the molecular basis underlying the elicitation of glyceollin, a phytoalexin (Yoshikawa et al., 1978a,b) produced by the expression of monogenic resistance in soybean (*Glycine max*) to incompatible races of *Phytophthora megasperma* f.sp. *glycinea* (Pmg).

Elicitors of Pathogen Origin That Function in Glyceollin Elicitation In Vivo

Isolated mycelial walls of many fungi possess potent elicitor activity to induce phytoalexin accumulation in plants. There are, however, several unresolved questions regarding the in vivo involvement of the wall-associated elicitors. A major detracting argument arises from the observation that active elicitor moieties can only be extracted from fungal walls by such severe treatments as autoclaving, acid, or alkaline treatment which are unlikely to exist in a biological environment. This raises the question of how normally insoluble elicitor molecules on or in fungal walls may come in contact with the corresponding recognitional molecules in plant cells during natural infection processes. Although unnaturally extracted elicitors are frequently used in many types of studies including biological and structural analysis, the possibility exists that such elicitors may not be those functioning in phytoalexin elicitation in fungus-infected plant tissues (Yoshikawa, 1983).

Our studies (Yoshikawa et al., 1981) with the soybean-Pmg system indicated that highly active carbohydrate elicitors of glyceollin production were

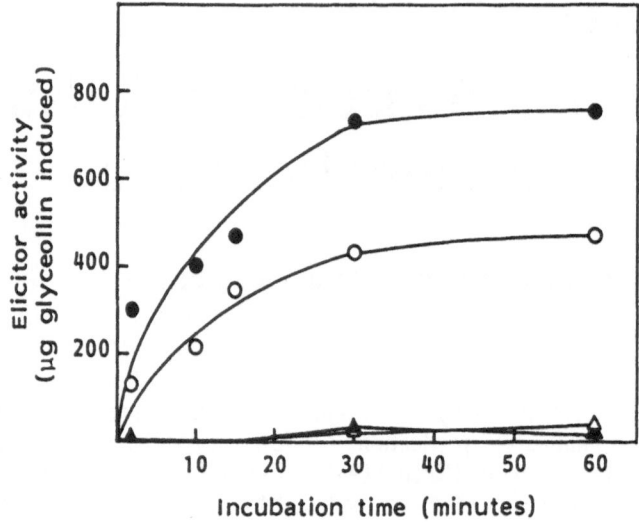

Figure 1. Time course of the release of a soluble elicitor after incubation of mycelial walls of *Phytophthora megasperma* f.sp. *glycinea* with soybean cotyledon tissues. The soluble fractions were obtained at indicated times of incubation, centrifuged and filtered, and diluted to give one-third (●) and one-ninth (o) of the relative concentration of the soluble fraction from the mycelial wall suspension that was not incubated with the tissue (▲). Elicitor activities of these soluble fractions and of a water control (△) were assayed in wounded cotyledons (Yoshikawa et al., 1981).

released into a soluble form from insoluble mycelial walls of the fungus by a factor contained in soybean tissues (Fig. 1), and provided a new insight into the in situ production of soluble elicitors which could be more efficiently recognized by plant cells. The elicitor release occurred as rapidly as 2 min after incubation of mycelial walls or actively growing hyphae with soybean tissues, suggesting that the process may be important as the earliest plant-pathogen interaction leading to the induction of glyceollin production.

The factor capable of releasing elicitors was purified from soybean cotyledons to apparent homogeneity and shown to possess ß-1,3-endoglucanase activity (Keen et al., 1983). Identity of the elicitor-releasing factor to the glucanase was further supported by the facts that ß-1,3-endoglucanase purified from the bacterial culture of *Arthrobacter luteus* possessed similar elicitor-releasing activity and that antibodies raised against the highly purified soybean glucanase inhibited both the elicitor-releasing and glucanase activities to a similar extent. Furthermore, pretreatment of soybean tissues with the antibodies partially inhibited glyceollin accumulation otherwise induced by mycelial walls, suggesting that the elicitor release indeed occurs in natural infection processes and the released elicitors are responsible for glyceollin elicitation in the infected tissues.

In addition to soybean, all twelve tested plant species representing diverse plant families were found to contain activity to release elicitors from cell walls of diverse groups of fungi, suggesting that the elicitor release is not unique to the soybean-Pmg interaction but may occur in various host-pathogen interactions.

Activity and Structure of the Glucanase-Released Elicitors

The glucanase-released Pmg elicitors active in glyceollin elicitation were heterogeneous in size, ranging >100,000 to 2000 Da when evaluated by gel filtration (see Fig. 2a). Activities of different size fractions of the released elicitors to induction of glyceollin accumulation in soybean tissues were at least 10 to 100 times higher, based on weight concentrations, than the previously reported elicitors extracted by autoclaving, acid, and alkaline treatments, and the released elicitors accounted for more than 90% of the total elicitor activity of the native mycelial walls (Table 1).

Tentative structures of the released elicitors were deduced by use of the sugar, ^{13}C-NMR, and enzymatic analysis. Hepta-ß-D-glucopyranoside (G7), the smallest elicitor-active molecule obtained by acid hydrolysis of Pmg cell walls (Sharp et al., 1984), was chemically synthesized and also used for structural comparison. ^{13}C-NMR and sugar analysis indicated the presence of ß-1,6- and ß-1,3-glucan linkages for various size fractions of the released elicitors. The released elicitors, as well as G7, were not hydrolyzed by ß-1,3-endoglucanase, suggesting the main chain to be ß-1,6-linked. In contrast, ß-1,3-exoglucanase partially degraded these elicitors, resulting in a complete loss of their elicitor activity. These results indicate that the released elicitors are composed of ß-1,6-linked main chains of different length and originally bound to fungal cell walls by ß-1,3-linked side chains (Fig. 2). Elicitors are thus released, upon infection, due to attack on the side chains by the host ß-1,3-endoglucanase, leaving 1 or 2 ß-1,3-linked glucose moieties on each side chain of the released elicitors.

Molecular Cloning of cDNA Encoding the Elicitor-Releasing Factor, ß-1,3-Endoglucanase, in Soybean

Soybean ß-1,3-endoglucanase thus appears to be a key host component involved in the earliest soybean-Pmg interaction leading to the induction of a plant

Table 1. Comparison of elicitor activity of various carbohydrate derived from cell walls of *Phytophthora megasperma* f.sp. *glycinea*.

Elicitor	Mass relative to total cell wall (% dry wt.)	Concn for half maximal elicitor activity (µg/ml)	Total elicitor activity (unit/g wall)
Total cell wall	100	9.0	111 x 103 (100%)
Soybean enzyme-released	2.5	0.2	125 x 103 (113%)
Autoclave-released	4.0	2.0	20 x 103 (18%)
Alkaline (NaOH)-released	2.8	5.2	5 x 103 (5%)
G7[a]	<0.1	9.0	<0.1 x 103 (0.1%)

[a]Hepta-ß-D-glucopyranoside, an elicitor reported by Sharp et al. (1984), was chemically synthesized.

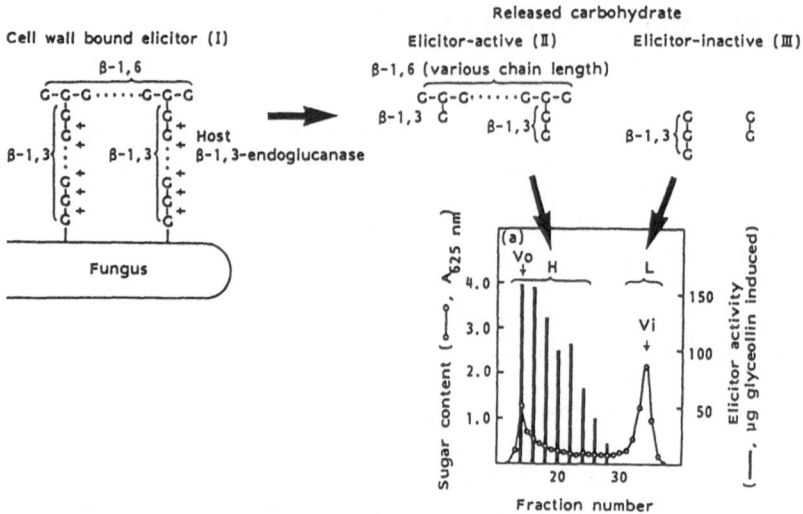

Figure 2. Tentatively proposed structures of elicitors bound to cell walls of *Phytophthora megasperma* f.sp. *glycinea* and its released forms due to attack by soybean ß-1,3-endoglucanase. Inserted figure (A) is an elution profile of total carbohydrates released by soybean glucanase on Sephadex G-100. ß-1,6-glucans of various chain length are bound to cell walls through ß-1,3-side chains (I). Upon infection, ß-1,3-side chains are attacked by endo-type soybean ß-1,3-glucanase, resulting in the release of elicitor-active ß-1,6-chains of various chain length with side chains of ß-1,3-linked one or two glucose moieties [II, correspond to fraction H in (A)] and di- or trimer of ß-1,3-glucans [III, correspond to fraction L in (A)] derived from the endoglucanase attack of ß-1,3-side chains of cell wall-bound elicitors. Small arrows indicate the sites for glucanase attack.

defense reaction, by releasing elicitor-active carbohydrates from mycelial walls. We therefore cloned and characterized cDNA encoding soybean ß-1,3-endoglucanase to further elucidate the role of this enzyme in the expression of disease resistance (Takeuchi et al., 1990).

Several cDNA clones for the glucanase gene were obtained by antibody screening of a lambda gtll expression library prepared from soybean cotyledons. Hybrid-selected translation experiments indicated that the cloned cDNA encoded a 36-kDa precursor protein product that was specifically immunoprecipitated with ß-1,3-endoglucanase antiserum. Nucleotide sequences of three independent clones revealed a single uninterrupted open reading frame of 1041 nucleotides, corresponding to a polypeptide 347 residues long (Fig. 3). The primary amino acid sequence of ß-1,3-endoglucanase as deduced from the nucleotide sequence was confirmed by direct amino acid sequencing of trypsin digests of the glucanase, indicating that the cloned cDNAs were indeed those for the soybean glucanase. The soybean ß-1,3-endoglucanase exhibited 53% amino acid homology to a ß-1,3-glucanase cloned from cultured tobacco cells and 48% homology to a ß-(1,3-1,4)-glucanase from barley. *E. coli* cells expressing the cloned full length cDNA (pEG488) synthesized protein positive to the glucanase antiserum which, upon a solubilization and reconstitution, possessed both the ß-1,3-endoglucanase and elicitor-releasing activities, firmly establishing that both the activities were due to ß-1,3-endoglucanase. Furthermore, tobacco

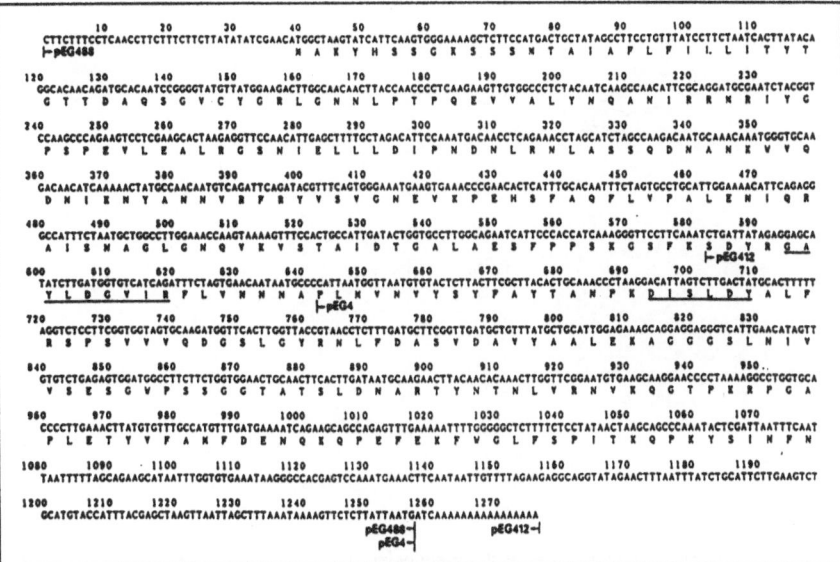

Figure 3. Nucleotide sequence and deduced amino acid sequence of a composite cDNA for soybean ß-1,3-endoglucanase mRNA obtained by combining pEG4, pEG412, pEG488. Nucleotides covered with each clone are indicated. Underlines below amino acid abbreviations indicate the portions of the sequence confirmed by amino acid sequencing of the peptides.

plants transformed by *Agrobacterium tumefaciens* carrying pEG488 also synthesized a glucanase antiserum-positive protein (Fig. 4). Experiments are now being conducted to see if the transgenic tobacco plants with higher levels of ß-1,3-endoglucanase activity enhanced disease resistance to several fungal pathogen including Pmg.

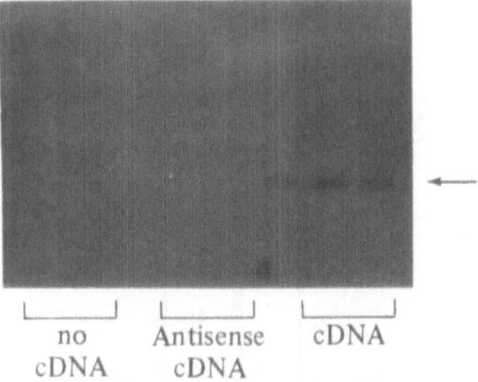

Figure 4. Synthesis of soybean ß-1,3-endoglucanase protein in tobacco plants transformed by *Agrobacterium tumefaciens* containing pEG488 as revealed by Western blot analysis using the soybean glucanase antiserum. Lanes of no cDNA, antisense cDNA, and cDNA refer to *A. tumefaciens* carrying the respective cDNA of pEG488. Arrow indicates the position of the soybean glucanase.

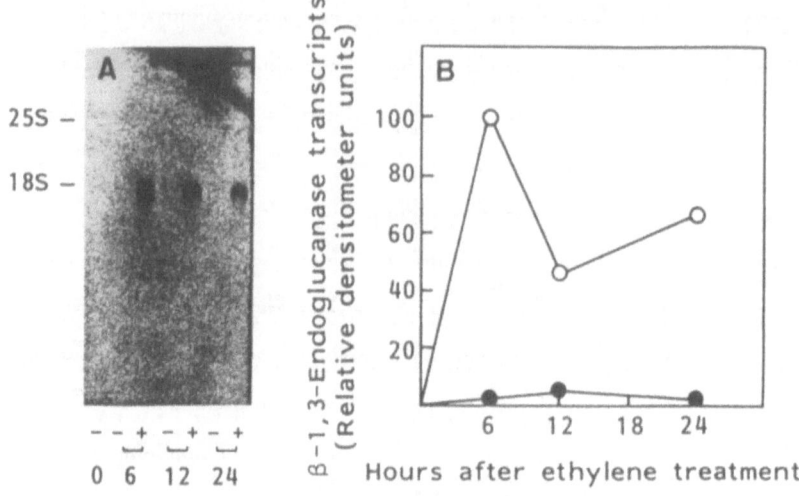

Figure 5. A. Northern blot analysis of ß-1,3-endoglucanase transcripts in cotyledons of soybean seedlings at the indicated times (hr) treated (+) or untreated (-) with 20 µl/1 ethylene. Blotted filters were hybridized with [³²P] labelled ß-1,3-endoglucanase cDNA, pEG488. B. Changes in amounts of ß-1,3-endoglucanase transcripts after ethylene treatment (o) or without ethylene treatment (●) as estimated by densitometry of the autoradiograph shown in (A).

Consistent with the contention that ß-1,3-endoglucanase plays a direct role in the induction of disease resistance and with the prediction that glucanase-enhanced plants may possess increased disease resistance are our observations that ethylene increased steady-state levels of the glucanase mRNA in soybean by 50- to 100-fold (Fig. 5), resulting in a 2- to 3-fold increase in the glucanase

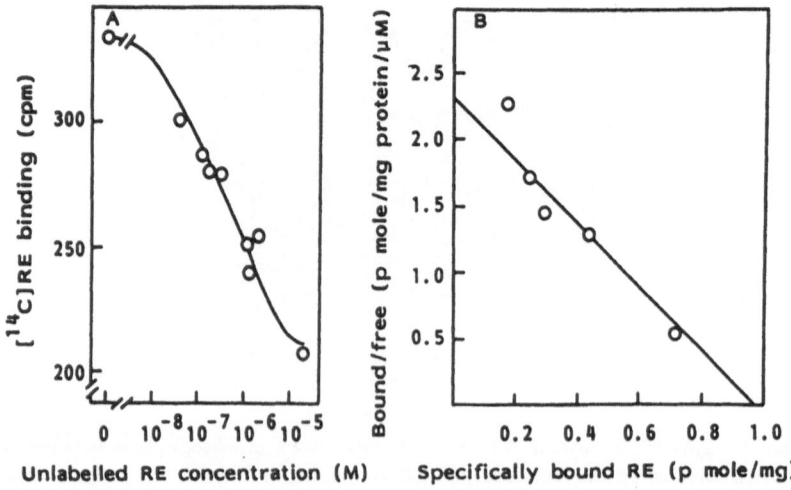

Figure 6. Effect of various concentrations of unlabelled soybean glucanase-released elicitor (RE) on [¹⁴C]RE binding to the isolated soybean cotyledon membrane fraction (A) and its Scatchard analysis

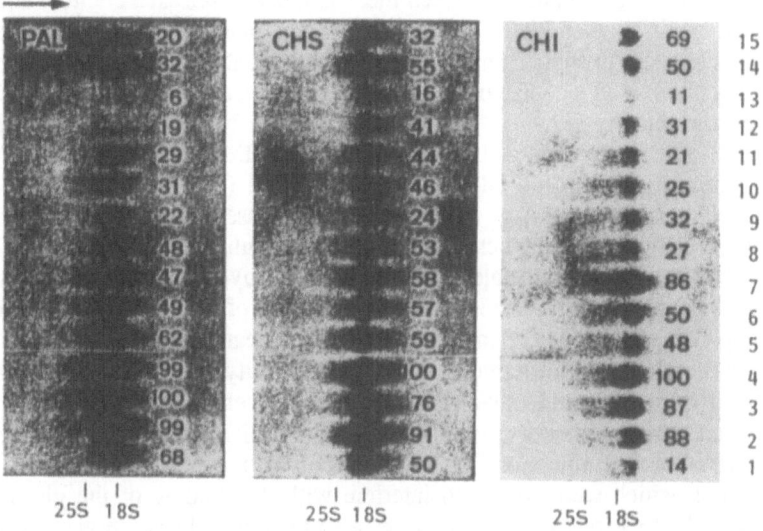

Figure 7. Northern blot analysis of transcripts for phenylalanine ammonia-lyase (PAL), chalcone synthase (CHS), and chalcone isomerase (CHI) in soybean cotyledons treated with various biotic and abiotic elicitors capable of inducing glyceollin accumulation. Blotted sheets were hybridized with [32P]labelled cDNA for the respective enzymes. Lanes 1 to 15 indicate the hybridization with RNA from cotyledons treated with various elicitors. 1, H_2O; 2, soybean glucanase-released elicitor (10 µg/ml); 3, unextracted mycelial wall (100 µg/ml); 4, Glutathione (5 mM); 5, Brilliant green (20 µg/ml); 6, Acridine red (20 µg/ml); 7, UV irradiation (10 min); 8, $CdCl_2$ (1 mM); 9, $CuSO_4$ (10 mM); 10, $K_2Cr_2O_7$ (3.3 mM); 11, $AgNO_3$ (3.3 mM); 12, $HgCl_2$ (1.5 mM); 13, $HgCl_2$ (10 mM); 14, Triton X-100 (5 mg/ml); 15, Nonidet P-40 (5 mg/ml). Values at right side of each band indicate the relative intensity of the bands measured by densitometry. Arrow indicates the direction of electrophoresis.

activity, and that ethylene-treated soybean hypocotyls showed partial resistance upon infection with a compatible race of Pmg, accompanied by higher levels of glyceollin production (Table 2).

A Specific Receptor on Soybean Membranes for the Glucanase-Released Elicitors

We previously demonstrated that membranes prepared from soybean cotyledons contained a specific binding site for an intercellular ß-1,3-glucan of Pmg, mycolaminaran, which possessed a weak elicitor activity (Yoshikawa et al., 1983). A further study was made to examine whether soybean membranes contain receptors specific for the glucanase-released elicitors which appear to play a crucial role in the elicitation of glyceollin accumulation in the fungus-infected soybean tissues.

A direct assay binding between the ^{14}C-labelled released elicitors and the isolated soybean membranes was used. Total binding of the ^{14}C-labelled released elicitors to soybean membranes was inhibited by addition of unlabelled released elicitors in a concentration dependent manner, suggesting the existence of specific binding sites. A Scatchard plot of the binding data disclosed the presence of a single class of binding sites having kDa value of 4×10^{-7} M and approximately 20 binding sites per cotyledon cell (Fig. 6). Preincubation of soybean membranes above 50°C or with proteases abolished the specific binding. A higher binding activity was observed with a membrane fraction rich in plasma membrane-associated ATPase obtained by an aqueous two polymer phase system. These results indicate the existence of heat-labile proteinaceous binding sites specific for released elicitors on soybean membranes, presumably plasma membranes. Furthermore, the binding activity of several carbohydrates obtained from Pmg and other sources, or from chemical modification of the glucanase-released elicitors, correlated with their elicitor activities in induction of glyceollin accumulation in soybean cotyledons. Mycolaminaran and laminaran, which were shown to interfere with the binding of the glucanase-released elicitors to soybean membrane, also inhibited glyceollin accumulation otherwise induced by the released elicitors. These observations support the contention that the observed binding site on soybean membranes is a receptor physiologically functioning in glyceollin elicitation.

Transcriptional Activation of the Genes for Glyceollin Biosynthetic Enzymes

Our previous study (Yoshikawa et al., 1978b) using transcriptional and translational inhibitors indicated that glyceollin production was mediated by de novo mRNA and protein synthesis. Further study directly measuring mRNA levels with the use of ^{32}P-labelled cDNAs for phenylalanine ammonia-lyase

Table 2. Effect of different concentrations of ethylene on ß-1,3-glucanase activity, growth of a compatible race (race 3) of *Phytophthora megasperma* f.sp. *glycinea*, and phytoalexin (glyceollin) production in soybean hypocotyls.

Concn of ethylene (μl/l)	Glucanase activity at inoculation[a] (DA$_{660}$/g hypocotyls)	Hyphal growth in host[b] (μm)	Glyceollin[b] (μg/g hypocotyls)
0	22.7	>1500	125.4
1	25.4	1250	155.1
5	44.5	970	386.0
20	45.0	927	401.8
50	40.5	932	348.9

[a]Soybean seedlings were pretreated with indicated concentrations of ethylene for 2 days before extraction of crude enzyme preparations for measurement of glucanase activity and inoculation of hypocotyls with the fungus.
[b]Rates of hyphal growth in the hypocotyls and glyceollin concentrations were measured 1 and 2 days, respectively, after inoculation.

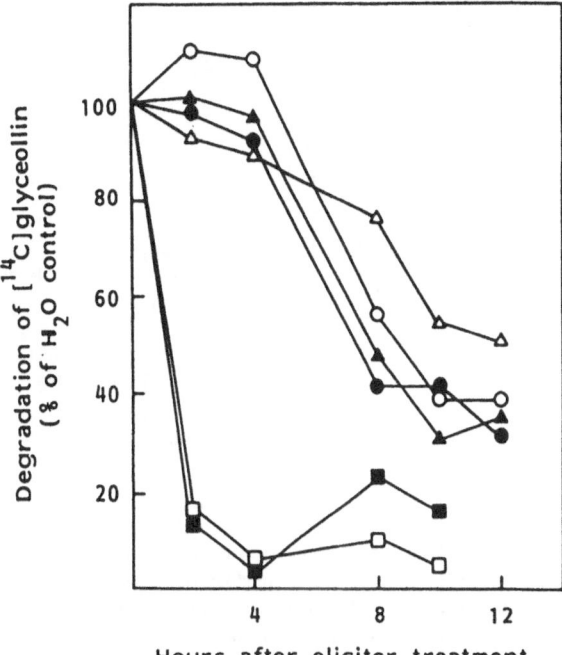

Figure 8. [14C] Glyceollin degrading activity of soybean cotyledons treated with various biotic and abiotic elicitors. Glyceollin degrading activity was measured by incubating elicitor-treated cotyledons with [14C] glyceollin for 1 hour at the indicated time. O, soybean glucanase-released elicitor (10 μg/ml); ●, unextracted mycelial wall (100 μg/ml); Δ, UV irradiation (10 min); ▲, Acridine red (20 g/ml); ☐, $HgCl_2$ (1.5 mM), ■, $AgNO_3$ (3.3 mM).

(PAL), chalcone synthase (CHS), and chalcone isomerase (CHI) demonstrated that the glucanase-released elicitors as well as the native fungal cell walls induced the gene transcription of each enzyme involved in glyceollin biosynthesis within 1 to 2-h after elicitor application (Fig. 7). Concomitant to the induction of glyceollin biosynthesis, glyceollin degrading activity present in soybean tissues was reduced after elicitor treatment (Fig. 8). It may therefore be deduced that these two biochemical actions of the elicitors, the induction of glyceollin synthesis and the reduction of glyceollin turn-over activity, synergistically lead to the massive glyceollin accumulation observed in the elicitor-treated soybean tissues.

Gene transcription for PAL, CHS, and CHI, however, was also induced by cutting of soybean cotyledons, an injury made for elicitor treatment, although the induction was not as strong as that observed in elicitor-treated tissues (Fig. 7). This was consistent with our previous observation (Yoshikawa, 1978;Yoshikawa et al., 1979) that glyceollin synthetic activity, as measured by the 14C-phenylalanine incorporation into glyceollin, was induced by the injury of soybean tissues. Glyceollin did not accumulate, however, to detectable levels by the injury alone due to the existence of a glyceollin turn-over system in soybean tissues. In contrast to the fungal cell wall-associated elicitors, abiotic elicitors such as heavy metal salts and detergents but not DNA-interacting

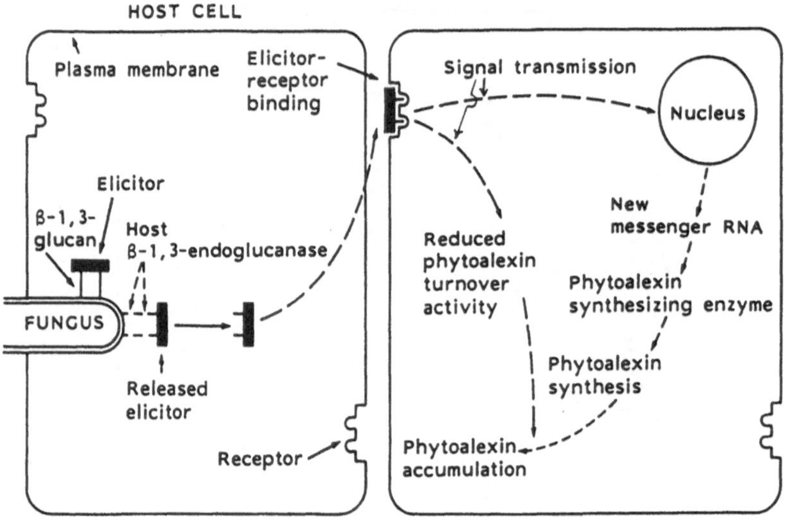

Figure 9. Schematic representation of possible sequence of biochemical events leading to plant defence reactions. The scheme exemplifies phytoalexin accumulation mechanism in the *Phytophthora megasperma* f.sp. *glycinea*-soybean interaction, but it may be applicable to other plant-pathogen systems. Contact of incompatible fungal races with host cells results in rapid release of phytoalexin elicitors from fungal cell wall surface, due to attack by ß-1,3-endoglucanase constitutively present in host cells. The released elicitors then interact with the complementary receptors on plant plasma membrane. This interaction generates second messengers which transmit the signal to the nucleus where de novo transcription is invoked. The resulting new messenger RNA leads to the synthesis of enzymes involved in phytoalexin biosynthesis and the phytoalexin is formed. Levels of phytoalexin accumulation are accelerated by simultaneous inhibition of phytoalexin degrading system, which may also result from the elicitor-receptor interaction. Compatible fungal races may either possess elicitors that, upon release, cannot interact with the host receptors, or produce "suppressors" that interfere with the elicitor-receptor interaction or inhibit one of the subsequent host metabolic processes leading to the phytoalexin accumulation.

agents, however, did not activate the gene transcription for glyceollin biosynthetic enzymes more than the cut-injured and water control (Fig. 7), but rapidly reduced glyceollin degrading activity (Fig. 8). These results at the level of gene transcription support our previous contention (Yoshikawa, 1978) that biotic and abiotic elicitors induce glyceollin accumulation with the different modes of action.

As summarized in this article (also see Fig. 9), researches in the last several years have disclosed the overall sequence of biochemical events that appears to be involved in the induction of plant defence reactions. The processes probably involve pathogen-associated molecule (elicitor)-plant receptor interaction, formation of signal transmitting substances, and gene activation, resulting in de novo biosynthesis of enzymes responsible for production of phytoalexins or other defence substances. At present, however, the scheme is still hypothetical in many respects and requires further evaluation of each process before molecular details of the expression of disease resistance are fully understood.

Acknowledgements

The authors are grateful to Dr. N.T. Keen for suggestions during the course of this research. The work was supported in part by a Grant-in-Aid for Cooperative Research (01304014) from the Ministry of Education of Japan to M.Y.

References

Keen, N.T., 1981, Evaluation of the role of phytoalexins, *In* Plant Disease Control (Staples, R.C., ed.), J. Wiley, New York, pp. 155-177.

Keen, N.T. and Yoshikawa, M., 1983, ß-1,3-Endoglucanase from soybean releases elicitor-active carbohydrates from fungus cell walls, *Plant Physiol.* **71**:460-465.

Sharp, J.K., McNeil, M., and Albersheim, P., 1984, The primary structure of one elicitor-active and seven elicitor-inactive hexa(ß-D-glucopyranosyl)-D-glucitols isolated from the mycelial walls of *Phytophthora megasperma* f.sp. *glycinea*, *J. Biol. Chem.* **259**:11321-11336.

Takeuchi, Y., et al., 1990, Molecular cloning and ethylene induction of mRNA encoding a phytoalexin elicitor-releasing factor, ß-1,3-endoglucanase, in soybean, *Plant Physiol.* **93**:673-682.

Yoshikawa, M., 1978, Diverse modes of action of biotic and abiotic phytoalexin elicitors, *Nature* **275**:546-547.

Yoshikawa, M., 1983, Macromolecules, recognition, and the triggering of resistance, *In* Biochemical Plant Pathology (Callow, J.A., ed.), J. Wiley, Chichester, pp. 267-298.

Yoshikawa, M., Keen, N.T., and Wang, M.C., 1983, A receptor on soybean membranes for a fungal elicitor of phytoalexin accumulation, *Plant Physiol.* **73**:497-506.

Yoshikawa, M. and Masago, H., 1982, Biochemical mechanism of glyceollin accumulation in soybean, *In* Plant Infection: The Physiological and Biochemical Basis (Asada, Y., ed.), Japan Sci. Soc. Press, Tokyo and Springer-Verlag, Berlin, pp. 265-280.

Yoshikawa, M., Matama, M., and Masago, H., 1981, Release of a soluble phytoalexin elicitor from mycelial walls of *Phytophthora megasperma* var. *sojae* by soybean tissues, *Plant Physiol.* **67**:1032-1035.

Yoshikawa, M., Yamauchi, K., and Masago, H., 1978a, Glyceollin: its role in restricting fungal growth in resistant soybean hypocotyls infected with *Phytophthora megasperma* var. *sojae*, *Physiol. Plant Pathol.* **12**:73-82.

Yoshikawa, M., Yamauchi, K., and Masago, H., 1978b, De novo messenger RNA and protein synthesis are required for phytoalexin-mediated disease resistance in soybean hypocotyls, *Plant Physiol.* **61**:314-317.

Yoshikawa, M., Yamauchi, K., and Masago, H., 1979, Biosynthesis and biodegradation of glyceollin by soybean hypocotyls infected with *Phytophthora megasperma* var. *sojae*, *Physiol. Plant Pathol.* **14**:157-169.

Summary of Discussion of Yoshikawa's Paper

The discussion was initiated by *Durbin* inquiring as to the MW of the three elicitors and whether they bind to the same receptor. *Yoshikawa* responded that the elicitors were very heterogeneous in size and a MW had not been confirmed. He further responded that in studies using labeled ligand and elicitor, fraction I which has a relatively large size shows greatest binding. Smaller fractions compete with fraction I for binding sites and thus may bind to the same site as the fraction I elicitor. *Durbin* noted that the higher MW elicitor fraction was more active than low MW elicitor fraction. *Yoshikawa* agreed. *Alexander* questioned the crossreactivity of the glucanase antibodies to proteins in transgenic tobacco plants or with proteins in infected tobacco plants. *Yoshikawa* replied that the antibodies did not crossreact with tobacco proteins as evidenced by Western blots, Ouchterlony plates, or rocket immunoelectrophoresis. The antibodies have not been tested against pathogen or elicitor treated tobacco. He further stated that tobacco plants transformed with their ß-1,3-endoglucanase gene had higher enzyme activity than transformed controls. The role of fungal ß-glucanase was questioned by *Yokoyama*. *Yoshikawa* answered that his model did not consider the fungal enzyme. However, fungal enzyme could be important. Fungal glucanases are required for hyphal tip growth and could be involved with spontaneous excretion of small amounts of elicitor from living hyphae. *Yokoyama* further questioned the role of ß-1,3-exoglucanases in susceptibility. *Yoshikawa* doubted the role of this enzyme in the interaction because they could obtain substantial elicitor from intercellular fluids. *Kunoh* asked if elicitors could not only induce phytoalexin synthesis but also affect phytoalexin turnover. *Yoshikawa* reminded that his model proposed a role for elicitors in phytoalexin turnover. Using ^{14}C-glyceollin they have shown that glycerollin degradation was reduced by elicitor treatments. *Bushnell* noted that the difference between resistant and susceptible cultivars did not lie in the amount of elicitor released. *Yoshikawa* agreed that both cultivars caused elicitor release but the released elicitor may have greater affinity for the incompatible cultivar. *Chumley* asked *Yoshikawa* to comment on the number of elicitor binding sites per cotyledon cell. *Yoshikawa* replied that there were very few (near 20) while the number of binding sites for ß-1,3-glucans were about 2000. Questioning was closed by *Yamada* who asked if ß-1,3-glucanases were induced during infection of both compatible and incompatible cultivars. *Yoshikawa* stated that they had not observed the induction of ß-1,3-glucanase activity in response to elicitors or pathogens.

Chapter 15
Genetic Fine Structure Analysis of a Maize Disease-Resistance Gene

Jeffrey L. Bennetzen, Scot H. Hulbert, and Philip C. Lyons

Many single genes have been identified in plants which determine resistance to specific pathogens. Many of these genes condition race-specific resistance, wherein particular alleles of a generally dominant resistance gene recognize and provide resistance to particular races of a pathogen species. In most cases, a unique series of race specificities is used to define a unique resistance gene allele. Similarly, a novel pattern of responses of a particular pathogen isolate on a series of resistance allele isolines is definitional for a unique pathogen race. Flor's pioneering studies determined that the incompatible (i.e., resistance) response generally was inherited as a monogenic dominant locus in both the host and pathogen (Flor, 1955, 1971). These dominant avirulence (*avr*) loci (Staskawicz et al., 1984) in the pathogen may be broadly envisioned as encoding antigenic determinants which the factor encoded by the resistance gene recognizes to determine the presence and nature of the invasive pathogen. This interaction between the gene products of the host resistance and pathogen avirulence loci is correlated with a complex series of biochemical events. Various host hydrolytic, microbicidal, and cell wall strengthening enzyme activities are induced and are commonly associated with hypersensitive necrosis at the site of pathogen introduction. The simplest model for the action of a race-specific or gene-for-gene resistance locus states that the resistance gene specifies a signal transduction function that recognizes the pathogen's presence and, in response, activates multiple biochemical and structural defense systems.

We would like to identify the genetic and molecular components and activities of a gene-for-gene disease resistance locus. The *Rp1* locus of maize determines race-specific resistance to *Puccinia sorghi*, a fungal leaf rust pathogen. We targeted this locus for detailed analysis due to its excellent prior characterization by Hooker and coworkers (Hooker et al., 1962; Lee et al., 1963; Hagan et al., 1965; Saxena et al., 1968; Wilkinson et al., 1968) and because of the potential for cloning any dominant gene in maize by transposon tagging.

Our preliminary experiments, and those of Pryor (1987), indicated that *Rp1* was meiotically unstable in an allele-specific manner (Bennetzen et al., 1988). Some alleles, such as $Rp1^G$, were inactivated at rates approaching 1 in 500, while other alleles were over ten-fold more stable. In order to better characterize the genetic components of the *Rp1* locus and to investigate possible mechanisms of its allele-specific instability, we have performed a fine structure genetic analysis of the *Rp1* region. Our data indicate that *Rp1* is probably not, in fact, a locus at all but consists of a series of tandemly linked genes with related structures and functions. Unequal crossing over between these tandem homologous regions could generate both the instability and "allelic" diversity observed at *Rp1*.

Table 1. Virulence phenotypes of *Puccinia sorghi* isolates on maize *Rp* differentials

Rp allele	Rust Isolate				
	1-4	HI1	IN1	GA1	TX1
Rp1-A	-	+	+	-	-
-B	I	-	+	-	-
-C	+	-	+	+	+
-D	-	+	-	-	-
-E	-	-	-	-	+
-F	-	+	+	-	-
-G	-	-	-	+	+
-H	+	+	+	+	+
-I	-	-	-	-	+
-J	+	+	+	+	+
-K	-	-	-	-	+
-L	+	-	+	+	+
-M	I	-	+	+	+
-N	+	-	+	+	+
Rp3-A	-	+	-	+	+
-B	-	+	-	+	+
-C	-	+	-	-	-
-D	-	+	-	+	+
-E	-	+	-	+	+
-F	-	+	-	+	+
Rp4-A	+	+	+	+	+
-B	+	+	I	+	+
Rp5	I	+	-	+	+
Rp6	+	+	+	+	+

(-) phenotype indicates pathogen was not virulent (incompatible interaction);
(+) phenotype indicates pathogen was virulent (compatible interaction);
(I) phenotype indicates pathogen caused intermediate reaction in host.

Collection and Characterization of Several Races of *Puccinia sorghi*

Hooker and coworkers identified fourteen separate *Rp1* alleles (A-N) which could be differentiated using their collection of *P. sorghi* isolates (Saxena et al., 1968). Each of the *Rp1* alleles (and several alleles of other *Rp* loci) was separately backcrossed into an R168 ("universal suscept") background ten or more generations and selected for *Rp* function. Although we were able to obtain these *Rp* differential isolines from Dr. W. Pedersen at the University of Illinois, the *P. sorghi* stocks purified and characterized by Hooker and colleagues have not survived storage. In order to perform a recombinational analysis of *Rp1*, a collection of rust races differentiating the *Rp1* alleles was required.

Our attempts to collect a variety of *P. sorghi* races in 1987 and 1988 were completely unsuccessful. Single spore isolates derived from several sites and sources in the midwestern United States, Connecticut, and Texas all had the same pattern of virulence and avirulence on our *Rp* differential lines as the previously isolated race 1-4 (Bennetzen et al., 1988) (Table 1). These data suggest that, often, very little *P. sorghi* variability is present in the continental USA. We did collect one novel *P. sorghi* biotype from Molokai, Hawaii, in the winter of 1988-1989 (H89W1, Table 1).

The summer of 1989 came after an unusually mild and wet spring that followed an unusually severe drought in 1988. A screen for *P. sorghi* races at the Purdue Agronomy Farm yielded three novel *P. sorghi* isolates, including I89S1 (Table 1). Since *P. sorghi* does not overwinter in Indiana, new populations must travel up from Mexico each growing season. It is likely that the unusual weather in 1988-1989 facilitated the arrival of the diverse population of rust races we detected in Indiana in the summer of 1989.

Our five identified *P. sorghi* races, or biotypes, are defined by their unique patterns of virulence and avirulence on our *Rp* differential lines (Table 1). Some of our *Rp* alleles ($Rp1^H$, $Rp1^J$, $Rp4^A$, $Rp4^B$, $Rp6$) do not provide resistance to any of these biotypes. We also are unable to distinguish some *Rp1* alleles, for instance, $Rp1^A$ from $Rp1^F$ or $Rp1^L$ from $Rp1^N$, with this small collection of rust races. We hope that two additional, apparently different, races of *P. sorghi*, which we collected this last summer but have not yet fully characterized, will remedy some of these deficiencies. However, it is possible that we have lost $Rp1^H$, $Rp1^J$, and/or the two *Rp4* alleles from the R168 background. Hooker and coworkers stored many of their differential isolines in an *Rp/rp* heterozygous state. In the absence of a *P. sorghi* resistance selection, the dominant *Rp* resistance allele may have been randomly lost during subsequent amplifications of the stocks. This was true of the $Rp1^A$, $Rp1^B$, and $Rp1^C$ stocks that we originally obtained, as shown by rust race resistance spectrum comparison of these stocks and the cultivars from which the resistance genes were derived. Moreover, our original $Rp1^A$, $Rp1^B$, and $Rp1^C$ stocks had the restriction fragment length polymorphism (RFLP) patterns of the R168 (*rp*) recurrent parent at the *Rp1*-linked NPI285, NPI422, and BNL3.04 sites (see below and Fig. 2). We believe that our $Rp1^H$ and $Rp1^J$ stocks may still carry functional *Rp1* determinants because they do have unique restriction fragment length polymorphism (RFLP) patterns at the NPI422 site.

Generation of a Fine Structure Recombinational Map of *Rp1* and *Rp5*

The simplest model for the nature of the *Rp1* allelic series would be multiple alleles of a single *Rp1* gene. However, *Rp1* could also be composed of several tandemly linked genes with a similar function but determining different rust race recognition specificities (Fig. 1). Hooker's laboratory (Hagan et al., 1965; Wilkinson et al., 1968) had already mapped two other *Rp* genes, *Rp5* and *Rp6*, within a few map units of *Rp1*. Ullstrup (1965) placed *Rpp9*, a gene specifying resistance to the southern leaf rust, *Puccinia polysora*, 1.6 map units from *Rp1*. Hence, ample evidence suggested that the short arm of chromosome 10 of maize contained a cluster of *Rp* genes.

In order to map the determinants of *Rp1* and *Rp5* function, we needed both rust races that could differentiate between *Rp* alleles and flanking genetic markers to allow us to score for a recombinational exchange associated with any genetic changes detected at *Rp1* or *Rp5*. We obtained RFLP probes previously mapped to the short arm of chromosome 10 in maize from Dr. David Hoisington at the University of Missouri (BNL3.04) and Dr. Tim Helentjaris at Native Plants, Inc. (NPI422 and NPI285). We mapped each of these markers relative to each other and to $Rp1^F$ (Fig. 2). We observed that marker BNL3.04 was centromere distal to $Rp1^F$ and that markers NPI422 and NPI285 were centromere proximal to $Rp1^F$. These *Rp1*-flanking markers were therefore sufficient to allow us to detect recombination at *Rp1*.

Several of the *Rp1* and *Rp5* allelic isolines in the R168 background could be distinguished by polymorphic gel blot hybridization patterns at BNL3.04, NPI422, and/or NPI285. Several of the alleles were indistinguishable at all (e.g., $Rp1^A$ and $Rp1^F$) or some of these three sites, however.

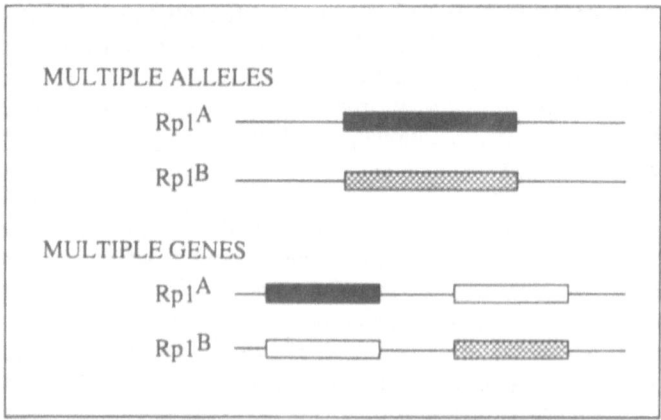

Figure 1. Alternative models for the nature of identified "alleles" of a plant disease resistance gene. The upper schematic represents a true allelic series where different alleles of *Rp1* carry related, but slightly different, genetic information at the same locus. The lower diagram portrays an alternative organization where different *Rp1* "alleles" actually have different, but tightly linked, genes that determine resistance to different *P. sorghi* races.

Our strategy to fine structure map *Rp1* alleles, and *Rp1* relative to *Rp5*, involved construction of a dominant *Rp/Rp* heterozygote that was then crossed as a female to an *rp/rp* tester line. The progeny were then scored for susceptibility to a rust race avirulent on both of the *Rp* alleles in the parental heterozygote. Since *Rp*-derived resistance is dominant, only those cases where a recombination between *Rp* determinants (Fig. 3) or some other *Rp* inactivation process had occurred would yield susceptible seedlings. In those cases where the two *Rp* alleles in the heterozygous parent differed at both distal and proximal RFLP sites, we could score to see whether the oocyte genetic contribution to this susceptible seedling had been generated through a recombinational event (Fig. 3). Moreover, the arrangement of flanking markers would indicate the relative order of the two *Rp* alleles. In the example shown in Figure 3, all susceptible seedlings due to recombination between $Rp1^E$ and $Rp1^F$ determinants would have the $Rp1^F$ pattern at the proximal NPI422 site and the $Rp1^E$ pattern at the distal BNL3.04 site. Figure 2 presents a summary of our data regarding the mapping of alleles of *Rp1* and *Rp5*, primarily relative to $Rp1^E$. The data set for these studies is fairly small, and therefore the accuracy of any given map distance is only within a few tenths of a map unit. The order of the alleles relative to $Rp1^E$ is unambiguous, however. Of 79 susceptible seedlings derived from a screen of 35,167 progeny of different *Rp/Rp* heterozygotes, all but two were recombinant at flanking markers. This indicates that any other *Rp* instability process active in these lines is relatively minor compared with intraregional recombination events. The two susceptible seedlings lacking recombination were from $Rp1^D/Rp1^F$ (4/5 recombinant, 1/5 non-recombinant) and $Rp1^E/Rp1^F$ (6/7 recombinant, 1/7 non-recombinant) test-crosses. We cannot guarantee that any of these "non-recombinant" seedlings arose by a non-recombinational process, however, since neither double cross-overs, sister chromatid exchange, intrachromatid recombination, nor gene conversion would have led to detected flanking marker exchange.

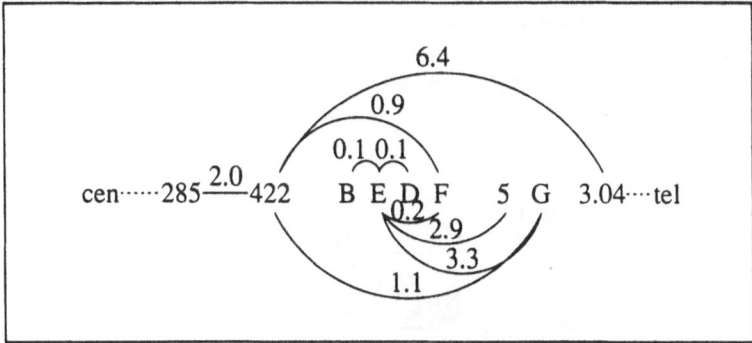

Figure 2. Fine structure recombinational map of *Rp1* and *Rp5* determinants on the short arm of chromosome 10 of maize. The data set for these populations is small, sometimes presenting approximate map distances based on as few as one detected recombination event. Hence, map distances (given in percent recombination) are generally only accurate within a few tenths of a percent. Map orders relative to $Rp1^E$ are consistent, however, as discussed in the text and outlined in Figure 3; 3.04, 422, and 285 are RFLP markers BNL3.04, NPI422, and NPI285, respectively; tel = telomere, cen = centromere.

The data indicated that most *Rp1* alleles cluster within a few tenths of a map unit of $Rp1^E$. Since recombination within a gene can occur at quite high frequencies in maize (Dooner, 1986), this region could contain several alleles of one *Rp* gene or several tandemly linked *Rp* loci. More detailed analysis of this region is under way via attempts to recover the doubly resistant progeny predicted by the multiple gene model (Fig. 3) (Saxena et al., 1968). At the time this study was initiated, however, we did not have a sufficient collection of rust biotypes to undertake this approach.

Our positioning of $Rp1^G$ placed it a good distance distal to all other *Rp1* alleles. By this criterion, $Rp1^G$ is most likely a separate *Rp* gene from *Rp1*, perhaps an allele of *Rp5* (Fig. 2).

A few of our recombinational fine structure results generated a quantitatively inconsistent map. For instance, $Rp1^G$ mapped 3.3 cM distal to $Rp1^E$, while $Rp1^E$ was placed about 0.7 cM distal to RFLP probe NPI422. Yet $Rp1^G$ mapped only 1.1 cM distal to NPI422 in a direct pairwise comparison (Fig. 2). Even with our small database, these results were incompatible at a statistically significant level. We believe this mapping incongruity is due to differential rates and/or modes of recombination in different *Rp/Rp* heterozygote pairs (see below).

Instability at *Rp1* in an $Rp1^A/Rp1^A$ "Heterozygote"

We constructed an $Rp1^A/Rp1^A$ homozygous line that was heterozygous at flanking RFLP sites by crossing two separate maize lines ("Golden King" and GG208R) which both carry the $Rp1^A$ allele (Lee et al., 1963). In crosses of this $Rp1^A/Rp1^A$ line as a female to an *rp/rp* tester, six susceptible seedlings were detected out of 5527 screened progeny. In all six cases, the susceptible seedling was recombinant at the flanking RFLP sites. These data suggest that the instability of *Rp1* homozygotes (Pryor, 1987; Bennetzen et al., 1988) is due to an unequal crossing over process (Fig. 4). Unequal crossing over, initially detected by Sturtevant (1925) at the *Bar* locus in Drosophila, requires tandemly arranged homologous sequences which can lead to mismatched pairing

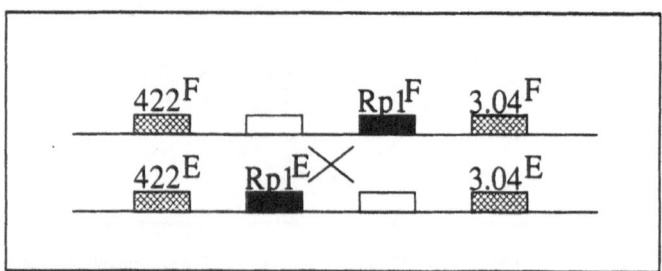

Figure 3. Recombination between *Rp1* allelic determinants or *Rp* genes at *Rp1* scored and positioned by analysis of flanking RFLP marker exchange. In the example shown here, a recombination event positioned at X would yield one chromosome with both $Rp1^E$ and $Rp1^F$ information flanked by 422^E (proximal) and 3.04^F (distal) markers and a chromosome without $Rp1^E$ or $Rp1^F$ determinants which carries 422^F (proximal) and 3.04^E (distal) markers. This indicates that $Rp1^E$ is proximal to $Rp1^F$.

and recombination. The result of such an unequal recombination event between tandem direct repeats is duplication of the sequences between mismatched regions on the one homologue and their reciprocal deletion on the other homologue (Fig. 4).

Our analysis of susceptible seedlings derived from an $Rp1^D/Rp1^F$ heterozygote test cross also suggested mispairing events. Of five susceptible seedlings found from 7184 progeny screened, all but one were recombinant. However, three of the recombinants indicated that $Rp1^F$ was the proximal allele, while one susceptible seedling indicated that $Rp1^D$ was the proximal allele. Unequal crossing over events could give rise to such an observation if more than one mode of unequal pairing could occur, as diagrammed in Fig. 5.

The Ramifications and Predictions of Unequal Crossing Over Between *Rp1* Determinants

Unequal recombination provides a simple explanation both of *Rp1* instability (Pryor, 1987; Bennetzen et al., 1988) and of the origin of a linked series of *Rp* genes. Once unequal crossing over has generated tandem duplications of an *Rp* locus, mutational/evolutionary events could act on one *Rp* gene without loss of the function of its linked twin. The generation of new race specificities could thereby proceed without any negative effect on the current resistance potential of the individual plant. Tandemly linked structural units also imply that a given *Rp* "allele" might actually consist of two or more functioning *Rp* genes. This would provide one explanation for the ability of some *Rp* alleles to act against a broader spectrum of rust biotypes than others (Table 1). Mapping studies of a given *Rp* allele and measures of its instability might thereby yield different results, depending upon the *P. sorghi* race used to detect loss of *Rp* function.

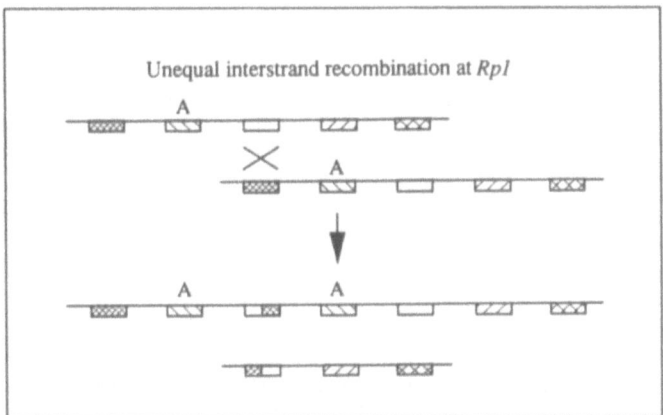

Figure 4. Unequal crossing over structures derived from mismatch pairing and their recombinational products. The boxes indicate different directly repeated tandem arrays of *Rp* information that could serve as regions of homology for mismatch pairing. After recombination at the site marked by an X, a duplication of $Rp1^A$ information on one chromosome and a reciprocal deletion of this same information on the other homologue are generated.

With instability, presumably due to unequal crossing over, as high as 0.2% with some *Rp1* alleles (e.g., *Rp1*^{G} and *Rp1*^{L}) (Bennetzen et al., 1988), separate *Rp1* homozygous lines would soon differ in the number of *Rp* duplications linked to *Rp1*. Those with more duplications would have more potential sites for unequal pairing and, therefore, a predicted higher rate of instability.

Different copy numbers of *Rp1* information would imply different overall physical sizes of *Rp1* "alleles." Hence, the inconsistent *Rp1* mapping data that we detected with different *Rp1/Rp1* heterozygotes, in reality, may accurately reflect different sizes of the *Rp1* region. However, the conclusion that unequal crossing over is a common phenomenon in the *Rp1* chromosomal domain casts serious doubts on the quantitative significance of any fine structure recombinational data in this area.

The six rust susceptible progeny derived from the test cross of an *Rp1*^{A}/*Rp1*^{A} line heterozygous at flanking RFLP markers were not only recombinant but also were recombinant in the same direction. In each case, the GG208R pattern was found at the distal BNL3.04 site, and the "Golden King" pattern was observed at the centromere proximal NPI422 site. Either of two models can explain these data. One model proposes that the two "*Rp1*^{A} alleles" are actually different *Rp1* alleles or genes, despite their identical pattern of resistance and susceptibility on our rust race collection (Table 1) and Hooker's. By this model, the GG208R "*Rp1*^{A} allele" would map proximal to the 'Golden King' "*Rp1*^{A} allele." A second model explains these results in terms of different numbers of *Rp* repeats at these two *Rp1*^{A} alleles. As shown in Fig. 6, there is only one unequal pairing format that leads to a susceptible recombinant with NPI422A information at the proximal RFLP site and BNL3.04A' at the distal site. Three mismatch' pairing modes would yield NPI422A' and BNL3.04A flanking markers in recombinant susceptible seedlings. It is easy to see how variations in the number and position of the regions of tandemly repeated homology in different *Rp1*^{A} alleles could significantly bias the direction of unequal crossing-over events.

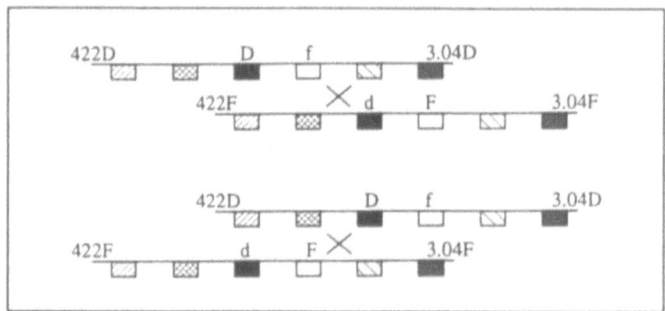

Figure 5. Two models of mismatch pairing and recombination between *Rp1* "alleles" that yield different predicted relative map positions. The upper diagram presents an unequal recombination structure that would yield a prediction that *Rp1*^{D} is proximal to *Rp1*^{F}, while the mismatch recombination event depicted in the lower half of the figure would lead to the conclusion that *Rp1*^{D} is distal to *Rp1*^{F}. D, F = dominant allelic information specifying *Rp1*^{D} or *Rp1*^{F} resistance properties, respectively. d, f = recessive information at *Rp1*^{D} or *Rp1*^{F} site, respectively, that does not specify resistance to the rust race screened.

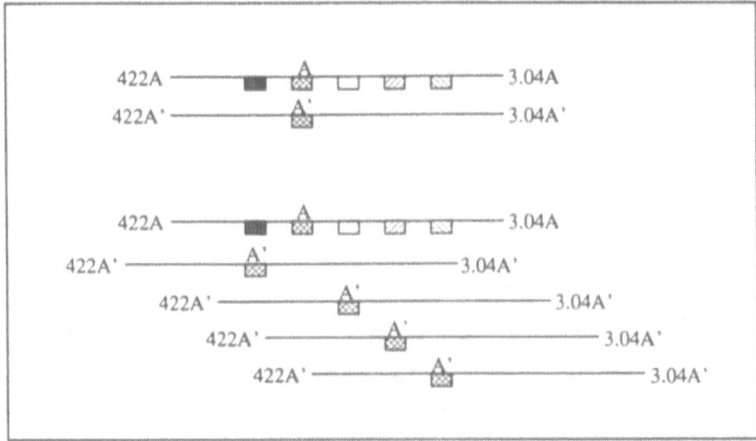

Figure 6. Model of mismatch pairing and unequal crossing over between two $Rp1^A$ alleles that differ in flanking sequences. The Rp homology boxes shown here suggest that one of the differences in these two $Rp1^A$ alleles ($Rp1^A$ and $Rp1^{A'}$) would be due to different outcomes of unequal crossing over events. The upper portion of the figure depicts the two $Rp1^A$ alleles. The lower portion of the figure shows the four different mismatch pairing structures that, resolved by recombination between $Rp1^A$ and $Rp1^{A'}$ determinants, would be guaranteed to yield one chromosome with $Rp1^A$ and $Rp1^{A'}$ information and the other homologue with neither $Rp1^A$ nor $Rp1^{A'}$ information. Of the four unequal pairing/recombination situations shown for the $Rp1^{A'}$ chromosome, three would indicate that $Rp1^{A'}$ is distal to $Rp1^A$, while one would indicate the opposite relative positions. Either or both of these conclusions might be incorrect, however, since $Rp1^A$ and $Rp1^{A'}$ are truly allelic in this scenario.

Future Prospects for *Rp1* Characterization

Our laboratory has targetted the molecular cloning of *Rp1* by transposon tagging as a major goal (Bennetzen et al., 1988). Once clones of *Rp1* are isolated, the structural rearrangements generated by the instabilities and recombination in this region will provide landmarks necessary for assembly of a physical map of the *Rp1* region. Analysis of any changes in *P. sorghi* race resistance specificities (Bennetzen et al., 1988) generated in these recombinants may also determine the position and number of functional *Rp* genes in any *Rp1* "allele." We believe that only such a combined genetic and molecular approach will yield a solid understanding of the structure, evolution, and function of this fascinating gene-for-gene system.

Acknowledgements

We would like to thank W. Pedersen for providing maize lines and T. Helentjaris and D. Hoisington for providing RFLP probes. This research was supported by NSF grant DMB-8552557.

References

Bennetzen, J.L., Qin, M.M., Ingels, S., and Ellingboe, A.H., 1988, Allele-specific and *Mutator*-associated instability at the *Rp1* disease-resistance locus of maize, *Nature* **322**:369-370.

Dooner, H.K., 1986, Genetic fine structure of the *bronze* locus in maize, *Genetics* **113**:1021-1036.

Flor, H.H., 1955, Host-parasite interaction in flax rust–its genetics and other implications, *Phytopathol.* **45**:680-685.

Flor, H.H., 1971, Current status of the gene-for-gene concept, *Annu. Rev. Phytopathol.* **9**:275-296.

Hagan, W.L., and Hooker, A.L., 1965, Genetics of reaction to *Puccinia sorghi* in eleven corn inbred lines from Central and South America, *Phytopathol.* **55**:193-197.

Hooker, A.L., and Russell, W.A., 1962, Inheritance of resistance to *Puccinia sorghi* in six corn inbred lines, *Phytopathology* **52**:122-128.

Lee, B.H., Hooker, A.L., Russell, W.A., Dickson, J.G., and Flangas, A.L., 1963, Genetic relationships of alleles on chromosome 10 for resistance to *Puccinia sorghi* in 11 corn lines, *Crop Sci.* **3**:24-26.

Pryor, A., 1987, The origin and structure of fungal disease resistance genes in plants, *Trends Genet.* **3**:157-161.

Saxena, K.M.S., and Hooker, A.L., 1968, On the structure of a gene for disease resistance in maize, *Proc. Natl. Acad. Sci. USA* **68**:1300-1305.

Staskawicz, B.J., Dahlbeck, D., and Keen, N., 1984, Cloned avirulence gene of *Pseudomonas syringae* pv. *glycinea* determines race-specific incompatibility on *Glycine max*, *Proc. Natl. Acad. Sci. USA* **81**:6024-6028.

Sturtevant, A.H., 1925, The effects of unequal crossing over at the Bar locus in *Drosophila, Genetics* **10**:117-147.

Ullstrup, A.J., 1965, Inheritance and linkage of a gene determining resistance in maize to an American race of *Puccinia polysodra*, *Phytopathol.* **55**:425-428.

Wilkinson, D.R., and Hooker, A.L., 1968, Genetics of reaction to *Puccinia sorghi* in ten corn inbred lines from Africa and Europe, *Phytopathol.* **58**:605-608.

Summary of Discussion of Bennetzen's Paper

Mills opened the discussion by noting that processing of the rat muscle gene at different processing sites within a single locus gave rise to structures leading to either smooth or striated muscle cells from a single gene. *Bennetzen* responded that such a model for generating different specificities through alternative processing is exciting. He then made an analogy with the regulation of the mammalian immunoglobulin loci, which involves a series of genes which undergo somatic generation of new functions, some of which are also associated with variable processing. He noted that his maize work had only confirmed one level of variability at the particular locus studied. *Alexander* asked in the most pessimistic analysis how far away the markers are from the $Rp1^F$ locus. Bennetzen replied that the two probes are inseparable from the Rp locus and have recombination events within this tight region. Some of the recombination events occur on either side of the Rp-flanking DNAs. By that definition those probes are within the $Rp1$ region. While he believes this is a rather small area, there is really no way to correlate numbers of recombination units with physical distance. *Chumley* mentioned that he was puzzled by the internal inconsistency with regard to the map distances between e and g. He said that he could accept that the distance from 422 to g could be the same as the distance from e to g, but could not see how it could be shorter. He wondered whether this represents differences between strains or a composite. *Bennetzen* noted that he worked with isogenic lines and having an isoline means that it's the same almost everywhere in the genome except for maybe 5 map units around the area which is of interest. This, in fact, does not represent a difference in a single gene, necessarily. From his data *Bennetzen* predicted that $Rp1^G$ may be one of the more unstable alleles. Rates of instability are 1 in 300 or 400 and rates of recombination are not that much higher. This implies that the $Rp1^G$ copy number between two different $Rp1^G$ lines should differ substantially. In a very short period of time some $Rp1^G$ lines will have several copies of that area due to unequal crossing over while others will be quite deficient in copy number for $Rp1^G$. We believe that doing any comparison with $Rp1G$ may be inaccurate because of its large size which would lead to more chances for equal or unequal crossing over. Another reason comparisons may be inaccurate is that unequal crossing over could lead to hyperrecombination events. Doubling the sequence in a given area leads to greater than a doubling of recombination events. *Bushnell* asked if this type of phenomenon occurs at other disease resistance loci where there are multiple alleles. *Bennetzen* indicated that he really did not know since so few disease resistance genes have been evaluated. In the few that have been studied, instability appears common. He knows of no cases where such genes had been found to be stable. *Keen* questioned whether vegetative recombination occurred and if chimeric plants or mesothytic reactions are indicative of vegetative recombination. *Bennetzen* noted that they had published that the gene was autonomous in nature. They occasionally get mesothytics, but don't know what that means. He also noted that on some homozygous resistant plants they sometimes see an area supporting the pathogen. This could be a chimera or it may have a physiological explanation. *Tsuge* questioned how many progeny are required for transposon tagging of a gene. *Bennetzen* replied that calculations have been made, but they are based on huge assumptions. Only the mutator (mu) system has been characterized at several loci and it generates mutations at 1/1000 to 1/100,000. He cautioned that the AC system has been frequently referred to as a great tagging molecule. However, its mutation rate has been characterized at one locus and that locus was linked to an AC element. *Mills* interjected that AC tends to jump to nearby loci from where it is residing. He questioned whether mu transposes to other chromosomes at high frequency in Bennetzen's system. *Bennetzen* closed by saying they are currently addressing that question. AC transposes to linked sites while mu elements are found in clusters. His thoughts were that mu elements did not excise when tranposed. Therefore, a mu element makes an extra copy that inserts nearby, and since no excision occurs you get a cluster effect.

Chapter 16
Recognition of Fungal Nonpathogens by Plant Cells at the Prepenetration Stage

Hitoshi Kunoh, Issei Kobayashi, and Naoto Yamaoka

Plant cells respond in discriminate cytological and physiological fashions to attacks by pathogenic and nonpathogenic microbes. Prior to these diverse responses, microbes and host plants must perceive and then recognize each other. Recognition thus has both a positive aspect, the ability to establish a recognizable interaction between certain hosts and microbes, and a negative aspect, the failure of encounters between inappropriate combinations of plant and microbe to lead to disease, detectable change, or advantage (Lippincott et al., 1984).

Preliminary inoculation of plants with a nonpathogen or an incompatible race of pathogen often elicits a series of biochemical changes which induce, at the interaction site, resistance to a compatible pathogen. Conversely, plants preliminarily inoculated with a pathogen become susceptible to a nonpathogen or an incompatible race of the pathogen. This induced resistance and susceptibility is thought to help the exploitation of molecular mechanisms of mutual recognition between microbes and plants (Ouchi et al., 1979). Ouchi and colleagues (1974a,b, 1982) extensively studied these phenomena at the cell and tissue levels employing mainly barley-*E. graminis*, and reported important findings: 1) susceptibility to an originally incompatible race of *E. graminis* or resistance to an originally compatible race was induced in barley leaves by prior inoculation of the compatible or incompatible race, respectively; and 2) completion of cellular conditioning toward susceptibility required 15 to 18-h while conditioning toward resistance required only 6-h. Based on their results, Ouchi et al. (1974a) proposed the term accessibility to connote the acceptance-rejection relations in parasite-plant interaction at the levels of tissues, cells, and molecules, and reserved the term susceptibility for parasite-host relations at the organ and whole plant levels. Thereafter, our group (1985a,b, 1986, 1988, 1989) has investigated these induced phenomena at the cellular level using a barley coleoptile cell-*E. graminis* and *E. pisi* system. In these studies, we used a modification of the concept of accessibility originally proposed by Ouchi et al. (1974a): the term accessibility was used to express the acceptance state of cells, and the term inaccessibility for the rejection state of cells. We consider that normal (uninfected) plant cells are in a state inaccessible to attacking agents, and that attack by pathogens leads to the development of a state accessible even to nonpathogens. On the other hand, attempted attack by nonpathogens enhances the normal level of the inaccessible state by directing cells toward a greater rejection state, which then may protect against pathogens which readily penetrate host cells under normal conditions (Kunoh et al., 1988).

To date, most studies on induced accessibility and enhanced inaccessibility deal with penetration and post-penetration stages of fungal development. Since cytoplasmic aggregation is thought to be the first visible response of host cells to a fungal attack (Aist et al., 1977; Bushnell et al., 1975; Zeyen et al., 1979), intracellular responses related to mutual recognition which must precede expression of accessibility or inaccessibility, if any, can easily be overlooked. However, several observations suggest that phenomena associated with initial stages of fungal contact on host surfaces play a critical role in host recognition and success of fungal infection (Hau et al., 1982; Longman et al., 1987; Staples et al., 1980). Thus, in this chapter emphasis is given to our recent findings concerning cytologically visible responses of host cells to germlings of a nonpathogenic fungus which are associated with recognition and following inaccessibility of host cells.

Growth of *E. graminis* and *E. pisi* Singly Inoculated on Barley Coleoptiles, and Host Cell Responses

Conidia of *E. graminis*, inoculated onto barley coleoptiles usually produce primary germ tubes between 1 and 2-h after inoculation and appressorial germ tubes between 3 and 4-h after inoculation. The latter germ tubes gradually elongate and swell, and finally develop into appressoria with a lobe at their apex. As appressoria attempt to penetrate host walls between 10 and 12-h after inoculation, host cytoplasmic aggregates are induced at the site. About 5 to 30 minutes later, clear, circular structures corresponding to penetration pegs or pores are visible in the lobes or host walls beneath the lobes. Therefore, these cytoplasmic aggregates are signals for the time of host wall traversal by *E. graminis*. Within 1 to 1.5-h after initiation of cytoplasmic aggregates, globose haustorial primordia become visible at the apices of penetration pegs. These gradually develop to form the typical haustoria with branched fingerlike projections. When *E. graminis* alone is inoculated onto coleoptiles, 70-80% of conidia succeed in producing haustoria, while the remaining conidia fail in penetration and a papilla becomes visible at the site of attempted penetration.

By contrast, when *E. pisi* is inoculated onto barley coleoptiles after 1 and 2-h, one appressorial germ tube arises from conidia of *E. pisi* without emerging short germ tubes equivalent to the primary germ tubes of *E. graminis*. By 4.5 to 6-h after inoculation, the appressorial germ tubes swell to form unique polymorphic appressoria with usually two to five short lobes. Host responses to penetration by this fungus are unique in that there are two stages of cytoplasmic aggregation (Kunoh et al., 1985b). Firstly, between 6 and 7-h after inoculation, small cytoplasmic aggregates appear beneath one of the short appressorial lobes; these last for 15 to 45 minutes, then vanish. Secondly, after an additional period ranging from 15 to 60 minutes, and sometimes longer, active cyto–plasmicaggregates develop beneath the same appressorial lobes. About 15 to 60 minutes later, clear, circular structures corresponding to the penetration

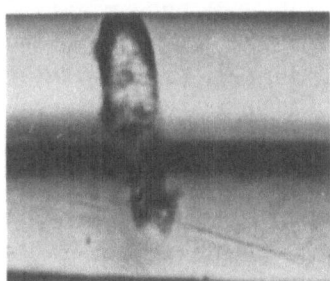

Figure 1. Several cytoplasmic strands below and appressorium of *Erysiphe pisi* in a barley coleoptile cell, 4.5-h before the actual penetration.

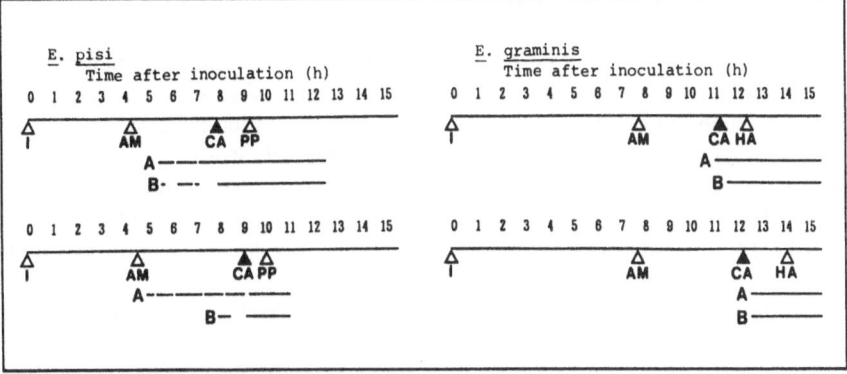

Figure 2. Variation of cytoplasmic streaming in barley coleoptile cells inoculated with either *Erysiphe pisi* or *E. graminis*, observed with a time-lapse video system. A: Prominent cytoplasmic streaming; B: several cytoplasmic strands visible below an appressorium; AM: maturation of an appressorium; CA: time of initiation of cytoplasmic aggregation; HA: time of appearance of a haustorial primordium; I: inoculation; PP: time of appearance of a penetration pore.

pegs or pores form suddenly in the lobes or host walls beneath the lobes. Therefore, for *E. pisi*, the second, rather than the first, cytoplasmic aggregate signals penetration. *E. pisi* never succeeds in producing haustoria in coleoptile cells when inoculated alone and papillae always form at the sites of attempted penetration.

Accessibility and Inaccessibility of Coleoptile Cells Induced by *E. graminis* and by *E. pisi*, Respectively

Kunoh et al. (1985a) reported the following concerning accessibility induced by *E. graminis* and inaccessibility enhanced by *E. pisi* in coleoptile cells, on the basis of time-interval between the initiation of cytoplasmic aggregates induced by these fungi in the same cells : 1) when *E. pisi* attempted penetration more than 60 minutes earlier than *E. graminis*, none of the *E. pisi* succeeded in penetration (% penetration of efficiency = rate of haustorium formation: PE) and PE of *E. graminis* was lowered to 28.6% on average; 2) when both fungi attempted penetration almost simultaneously (within 30 minutes of each other), PE of *E. graminis* recovered to 55.8% on average; 3) if *E. graminis* attempted peneration 60 minutes or more earlier than *E. pisi*, the mean PE of *E. pisi* was enhanced to 29.2% and that of *E. graminis* to 75.0%. These observations definitively show that conditioning coleoptile cells either toward accessibility or inaccessibility is closely associated with timing of attempted penetration by *E. graminis* and *E. pisi* in the same cells. However, these results do not necessarily mean that the cells recognize the presence of both fungi at the time of their penetration.

Discriminative Responses of Coleoptile Cells to Both Fungi at the Prepenetration Stage

Assuming that the cells might recognize the presence of both fungi before their penetration, we observed the cytoplasmic responses of barley coleoptile cells from inoculation of conidia through penetration with a time-lapse video system. When *E. pisi* was inoculated onto a coleoptile cell, cytoplasmic streaming became suddenly active below an appressorium 4 to 5-h before actual penetration occurred (Figs. 1, 2). On the contrary, when *E. graminis* was inoculated, the pattern of cytoplasmic streaming was not altered until the time of penetration (Fig. 2). In order to evaluate the velocity of cytoplasmic streaming, a silicon oil droplet was injected into a coleoptile cell inoculated with *E. pisi*, then the movement of oil droplet was monitored with a video system. As indicated in Fig. 3, oil droplets moved more rapidly when appressoria of *E. pisi* matured (defined below) on the cell surface. The number of cytoplasmic strands in a single coleoptile cell which had been inoculated with either living or freeze-killed *E. pisi* or *E. graminis* was estimated from four photographs of a single coleoptile cell taken at 10 μm intervals in depth (Fig. 4). The number of cytoplasmic strands in the inoculated cells also increased when appressoria of *E. pisi* matured on the cell surface, but no change was detected when a freeze-dried conidium of this fungus was placed on a coleoptile cell (Fig. 5). As mentioned above, 70-80% of *E. graminis* conidia usually succeed in producing a haustorium. In these cases, the number of cytoplasmic strands did not change until attempted penetration (Fig. 5). On the contrary, when *E. graminis* failed in penetration, that number had increased at the time of appressorial maturation (defined below), 4 to 5-h before the time of actual penetration (Fig. 5). These results suggest that whether the penetration attempt of both fungi will succeed or fail in the subsequent stages may have been already determined by the time appressoria matured on the surface of coleoptile cells. Thus, appressorial maturation may be one of the key phenomena for the recognition of both fungi by the host cell.

Definition of Maturation of *E. graminis* and *E. pisi* Appressoria

Within 0.5 to 1-h after appressorial germ tubes of *E. graminis* ceased elongation, a lobe always arose near the apex. From this time until the time of initiation of cytoplasmic aggregation in coleoptile cells below the lobe, no

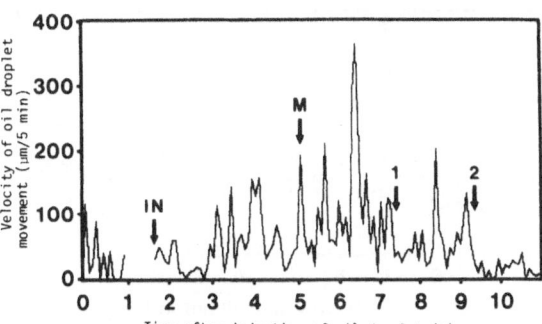

Figure 3. Varied velocity of the movement of an injected silicon oil droplet in an barley coleoptile cell inoculated with *Erysiphe pisi*. IN: Inoculation (transfer); M: time of maturation of an appressorium; 1 & 2: time of initiation of 1st and 2nd cytoplasmic aggregation.

morphologically visible changes occurred in the appressoria. Thus, the emergence of the lobe was defined as a sign of maturation of the *E. graminis* appressoria. On the other hand, after appressorial germ tubes of *E. pisi* ceased elongation, several short lobes usually developed at the swollen apex of the germ tubes. Within 2 to 2.5-h after the development of lobes, a circular to semicircular structure as observed through the surface suddenly appeared in one of the lobes (never more than one). Immediately after the appearance of such a structure, germlings were fixed for scanning electron microscopy (SEM) and their contact side was subjected to micromanipulation at the SEM level. It was revealed that the contact side of only the lobe with the above structure was flat

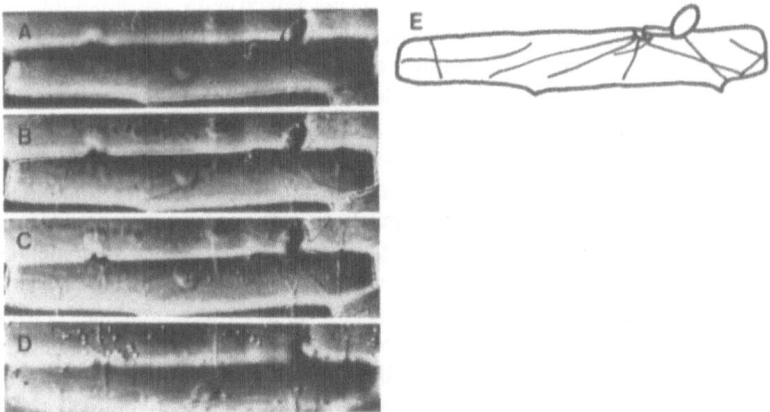

Figure 4. Micrographs of a coleoptile cell inoculated with an *Erysiphe graminis* conidium, taken at 10 μm intervals in depth from the cell surface (A-D). Cytoplasmic strands visible in frames A-D are totally drawn in frame E. The number of cytoplasmic strands is estimated as 10. Number at the upper left corner in each of frames A-D represents the depth from the cell surface in μm.

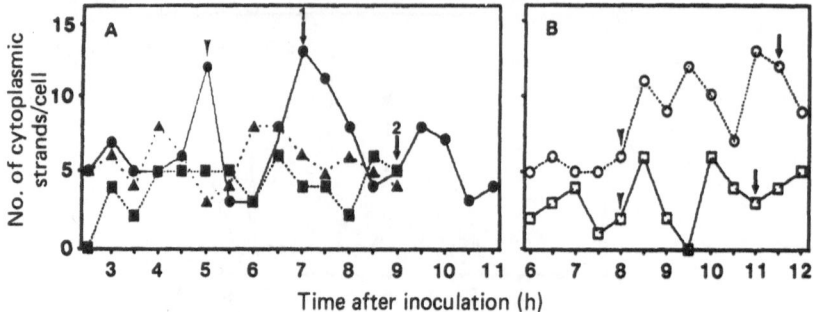

Figure 5. Variation of number of cytoplasmic strands in a barley coleoptile cell inoculated with *Erysiphe pisi* (A) and *E. graminis* (B) with the time after inoculation. ▼Time of maturation of appressorium.↓Time of initiation of cytoplasmic aggregates (1 & 2 = 1st & 2nd). *E. pisi*: ● inoculated with a living conidium; ▲ inoculated with a freeze-killed conidium; ■ not inoculated (control). *E. graminis*: ○ penetration failed; □ penetration succeeded.

but no penetration sign was visible on both the flat side and a track of the lobe on the coleoptile surface. Thus, the appearance of a circular to semicircular structure seemed to reflect the attachment of the lobe to the cell surface. From the time of the appearance of this structure to the time of initiation of cytoplasmic aggregation in a coleoptile cell, there was no visible change in appressoria. Therefore, the appearance of such a structure is defined as a sign of maturation of the *E. pisi* appressoria.

Inaccessibility of Coleoptile Cells Induced by *E. pisi* at the Prepenetration Stage

The incipient responses of host cells to the appressorial maturation of *E. pisi* raised a question whether these responses are linked with inaccessibility of the cells at the prepenetration stage. Experiments as indicated in Fig. 6 gave clues to answer this question. Germlings of *E. pisi* were removed from the cell surfaces by micromanipulation, 1) when an apex of an appressorial germ tube initiated swelling (before maturation), 2) when appressoria matured, or 3) 2 h after appressoria matured. Then germlings of *E. graminis*, which had been inoculated onto other coleoptiles, were transferred by micromanipulationonto the same cells from which *E. pisi* had been removed so that appressoria of transferred germlings could attempt to penetrate the cells. The degree of inaccessibility enhanced in these cells was evaluated through PE of transferred *E. graminis*. As indicated in Table 1, inaccessibility was not enhanced when *E. pisi* was removed before appressorial maturation. On the contrary, inaccessibility was enhanced in the cells when *E. pisi* had been removed at the time of appressorial maturation: when time intervals between removal of *E. pisi* and penetration of *E. graminis* were 6-h or longer, PE of the latter was suppressed by approximately 50%. Similar results were obtained when *E. pisi* had been removed 2-h after appressorium maturation: when time intervals between removal of *E. pisi* and penetration of *E. graminis* were 4-h or longer, PE of the latter was also suppressed by 50%. These results clearly demonstrate

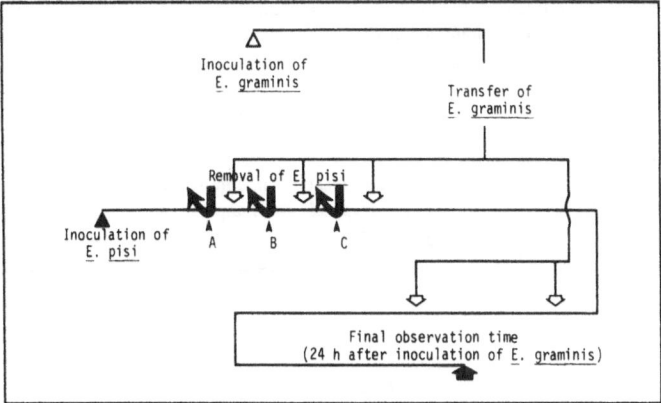

Figure 6. Time course of inoculation of *Erysiphe pisi* and *E. graminis*, removal of the former, and transfer of the latter. A: at the time of initiation of swelling of an appressorial germ tube apex; B: at the time of maturation of an appressorium; C: 2-h after the time of maturation of an appressorium.

that whether inaccessibility could be enhanced in the subsequent stages in a coleoptile cell was determined when *E. pisi* appressorium attached to the cell surface: 4 to 5-h prior to the actual penetration.

Capability of *E. graminis* to Suppress Inaccessibility of Coleoptile Cells

The above results led us to assume that after maturation of the *E. pisi* appressoria, 6-h may be required to enhance inaccessibility to a sufficient level to suppress the *E. graminis* penetration by about 50%, regardless of timing of removal of *E. pisi*. Since we evaluated the PE of transferred *E. graminis* 24-h after its inoculation, we were unable to determine when enhanced inaccessibility was initiated in cells between the times of penetration of *E. graminis* and final observation. However, because the haustorial primordia of *E. graminis* are produced at apices of the penetration pegs, normally 1 to 1.5-h after the initiation of cytoplasmic aggregates, success or failure of the penetration attempt of this fungus can be judged at this time. Taking this fact into consideration, we assumed that inaccessibility might be enhanced 7 to 7.5-h at the latest after the *E. pisi* appressoria matured (assumption 1) (Fig. 7). However, in the experiments described above, timings of transfer of *E. graminis* onto coleoptile cells varied depending on time intervals required between removal of *E. pisi* and penetration of *E. graminis*: when short time intervals were required, *E. graminis* was transferred onto the cell before *E. pisi* was removed, and conversely when longer time intervals were required, the *E. graminis* transfer was made after *E. pisi* had been removed. Therefore, it was inappropriate to rule out the possibility that such timings of the *E. graminis* transfer might interfere with the enhancement of inaccessibility.

Table 1. Penetration efficiency of *Erysiphe graminis* transferred onto barley coleoptile cells from which preinoculated *E. pisi* was removed at various stages of appressorial development

Time intervals between removal of *E. pisi* and initiation of penetration of *E. graminis*	*E. pisi* removed at		
	immature stage of appressorial germ tubes	mature stage of appressoria	2 h after maturation of appressoria
0.00 - 1.75 h			61.8%
2.00 - 3.75		61.3%	(34/55)
4.00 - 5.75		(33/56)	
6.00 - 7.75	62.2%		
8.00 - 9.75	(56/90)		
10.00 - 11.75		32.2%	33.0%[*]
12.00 - 11.75		(28/87)	(33/100)

[*] Significantly different from penetration efficiency of control (64.0%)

Focusing on timing of the *E. graminis* transfer, the data in Table 1 were rearranged. Figure 8 demonstrates that inaccessibility was enhanced only when transfer of *E. graminis* was made after appressoria of *E. pisi* matured, regardless of timings of removal *E. pisi*. Coleoptile cells were never conditioned toward inaccessibility when *E. graminis* was transferred onto the cells before the *E. pisi* appressoria matured. These results lead us to assume that *E. graminis* might be capable to suppress the enhancement of inaccessibility triggered by *E. pisi* and that this suppressive factor (suppressor) may be ineffective once the *E. pisi* factor (inducer) comes into play (assumption 2) (Fig. 7).

Which is a more plausible assumption? Clues to answer this question were given in the following experiment (Fig. 9). Several *E. pisi* conidia were inoculated onto coleoptiles, and then a single coleoptile cell into which an appressorium of *E. pisi* would have penetrated was selected after incubation. Before the *E. pisi* appressorium matured, a germling of *E. graminis* (G1) was

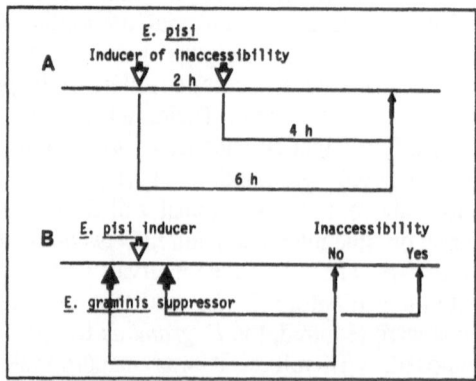

Figure 7. Interaction of an inducer of *Erysiphe pisi* and a suppressor of *E. graminis*. Assumption 1 (A): Six hours after maturation of an appressorium of *E. pisi* (its inducer given to a coleoptile cell) are required before inaccessibility is expressed. Assumption 2 (B): The suppressor of *E. graminis* cancels the effect of an inducer of *E. pisi* but not vice versa.

transferred onto the cell. After the 3-h incubation, another *E. graminis* (G2) was transferred onto the same cell, then G1 was removed. After the subsequent 20-h incubation, PE of G2 was determined.

Such a complicated experiment was designed on the basis of the following consideration. As mentioned above, *E. graminis* normally penetrates 9 to 11-h after inoculation. Thus, transferred *E. graminis* should initiate penetration within 4 to 5-h, because transfer is made 5 to 5.5-h after inoculation. According to assumption 1, coleoptile cells which are inoculated with *E. pisi* conidia should be in an inaccessible state at latest 7 to 7.5-h after appressoria of *E. pisi* matured on these cells. Therefore, if *E. graminis* is transferred before an appressorium of *E. pisi* matured, and if *E. graminis* might have a suppressor, the enhancement of inaccessibility may be suppressed by the G1-G2 system when G2 initiates penetration. If PE of G2 is lowered, assumption 1 should be more plausible.

As indicated in Fig. 9, PE of G2 was not lowered, comparing with that of control. Therefore, these results strongly support the notion that assumption 2 is more plausible. Since the suppressor seems effective soon after transfer of *E. graminis* was made, this undetermined factor is released plausibly before the appressorium matures.

Concluding Remarks

Our series of cytological studies demonstrate that coleoptile cells recognize the presence of *E. pisi* when its appressorium is attached to the cell surface. The most interesting result is that responses of coleoptile cells at the prepenetration

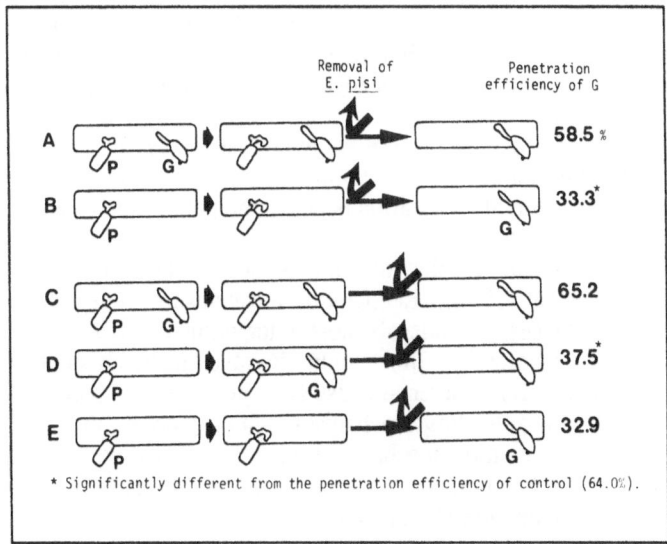

Figure 8. Effects of timing of *Erysiphe graminis* transfer on the penetration efficiency of *E. graminis*. A,B: *E. pisi* (P) was removed at the time of maturation of its appressorium. *E. graminis* (G) was transferred prior to (A) and after (B) removal of *E. pisi*. C-E: P was removed 2 h after maturation of its appressorium. G was transferred prior to (C) and after (D,E) maturation of P's appressorium. Transfer of G was made prior to (D) and after (E) removal of P.

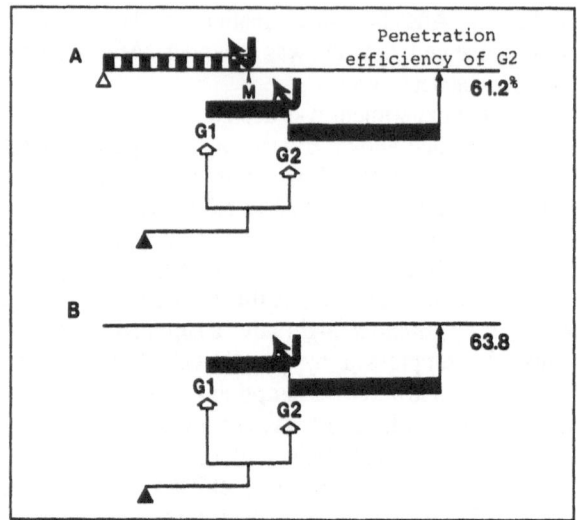

Figure 9. Time course of inoculation of *Erysiphe pisi* (P) and *E. graminis* (G1, G2), removal of P and G1, and transfer of G1 and G2, and effects of this process on the penetration efficiency of G2. A: P was removed at the time of maturation of its appressorium. G1 was transferred and incubated for 3-h, then removed, followed by the immediate transfer of G2. B: P was not inoculated, but transfer of G1 and G2 and removal of G1 was same as those in A.▲▲: inoculation of P and G, respectively;↘: removal of P and G1; ♠ : time of maturation of P's appressorium.

stage were essentially the same when *E. pisi* was inoculated or when *E. graminis* failed in penetration. When *E. graminis* succeeded in penetration, any dynamic response of cytoplasm did not occur until the fungus actually attempted penetration. This result can be interpreted in several ways: a suppressor of *E. graminis*, 1) may block recognition of its presence by coleoptile cells, 2) may interfere with the steps between the recognition and expression of inaccessibility, or 3) may more actively suppress the expression of inaccessibility. Further studies are required for elucidating the biochemical and molecular basis of mutual recognition between the pathogenic or nonpathogenic fungus and the host cells.

Our experimental results mentioned above support Lippincott and Lippincott's statement (1984), "Recognition mechanisms seem to have evolved to enhance the success of microbe-host interaction once they are in close proximity. In succession, this could involve tactic responses of the microorganism, specific adherence substances produced by both microbe and host, and sensing or activating mechanisms which promote microbe and host responses. The recognition mechanisms may function directly as a sensing device, initiating metabolic changes in either the microorganism, host, or both which contribute to the ultimate relationship."

Acknowledgements

This work was supported in part by Grants-in-Aid for Cooperative Research Nos. 60304022 (1985, 1986), 62304015 (1987, 1988) and 01304014 (1989, 1990) from the Ministry of Education, Science and Culture of Japan.

References

Aist, J.R. and Israel, H.W., 1977, Papilla formation: timing and significance during penetration of barley coleoptiles by *Erysiphe graminis hordei*, *Phytopathology* **67**:455-461.

Bushnell, W.R. and Bergquist, S.E., 1975, Aggregation of host cytoplasm and the formation of papillae and haustoria in powdery mildew of barley, *Phytopathology* **65**:310-318.

Hau, F.C. and Rush, M.C., 1982, Preinfectional interactions between *Helminthosporium oryzae* and resistant and susceptible rice plants, *Phytopathology* **72**:285-292.

Kunoh, H., et al., 1985a, Induced susceptibility and enhanced resistance at the cellular level in barley coleoptiles. I. The significance of timing of fungal invasion, *Physiol. Plant Pathol.* **27**:43-54.

Kunoh, H., Aist, J.R. and Hayashimoto, A., 1985b, The occurrence of cytoplasmic aggregates induced by *Erysiphe pisi* in barley coleoptile cell before the host cell walls are penetrated, *Physiol. Plant Pathol.* **26**:199-207.

Kunoh, H., et al., 1986, Induced susceptibility and enhanced resistance at the cellular level in barley coleoptiles. II. Timing and localization of induced susceptibility in a single coleoptile cell and its transfer to an adjacent cell, *Can. J. Bot.* **64**:889-895.

Kunoh, H., et al. 1988, Induced accessibility and enhanced inaccessibility at the cellular level in barley coleoptiles. III. Timing and localization of enhanced inaccessibility in a single coleoptile cell and its transfer to an adjacent cell, *Physiol. Mol. Plant Pathol.* **33**:81-93.

Kunoh, H., et al. 1989, Induced accessibility and enhanced inaccessibility at the cellular level in barley coleoptiles. V. Duration of stimulus by a non-pathogen in relation to enhnaced inaccessibility, *Physiol. Mol. Plant Pathol.* **35**:507-518.

Lippincott, J.A. and Lippincott, B.B., 1984, Concepts and experimental approaches in host-microbe recognition, *In* Plant-Microbe Interactions: Molecular and Genetic Perspectives, vol. I (Kosuge, T. and Nester, E.W., eds.), Macmillan Pub. Co. Inc. New York, pp. 195-214.

Longman, D. and Callow, J.A., 1987, Specific saccharide residues are involved in the recognition of plant root surfaces by zoospores of *Pythium aphanidermatum*, *Physiol. Mol. Plant Pathol.* **30**:139-150.

Ouchi, S., et al., 1974a, Induction of accessibility and resistance in leaves of barley by some races of *Erysiphe graminis*, *Phytopathol. Z.* **79**:24-34.

Ouchi, S., et al., 1974b, Induction of accessibility to a non-pathogen by preliminary inoculation with a pathogen, *Phytopathol. Z.* **79**:142-154.

Ouchi, S., et al., 1979, The induction of resistance or susceptibility, *In* Recognition and Specificity in Plant-Parasite Interactions (Daly, J.M. and Uritani, I., eds.), Japan Sci. Soc. Press, Tokyo/Univ. Park Press, Baltimore, pp. 49-65.

Ouchi, S. and Oku, H., 1982, Physiological basis of susceptibility induced by pathogens, *In* Plant Infection: The Physiological and Biochemical Basis (Asada, Y. et al., eds.), Japan Sci. Soc. Press, Tokyo/Springer-Verlag, Berlin, pp. 117-136.

Staples, R.C. and Macko, V., 1980, Formation of infection structures as a recognition response in fungi, *Exp. Mycol.* **4**:2-16.

Zeyen, R.J. and Bushnell, W.R., 1979, Papilla response of barley epidermal cells caused by *Erysiphe graminis*: rate and method of deposition determined by microcinematography and transmission electron microscopy, *Can. J. Bot.* **57**:898-913.

Summary of discussion of Kunoh's paper

At first *C.P. Vance* and subsequently *Dr. W.R. Bushnell* asked whether the primary germ tube induced or affected the inaccessibility. *Kunoh* replied that he has no evidence to indicate the effect of the primary germ tube. *Durbin* asked about the morphological development of *E. pisi* on barley and its natural host, pea. *Kunoh* answered that the timings of morphogenesis were the same both on pea and barley. *Bushnell* raised a question as to the time required for the induction of enhanced inaccessibility. *Kunoh* answered that the minimum time required for the induction should be 6-h on the basis of data shown in Table 1. *Bushnell* further suggested that there is a report indicating the positive effect of the primary germ tube on subsequent infection. *Kunoh* agreed that inaccessibility is induced 4-h after primary germ tube formation, but it is weak and disappears later on. *Bushnell* thought that most of the primary germ tubes in *Kunoh's* diagram are extended to neighboring cells, hence might not have had a chance to induce inaccessibility in the cells that were challenged later on. *Kunoh* commented that all the conidia were transferred carefully to the target cells and he did not pay attention to the effect of the primary germ tubes. *Bushnell* added that cytological responses to primary germ tubes and penetration hyphae resemble each other suggesting that inaccessibility is induced by both, but at different times. He also asked about the correlation between the enhanced inaccessibility and the development of papilla. *Kunoh* replied that the enhanced inaccessibility is always associated with papilla formation, but no experiment was done to analyze the correlation statistically.

Alexander asked if we can apply other chemicals like protein synthesis inhibitor in the system to biochemically characterize the processes involved in the induction of accessibility, and extended questions to the essentiality of single cell layers for this type of morphological study. *Kunoh* replied that we could use any type of chemical in sustaining solution containing calcium chloride and that single cell layer is essential for detailed observation of cytological changes. He added that he tried similar experiments with rice plant seath, but activity was very low, and that he was unable to get single cell layers with corn and sorghum. *Van Alfen* raised a question about the physiological changes during the experiments, especially about the availability of ATP in the detached single cell layer during the extended period of incubation. He then asked if there is a way to compare the physiological state in a single layer cell with that intact tissues. *Bushnell* commented that disease development on coleoptile is close to that on leaves for about 2 days and cytoplasmic streaming looks normal for 2 days or more, and the single layer cells will live for about 1 week. *Yokoyama* asked about cytological responses in resistant cultivars. *Kunoh* replied that *E. graminia* produces haustonia even in the coleoptile cells of immune type barley, like *H. nigrum* at about 20-30% frequency, hence he had to use *E. pisi* as the inaccessibility inducer. *Chumley* asked about the timing where removal of appressoria become impossible and extended the question about the differences between *E. pisi* and *E. graminis* in the timing. *Kunoh* replied that the attachment of *E. pisi* to barley surface is not as tight as one would assume. *Chumley* mentioned that the attachment of *Pyricularia oryzae* is very tight and becomes impossible to remove after appressorium maturation, but could easily be removed from barley leaves. *Kunoh* added that it is easy to remove *E. graminis* from barley leaf surface that is ordinarily covered by wax crystals. *Yamada* stated that the induced accessibility could be explained by the presence of suppressor and the enhanced inaccessibility by elicitors. *Kunoh* agreed with this concept.

Chapter 17
Role of Phytoalexins in Host Defense Reactions

Shigeyuki Mayama, Ana Paula Ayres Bordin, Toshinobu Morikawa, and Toshikazu Tani

Phytoalexins are antimicrobial compounds which are synthesized by and accumulate in plants at the infection sites of pathogenic microorganisms. It has been indicated that phytoalexins could be responsible for inhibition of the growth of pathogens in plants and regulation of the plant-parasite interactions (Ebel, 1986; Keen, 1982; Mansfield, 1982). It does not imply, however, that the production of phytoalexins is the only response of the plant in functional defense. Besides the production of low molecular phytoalexins, the activation and accumulation of preformed compounds (Tani et al., 1982), lignification (Asada et al., 1979) and hydroxyproline-rich protein deposition in cell walls (Hammerschmid et al., 1984) are also induced at the sites under stress of infection and are involved in defense reactions of some plants.

Phytoalexin accumulation is closely associated with the hyper sensitively formed necrosis in the host cells. Substantial evidence indicates that rapid and large accumulation of phytoalexin is followed by the host cell necrosis and these dead host cells function as reservoirs for the accumulation of phytoalexin (Bailey et al., 1980; Sato et al., 1977). The significance of phytoalexin in disease resistance has been evaluated from various points of view such as the localized concentration at infection sites (Yoshikawa et al., 1978; Mansfield, 1982), the promotion or inhibition of fungal growth by regulating the accumulation (Mayama et al., 1982a), and the relationship between fungal pathogenicity and phytoalexin tolerance (Tegtmeier et al., 1982).

The specific gene-for-gene interactions between plant cultivars and fungal races have been extensively studied in soybean-*Phytophthora megasperma* and potato-*Phytophthora infestans* systems in relation to phytoalexin production (Keen, 1982; Yoshikawa et al., 1986; Doke, 1987). Major research is now focused on the mechanisms of recognition and gene expression of phytoalexin production (Keen, 1982; Doke, 1987; Lamb, et al., 1989). It has been found that the expression of single gene mediated crown rust resistance in oats is highly correlated with the production of avenalumin, a phytoalexin produced by oats (Mayama et al., 1982b; Mayama, 1983). In the present study, to establish the genetical basis of phytoalexin for resistance, the genetic behavior of avenalumin accumulation and resistance expression was analyzed in the progenies of various crosses between several oat cultivars possessing the *Pc-2* gene which seems to confer both resistance to some races of *Puccinia coronata* f.sp. *avenae* and susceptibility to *Helminthosporium victoriae*. A recognition hypothesis for gene-for-gene interactions in crown rust of oats is discussed.

Avenalumin Accumulation and Resistance Expression

Avenalumins which are styrylbenzoxazinone, highly hydrophilic nitrogen-containing phenolics accumulate in incompatible oat-parasite interactions (Mayama, et al., 1982a,b). The relationship between the degree of expression of resistance and the production of avenalumin I was investigated in the interactions of two physiological races of *Puccinia coronata avenae* and 21 oat Pc lines, each carrying a different single major gene for resistance (Mayama, 1981). More rapid and abundant accumulation of avenalumins was found in the more incompatible interactions where fungal growth was more rapidly restricted. Large accumulation of avenalumins coincided with the time of detection of retardation of hyphal growth within leaf tissues. Very little accumulation of avenalumins was found in any compatible interaction. The amount of accumulated avenalumins was linearly correlated with infection rate through stomata up to 50% infection, and no avenalumins were detected in areas apart from the infected area.

When the slightly-infected leaves of Shokan 1 were treated with cellulase and pectinase 36-h after inoculation, the cells at infection sites remained undigested while those cells apart from the infection sites were released and became protoplasts (Fig. 1). The several-layered cells around the fluorescent collapsed cells at each infection site exhibited resistance to the wall splitting enzymes. If leaves were heavily inoculated, the whole inoculated region of each leaf turned resistant, whereas the cells in the adjacent uninoculated regions were completely released from the leaves by forming protoplasts. The data indicate that the undigested region around the infected cells was under stress because of infection, and that resistance to the enzymes could be a result of accumulation of stress compounds in the cell wall. Avenalumin production could occur at the affected cells around the hypersensitively collapsed cells since a linear increase of avenalumin accumulation in parallel with stomatal infection slowed down when the cells affected at infection sites exceeded in the leaves over 50% stomatal infection.

The enhanced syntheses of p-coumaric acid and ferulic acid which were direct precursors of avenalumin I and II respectively, were confirmed in GC MS and HPLC analysis. In race 226-inoculated Shokan 1 leaves, the peaks corresponding to p-coumaric acid and ferulic acid were much higher than those in race 203-inoculated and uninoculated. However, the amounts of the phenolic acids detected in the leaves were considerably lower than those of avenalumins (Table 1). Thus, avenalumins are considered as the major induced antifungal compounds in the infected oat leaves, although the phenolic acids also have antifungal activity.

The rapid accumulation of avenalumins in race 226 crown rust-infected Shokan 1 oat leaves which occurred at 20°C was greatly inhibited when infected plants were grown at 25° to 35°C. At 30°C, avenalumin accumulation was almost totally inhibited and much greater hyphal growth occurred than at 20°C. When infected plants were grown at 20°C for 36-h and then transferred to 30°C, most of the previously accumulated avenalumins disappeared rapidly within 48-h and hyphal growth continued, even though the hypersensitive collapsed cells remained.

Table 1.   Amounts of avenalumins I and II and the precursor phenolic acids in Shokan 1 oat leaves inoculated with incompatible race 226 of *Puccinia coronata* f.sp. *avenae*.

	μg/g fresh weight	
	Time after inoculation (h)	
	36	42
Avenalumin I	183	205
Avenalumin II	152	163
p-Coumaric acid	3	3
Ferulic acid	3	3

Treatment of plants with α-aminooxyacetate, a potent competitive inhibitor of phenylalanine ammonialyase, greatly inhibited the rapid accumulation of avenalumins in normally incompatible interactions. This was correlated with an increase in susceptibility of the host plants. Significant inhibition of avenalumin accumulation was observed 48-h after inoculation when inoculation was followed by application of the inhibitor at 200 μM, but much less inhibition of hypersensitive host cell collapse occurred in the treated leaves.

These facts suggest that rapid and large accumulation of the avenalumins is crucial for resistance expression in oat leaves against an incompatible crown rust race.

Segregation of Avenalumin Accumulation and Resistance Gene

Four cultivars of oats (*Avena sativa* L.), Victoria, Shokan 1, CW-491-4 (Pc-38) and Kanota were used in the genetical analysis. Victoria with *Pc-2* gene and Shokan 1 are resistant to race 226 of *P. coronata* Cda. f.sp. *avenae* Fraser et Led. Pc-38 and Kanota are susceptible to the same race. Intercrosses were made among the four cultivars and F_1, F_2 and F_3 progenies were obtained. In addition, two cultivars Iowa X469 (*Pc-2Pc-2*) and Iowa X424 (*pc-2pc-2*) which are near-isogenic lines differing at the *Pc-2* gene, and their F_1 hybrid were used. Plants were grown in vermiculite in a growth chamber at 20°C with a 16-h light period. The uredospores of race 226 were harvested and the inoculation was carried out as described previously. The resistance degree of each plant against race 226 was determined by its final infection phenotype and the length of infection hyphae, which was measured by staining with calcofluor 72 or 96-h after inoculation.

Avenalumin I was quantitated in four F_1 hybrids and the four parental oat cultivars. Shokan 1, whose resistant gene has not been identified, was highly resistant to race 226 and accumulated avenalumin extensively. The cultivar Pc-2 was moderately resistant to race 226 and accumulation of avenalumin in the

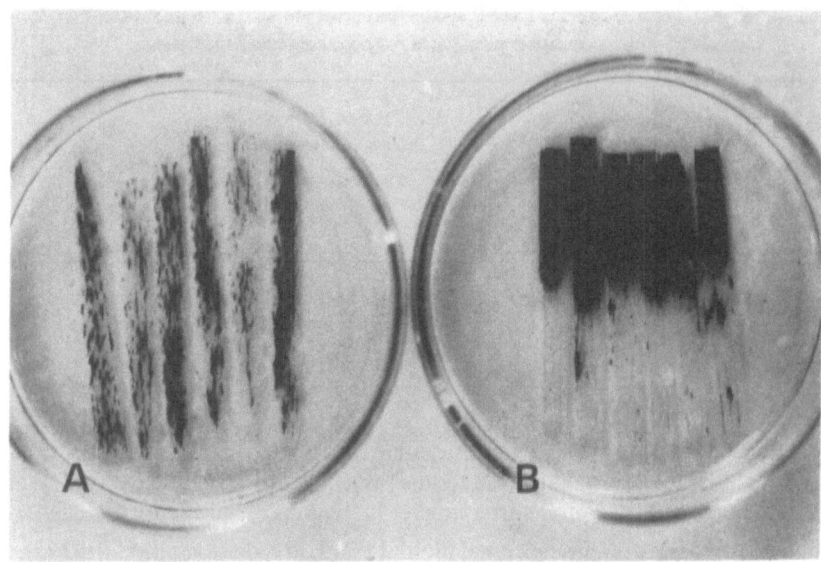

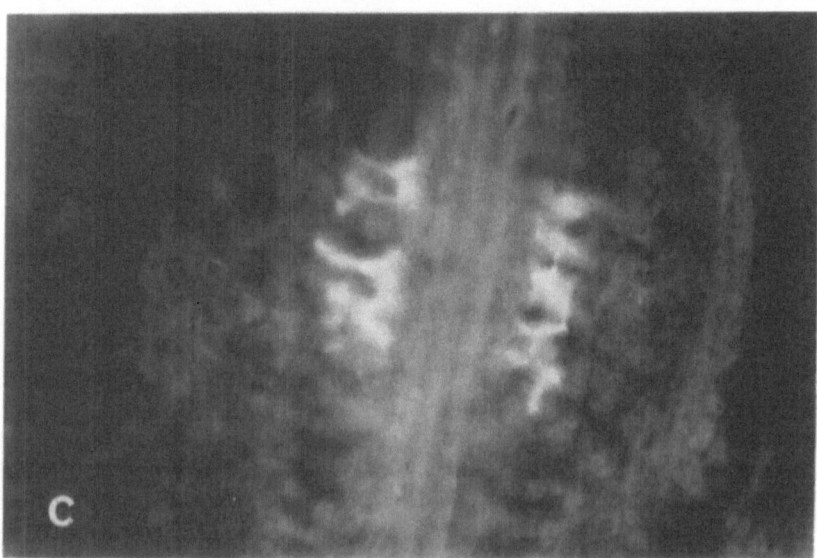

Figure 1. Effect of the hypersensitive cell response at crown rust-infected cells on the neighboring cells as assessed by resistance to wall splitting enzymes. The cells localized at infection sites are undigested by wall-splitting enzymes as shown in (A) slightly and (B) heavily inoculated leaves. The cells unaffected by the hypersensitive fluorescent collapsed cells (C) were released by forming protoplast.

cultivar was about one-third of that found in Shokan 1. Little accumulation of avenalumin I was found in Pc-38 and Kanota which are susceptible to the race. F_1 hybrids of resistant Shokan 1 and *Pc-2* crossed with the susceptible cultivars showed intermediate accumulation of avenalumin as compared with those found in each parental resistant plant. Infection types of the F_1 hybrids were resistant. However, infection hyphae of the rust fungi in the hybrids grew better than those in the parental resistant plants, and small uredia were slightly formed. Segregation for resistance and avenalumin accumulation in F_2 plants of Shokan-1 x Pc-38 and Shokan 1 x Kanota significantly fitted the expected ratio 3:1 (Table 2). The relationship between length of infection hyphae and avenalumin accumulation at 96-h after inoculation was carefully examined with the upper leaf portion of each plant of 194 F_2 plants of Shokan 1 x Pc-38 (Fig. 2). Fifty-five plants in which infection hyphae were more than 1200 μm long were completely susceptible since they produced abundant sporulation and no accumulation of avenalumin. About 55 plants were highly resistant, since the infection hyphae grew no longer than in Shokan 1 and thus appeared to be homozygous in the resistance gene. The amounts of avenalumin accumulation in that group were significantly higher than in the remaining 84 plants, which showed intermediate lengths of infection hyphae and thus were considered as heterozygous. Avenalumin accumulation was inversely related to the length of infection hyphae. The segregation was significantly fitted to the expected ratio. The data clearly indicate that resistance in Shokan 1 against race 226 is controlled by a single partially dominant gene.

Avenalumin Production and Expression of the *Pc-2* Gene

It is currently believed that both victorin sensitivity and crown rust resistance are controlled by the single *Pc-2* locus or possibly by adjacent loci. Victorin sensitivity and avenalumin production in F_1, F_2, and BC_1 progenies of Kanota x *Pc-2* were analysed. Victorin D was isolated from culture fluid of isolate Hv-1 of *H. victoriae* Meehan and Murphy and used as elicitor of avenalumins as described previously. Three millimeter leaf segments were obtained from the terminal portions of primary leaves and placed in serial dilutions (1 to 1000 ng/ml) of victorin D, which were kept in wells of a microplate for 24-h at 25° C under continuous illumination. The assay endpoint was taken at that lowest dilution of toxin which still showed clear necrosis in the edge of segments. It enabled us to compare the sensitivities of individual plants and to distinguish genotypes of *Pc-2* gene. It was found that the microplate bioassay for victorin sensitivity described above could distinguish homozygous, heterozygous, and recessive genotypes of *Pc-2*. The segregation ratio of the *Pc-2* gene in F_2 progenies, which was determined with the tip portion of each leaf, fitted the expected ratio of 1 sensitive (28) : 2 intermediate (64) : 1 resistant (28) (Table 3). Resistance of the same F_2 plants to race 226 was then examined by inoculating the uredospores on the remaining leaves. All plants that were either homozygous or heterozygous in the *Pc-2* gene were resistant to this race whereas the recessives are susceptible, except several plants which showed resistance to race 226. Resistance of these plants (as shown in the parentheses)

Table 2. Segregation for resistance to crown rust in F2 populations of Shokan 1 X Pc-38 and Shokan 1 X Kanota

Cross line	Total	Resistant	Susceptible	P
Shokan 1 X Pc-38	194	139	55	0.1-0.2
Shokan 1 X Kanota	246	174	72	0.1-0.2
(Expected ratio)	(3)	(1)		

may be due to an unknown resistance gene which had segregated from the Pc-2 gene. Cosegregation of victorin sensitivity and crown rust resistance in F_3 plants proved that the microplate bioassay was accurate enough to distinguish between Pc-2 genotypes. It was found that the Pc-2 cultivar, Victoria, has an unknown resistance gene which was linked to the Pc-2 gene.

Specific Elicitor of Avenalumin Production

If victorin sensitivity and crown rust resistance are both conferred by the Pc-2 locus, then victorin may elicit a hypersensitive reaction just like a race-specific elicitor which may be present in race 226 of the crown rust fungus. As the data shown above also predicted, it was found that victorin functions as a specific elicitor of avenalumin production only in Pc-2 oats. Several substances known to elicit phytoalexins in other plants were tested; mycolaminaram, glucan, glucomannan, and endogenous pectates could elicit avenalumin production, but they were all cultivar non-specific. In contrast, avenalumin I did not accumulate in the toxin-insensitive cultivars at any tested concentration of victorin, but considerable accumulation occurrred in the sensitive leaves exposed to victorin. In the sensitive plants, little electrolyte leakage was detected at the optimum victorin concentration for avenalumin production. However, avenalumin production progressively decreased as electrolyte loss occurred at higher concentrations. Avenalumin production was prevented by cyclohexamide treatment (5 μg/ml) but not electrolyte leakage. The sensitivities of F_1 hybrids of Kanota x Pc-2 and Iowa X469 x Iowa X424 to victorin for avenalumin induction were one-third to one-tenth of those of Pc-2 homozygous plants (Fig. 3), suggesting that there was a gene-dosage effect on the receptor of victorin.

Concluding Remarks

The present study of genetical analysis of phytoalexin accumulation and resistance expression clearly shows that the genetic behavior of avenalumin accumulation is a codominant single factor inheritance, closely linked with genes for resistance to crown rust of oats. Phytoalexins are antimicrobial secondary metabolites which are generally induced in plant cells by any stress factors; however, the fact of phytoalexin segregation in parallel with resistance in crown rust of oats clearly indicates that recognition controlled by the gene

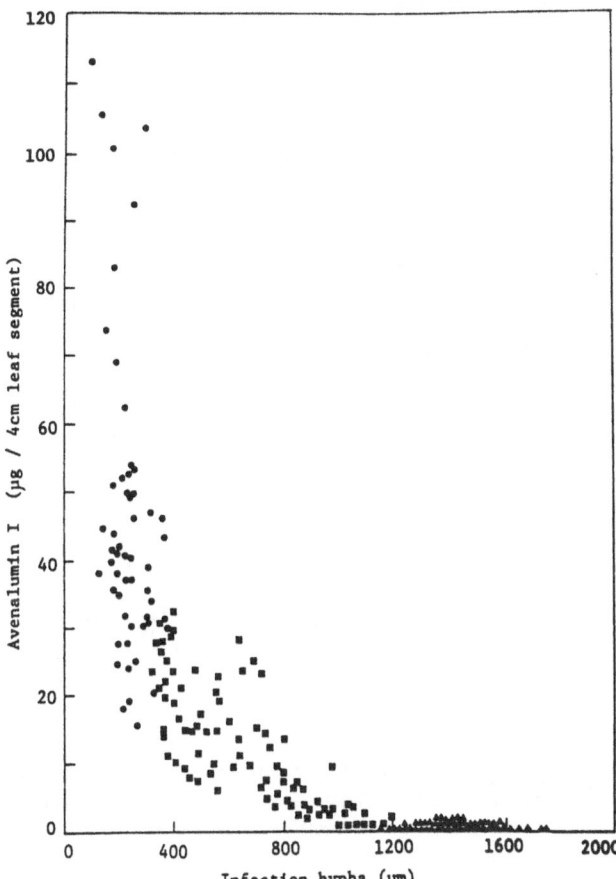

Figure 2. Segregation for resistance to crown rust race 226 and accumulation of avenalumin I in F_2 population of Shokan 1 x Pc-38. Plants with infection hyphae over 1200 μm in length 3 days after inoculation (shown by triangles) were completely compatible and formed abundant spores thereafter.

Table 3. Segregation for resistance to crown rust, avenalumin accumulation, and sensitivity to victorin in F_2 population of Kanota X Victoria (*Pc-2*).

Treatment	Total	No. of Plants			χ^2	p
		Pc-2Pc-2	Pc-2pc-2	pc-2pc-2		
Victorin	120	28	64	28	0.533	0.5-0.7
Race 226	120	28	64	26(2)	0.915	0.5-0.7
Victorin	104	22	56	26	0.923	0.5-0.7
Race 226	104	22	56	20(6)	2.082	0.3-0.5

The numbers in parentheses show the plants resistant to race 226 while the other recessives were susceptible, indicating the presence of an unknown resistance gene which segregated from the *PC-2*.

products for resistance and avirulence is functionally dominant to elicit phytoalexin production. The elucidation of an elicitation mechanism of avenalumin would be crucial to understand the specific parasitism in crown rust of oats.

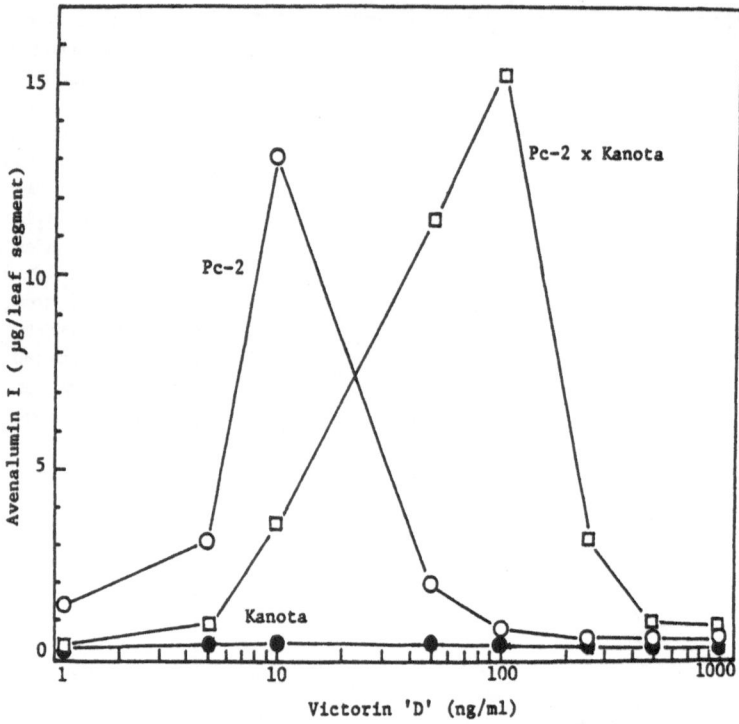

Figure 3. Differential induction of avenalumin accumulation by victorin in Victoria (*Pc-2*) and its F₁ hybrids. The homozygous *Pc-2* line (O) was about 10 times more sensitive to victorin than the hybrid (☐) in the elicitation of avenalumin I. (These data demonstrate partial dominance. If the trait were fully dominant, the heterozygote and homozygote would have the same phenotypes.)

Victorin functions as a specific elicitor of avenalumin accumulation in oats carrying the *Pc-2* gene. The elicitor data therefore support the possibility that avenalumin accumulation is a physiologically important consequence of victorin activity in *Pc-2* oat tissues. The differential sensitivity for avenalumin accumulation in *Pc-2* homozygous and heterozygous plants indicates that there was a gene-dosage effect on avenalumin production by affecting the quantity of receptor molecules of the elicitor, victorin. There is recent evidence that a receptor protein for victorin C may have been isolated (Wolpert et al., 1989). It would be interesting, therefore, to examine whether the gene-dosage effect is found in the receptor protein as suggested by the present study. Based on the facts for avenalumin elicitation by victorin in *Pc-2* oats, it is suggested that specific elicitor-receptor molecules could be involved in the interaction between oats and rust races. Thus, the genes responsible for the formation of elicitors and their receptors are considered as avirulence and resistant genes respectively.

Summary

Avenalumins, phytoalexin of oats, accumulate in incompatible interactions between oat cultivars and crown rust races. Genetical behavior of avenalumin accumulation was analysed by making diallele crosses among four hexaploid oats (Victoria and Shokan 1 : resistant to race 226; CW-491-4 (Pc-38) and Kanota : susceptible to race 226). In F_1 hybrids, avenalumin I accumulation was about one-third to one-tenth of those of resistant parents. Segregation for resistance and avenalumin accumulation in 194 F_2 plants of Shokan 1 x Pc-38 significantly fitted the expected ratio. Victorin functions as a specific elicitor of avenalumin production only in the line with the *Pc-2* gene. Avenalumin accumulation and sensitivity to victorin were examined in F_1, F_2 and BC_1 progenies of Kanota (*pc-2pc-2*) x Victoria (*Pc-2Pc-2*). The sensitivity of F_1 (*pc-2Pc-2*) to induce avenalumin by victorin was one-third to one-tenth of that of Victoria, suggesting that there was a gene-dosage effect on the receptor of victorin. The segregation ratio of victorin sensitivity and avenalumin accumulation in 120 F_2 plants fitted the expected ratio. Segregation in F_3 progenies proved that the bioassay method used for classification of F_2 plants was sensitive enough to distinguish each genotype of *Pc-2*. The genetic behavior of avenalumin accumulation is inherited as a codominant single factor, closely linked with genes for resistance to crown rust.

References

Asada, Y., et al., 1979, Induction of lignification in response to fungal infection. *In Recognition and Specificity in Plant Host-Parasite Interactions* (Daly, J.M. and Uritani, I, eds.), Japan Sci. Soc. Press, Tokyo and Univ. Park Press, Baltimore, pp. 99-112.

Bailey, J.A., et al., 1980, The temporal relationship between host cell death, phytoalexin accumulation and fungal inhibition during hypersensitive reactions of *Phaseolus vulgaris* to *Colletotrichum lindemuthianum*, *Physiol. Plant Pathol.* 17:329-339.

Doke, N., et al., 1987, Biochemical basis of triggering and suppression of hypersensitive cell response. *In* Molecular Determinants of Plant Diseases (Nishimura, S., et el., eds._, Japan Sci. Soc. Press, Tokyo and Springer-Verlag, Berlin, pp. 235-251.

Ebel, J., 1986, Phytoalexin synthesis: The biochemical analysis of the induction process, *Annu Rev. Phytoapthol.* 24:235-264.

Hammerschmidt, R., et al. 1984, Cell wall hydroxyproline enhancement and lignin deposition as an early event in the resistance of cucumber to *Cladosporium cucumerinum*, *Physiol. Plant Pathol.* 24:43-47.

Keen, N.T., 1982, Phytoalexins--Progress in regulation of their accumulation in gene-for-gene interactions. *In* Plant Infection (Asada, Y., et al., eds.) Japan Sci. Soc. Press, Tokyo/Springer-Verlag, Berlin, pp. 281-299.

Lamb, C.J., et al., 1989, Signals and transduction mechanisms for activation of plant defenses against microbial attack, *Cell* 56:215-224.

Mansfield, J.W., 1982, The role of phytoalexins in disease resistance *In* Phytoalexins (Bailey, J.A. and Mansfield, J.W., eds.), Blackie, Glasgow and London, pp. 253-288.

Mayama, S., et al., 1981, The production of phytoalexins by oat in response to crown rust, *Puccinia coronata* f. sp. *avenae, Physiol. Plant Pathol.* **19**:217-226.

Mayama, S., et al., 1982a, The role of avenalumin in the resistance of oat to crown rust, *Puccinia coronata* f. sp. *avenae, Physiol. Plant Pathol.* **20**:189-199.

Mayama, S., et al., 1982b, Effects of elevated temperature and α-amino-oxyacetate on the accumulation of avenalumins in oat leaves infected with *Puccinia coronata* f. sp. *avenae, Physiol. Plant Pathol.* **20**:305-312.

Mayama, S., 1983, The role of avenalumin in the resistance of oats to crown rust. *Mem. Fac. Agr. Kagawa Univ.* **42**:1-64.

Mayama, S., et al., 1986, The purification of victorin and its phytoalexin elicitor activity in oat leaves, *Physiol. Mol. Plant Pathol.* **29**:1-18.

Sato, N. and Tomiyama, K., 1977, Relation between inhibition of intra-cellular hyphal growth of *Phytophthora infestans* and rishitin concentrations in infected potato cells, *Annu. Phytopathol. Soc. Japan* **43**:598-600.

Simons, M.D., 1970, Crown rust of oats and grasses. *Am. Phytopathol. Soc. Monograph No. 5.*

Tani, T. and Mayama, S., 1982, Evaluation of phytoalexins and pre-formed antifungal substances in relation to fungal infection. *In* Plant Infection (Asada, Y., et al., eds.), Japan Sci. Soc Press, Tokyo and Springer-Verlag, Berlin, pp. 301-314.

Tegtmeier, K.J. and VanEtten, H.D., 1982, The role of pisatin tolerance and degradation in the virulence of *Nectoria haematococca* on peas; a genetic analysis. *Phytopathogy* **72**:608-612.

Wolpert, T.J., et al., 1985, Structure of victorin C, the major host-selective toxin from *Cochliobolus victoriae. Experientia* **41**:1524-1529.

Wolpert, T.J. and Macko, V., 1989, Specific binding of victorin to a 100-kDa protein from oats. *Proc. Natl. Acad. Sci. USA* **86**:4092-4096.

Yoshikawa, M., et al., 1978, Glyceollin: its role in restricting fungal growth in resistant soybean hypocotyls infected with *Phytophthora megasperma* var. *sojae, Physiol. Plant Pathol.* **12**:73-82.

Summary of Discussion of Mayama's Paper

The discussion opened with *Hammerschmidt* questioning whether avenalumins were incorporated into the cell wall of oats and thereby contributing to the resistance to cell wall degradation. *Mayama* replied that he did not have specific evidence for any incorporation but several studies have shown incorporation of phenolics into plant cell walls during disease resistance. *Macko* inquired whether abiotic treatments could induce avenalumins. *Mayama* responded that abiotic treatments resulted in avenalumin accumulation comparable to that observed with pathogens. *Keen* then queried whether incompatible rust races on the Pc2 cultivar produced an analog of victorin. *Mayama* indicated that this was a novel thought but he had not evaluated this phenomenon. *Essenberg, Macko,* and *Mayama* then discussed the rationale for conducting a competition experiment by the coincubation of reduced victorin and victorin with inoculation of an incompatible race. Such an experiment might result in less injury and reduced avenalumin accumulation. Although that specific experiment had not been done, *Mayama* noted that incubation of putrescine at inoculation cites suppressed avenalumin formation even after victorin treatments. *Yoshikawa* commented that oats must have a second resistance mechanism since *Helminthosporium* was not affected by avenalumin. *Mayama* reminded the audience of previous studies showing that *Helminthosporium* resistance was due to a highly toxic preformed compound, 26-desglucoavenacoside, which was activated upon inoculation. *Yoshikawa* asked if inhibitors of transcription and translation inhibited avenalumin accumulation. *Mayama* replied that such inhibitors did suppress avenalumin accumulation but were difficult to interpret with respect to the significance of toxins and phytoalexins in disease resistance. *Bushnell* questioned whether susceptible infected oat cells became resistant to cell digestion during infection. This does occur, *Mayama* responded, but much later in disease development. The participants then took part in a general discussion of whether elicitors could also be characterized as toxins. This discussion stemmed from the observation that the elicitor for resistance to *Puccinia coronata* may act at the same time as victorin. There was no consensus regarding this idea. *Durbin* closed the discussion by reminding the audience that elicitors may act as not only inducers of phytoalexins but as effectors of turnover.

Chapter 18
Regulation of Nodule Gene Expression in Plant-Controlled Ineffective Alfalfa

M.A. Egli, C.P. Vance, and R. J. Larson

Legume root nodules are morphologically and biochemically unique organs. Their formation involves infection of the host plant by *Rhizobium* bacteria, differentiation of root cortical cells to give rise to a nodule meristem, synthesis of new vascular elements, and development of infected and uninfected nodule cells, both of which undergo alterations in ultrastructure and metabolism (Long, 1989; Schubert, 1986; Vance et al., 1988). These processes result in the formation of organs that provide an ecological niche for N_2-fixing *Rhizobium* symbionts, and that can assimilate NH_4^+ derived from N_2 fixation into amides or ureides that are ultimately exported to other organs. This complex developmental sequence requires controlled coordinated expression of both bacterial and plant genes. (Downie et al., 1988; Long, 1989; Rolfe et al., 1988). While substantial progress has been made in identifying microbial genes and gene products contributing to nodulation and N_2 fixation (currently over 26 such genes), comparatively little progress has been made in understanding the contribution of plant genes to symbiosis (Govers et al., 1987; Vance et al., 1988; Verma et al., 1987). Although plant genetic control of symbiosis has been documented through classical studies of several legume species (Djordjevic et al., 1987; La Rue et al., 1985) the physiological and biochemical manifestation of these genes are poorly understood.

To date some 53 genes across 10 legume species have been identified as affecting nodulation and N_2 fixation (Table 1). These genotypes were obtained either as spontaneous variants from normal populations or by ethyl methanesulfonate (EMS) and γ-irradiation mutagenesis (La Rue et al., 1985; Vance et al., 1988). Most are inherited as recessive traits and involve a single gene. The identified genes condition four major phenotypes: 1) nonnodulating; 2) ineffective (little or no N_2 fixation) tumor-like nodules; 3) ineffective early-senescing nodules; and 4) supernodulation where plants form several-fold more nodules than normal.

Evidence for the importance of plant genes to N_2 fixation has also come from the identification of plant synthesized, nodule-specific proteins (nodulins) and nodule-enhanced proteins (Delauney et al., 1988). Nodulins and nodule-enhanced gene products are named by the prefix N followed by the protein molecular weight (i.e., N-40 refers to a nodulin of 40 kDa). Nodulins are

Table 1. Host plant genes affecting nodulation and N_2 fixation

Species	Number of genes	Comments
Trifolium pratense L.	7	Naturally occurring; condition non-nodulation and ineffective nodulation; nodules vary from early senescencing to tumor-like.
Pisum sativum L.	16	Naturally occurring and EMS mutagenesis; condition non-nodulation, ineffective, and supernodulation, and nodulation in presence of NO_3^-; some traits temperature sensitive.
Medicago sativa L.	7	Naturally occurring; condition non-nodulation and ineffective nodulation; nodules vary from early senescencing to tumor-like.
Glycine max L. Merr.	9	Naturally occurring and EMS mutagenesis; condition non-nodulation, ineffective, and supernodulation, and nodulation in the presence of NO_3^-.
Trifolium incarnatum L.	1	Naturally occurring; condition ineffective nodulation.
Arachis hypogeae L.	2	Naturally occurring; conditions non-nodulation.
Cicer arietinum L.	5	Derived by g-irradiation; condition non-nodulation, reduced nodulation, and ineffective nodulation; some traits temperature sensitive.
Vicia faba L.	2	Naturally occurring, condition ineffective nodulation.
Phaseolus vulgaris L.	3	EMS mutagenesis; condition non-nodulation, super-nodulation, and nodulation on presence of NO_3^-.
Melilotus alba L.	1	EMS mutagenesis; conditions non-nodulation.

distinguished from bacterial proteins because they are synthesized from poly A^+RNA on 80S ribosomes (Van Kammen, 1984). Complementarity of nodulin mRNA to host genomic DNA is additional evidence for the plant origin of nodulins.

From 9 to 30 nodulins or nodule-enhanced polypeptides and mRNAs have been identified in mature nodules from various legume species (Auger et al., 1981; Govers et al., 1985; Lang-Unnasch et al., 1985). Contributions of these host plant gene products to symbiosis can be grouped into several types of functions including: recognition, root hair invasion, infection thread formation, meristem initiation, vascular bundles, supply of reduced carbon to bacteroids, assimilation of NH_4^+, synthesis of exportable forms of nitrogen, maintenance of low nodule (O_2), separation of bacteroids from host cytoplasm (peribacteroid membrane), and possibly suppression of host defense reactions. Approximately 15 nodule-specific/enhanced host gene products have been identified (Table 2). Nodulin expression appears to be induced in response to specific steps in nodule morphogenesis. Studies of nodule development indicate that expression of most nodulins occurs at or slightly before the onset of N_2 fixation (Dunn et al., 1988; Lullien et al., 1987; Vance et al., 1985). However, reduced nodulin expression in bacterially induced ineffective nodules suggest that other factors, perhaps products of N_2 fixation may also regulate nodule gene expression.

Table 2. Nodule specific/enhanced host genes and their products.

Nodulin	Isolated	Gene subcellular location	(kDa)	MW Function
Leghemoglobin	Yes	Infected cell (cytoplasm)	16	Oxygen carrier
Sucrose synthase	Yes	Unknown	90-100	Carbon metabolism
Uricase	Yes	Uninfected cell (peroxisome)	35	N assimilation (ureides)
Glutamine synthetase	Yes	Infected cell (cytoplasm, plastids)	37, 38, 44	N assimilation
Choline kinase	No	Infected cell (peribacteroid membrane)	60	Membrane structure
Phosphoenolpyruvate carboxylase	Yes	Infected cell (cytoplasm)	101	N assimilation and organic acids
Glutamate synthase (NADH)	No	Infected cell (cytoplasm)	200	N assimilation
Aspartate aminotransferase	Yes	Infected cell (cytoplasm, plastids, peribacteroid membrane)	45, 40	N assimilation, carbon metabolism
Nodulin 23 and 24	Yes	Infected cell (peribacteroid membrane)	23-30	Unknown
(Hydroxy) proline rich glycoprotein	Yes	Unknown	75	Early nodule development
Malate dehydrogenase	No	Infected and uninfected cell	30-37	Organic acid and carbon metabolism
Xanthine dehydrogenase	No	Infected cells (plastids)	---	N assimilation (ureides)
Pyrroline carboxylate reductase	Yes	Infected cell	28	Proline biosynthesis

While numerous studies have documented the role of the *Rhizobium* genome in nodulin expression, there are few reports describing the effect of the host plant genome on nodulin expression. We report here the effects of alfalfa genes on nodulin gene expression. We compare the expression of plant genes in normal N_2 fixing alfalfa cv. "Saranac" ("Sar") to that of ineffective (Fix⁻) nodules formed by plants which are nulliplex for the in_1 gene in the "Saranac" background (in_1Sa). Mature in_1Sa nodules contain *Rhizobium* infected cells, but they fix little N_2 and senesce earlier than do effective nodules (Egli et al., 1989; Vance et al., 1983).

Nodule Development and Nitrogenase Activity

Alfalfa seeds inoculated with *R. meliloti* germinated by d3, when roots were first collected. The first sign of infection (infection threads and microscopic nodules) was seen on d6. From d7 to d10, the diameter of the largest nodules increased from about 0.5 to 1.0 mm. On d10, the largest "Sar" noduleswere faint pink and the first trifoliates of both "Sar" and in_1Sa plants were

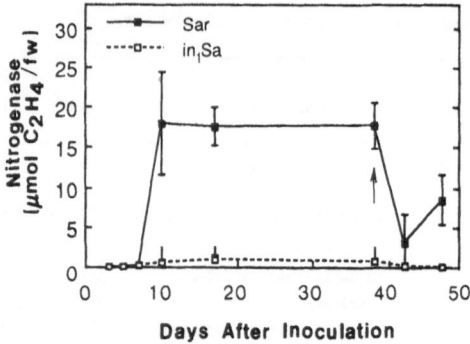

Figure 1. Nitrogenase activity, throughout development and senescence of effective "Saranac" and ineffective *in*1Sa root nodules. Senescence was induced on d38 by defoliation.

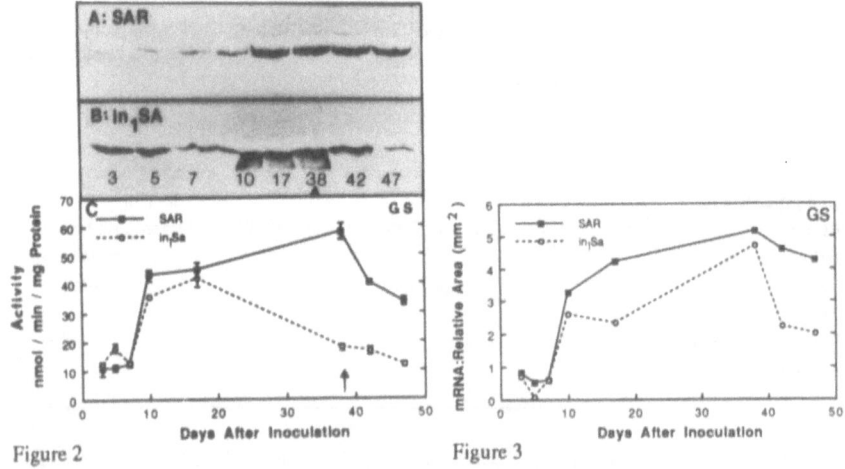

Figure 2

Figure 3

Figure 2. Expression of GS polypeptides (A and B) and GS enzyme activity (C) throughout development and senescence of effective "Saranac" and ineffective *in*1Sa root nodules.

Figure 3. Accumulation of GS mRNAs throughout development and senescence of "Saranac" and *in*1Sa root nodules.

partially expanded. By d17, "Sar" nodules were pink and foliage was dark green, while *in*1Sa nodules were white or faint pink, and foliage was pale green and appeared N-deficient. On d38, the fresh weights of "Sar" and *in*1Sa were 4.31 and 1.53 g, respectively. Proximal ends of *in*1Sa nodules were greenish and senescent. Defoliation on d38 caused extensive senescence of the proximal ends of effective "Sar" nodules by d42. Regrowth of "Sar" foliage began by d47.

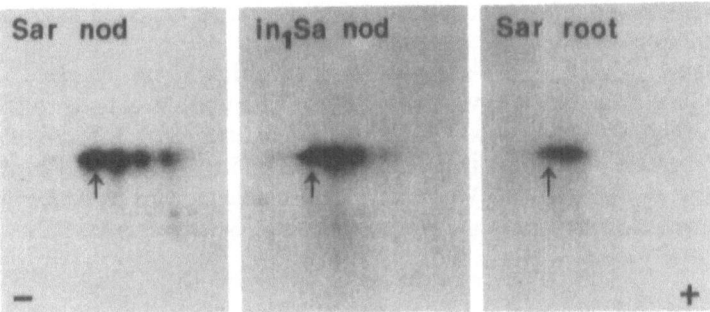

Figure 4. Immunoprecipitation of GS polypeptides from in vitro translations of "Saranac" nodule and root and in_1Sa nodule poly A$^+$RNA. Arrow indicates nodule enhanced GS polypeptide.

"Sar" nodule nitrogenase increased from almost nil on d7 to a maximum specific activity on d10 and remained constant through d38 (Fig. 1). Nodule nitrogenase of in_1Sa also increased from d7 to d10, but the maximum activity of in_1Sa nodules was 5% or less than that of "Sar." Four d after defoliation (d42), nitrogenase of both "Sar" and in_1Sa decreased 80% relative to d38 rates. Effective "Sar" nodule nitrogenase had partially recovered by d47 while activity of in_1Sa was almost nondetectable.

Expression of Glutamine Synthetase (GS)

In vitro GS activity of effective "Sar" nodules increased dramatically from d7 to d10 and continued to increase through d38 (Fig. 2C). By comparison, GS activity of in_1Sa nodules increased from d7 to d17, being comparable to that of effective nodules early in development. However, after d17 GS activity of in_1Sa nodules decreased and was only 25% that of effective nodules by d38. Defoliation resulted in a decrease in GS activity in both genotypes. Nine days after defoliation GS activity of "Sar" was about 50% that of d38 nodules, but was still threefold greater than GS activity of in_1Sa nodules.

Western blots showed that increased GS activity was accompanied by increased expression of GS polypeptides (Fig. 2A, B). The density of the GS polypeptide bands increased dramatically from d7 to d10 and remained relatively constant until 9d after defoliation. By this time the GS band had decreased slightly in effective nodules and was barely detectable in ineffective nodules. It should be noted that the GS blots of in_1Sa nodules contained 20 µg protein/lane while those of effective "Sar" contained 10 ug protein/lane. Thus, the visual comparison of GS polypeptides between effective and ineffective nodules becomes even more disparate with this in mind.

Northern blots showed that GS mRNA accumulation in nodules increased dramatically from d7 to d10 and continued to gradually increase to d38 (Fig. 3). The amount of GS message in nodules during the first 38 days of development was generally comparable between effective and ineffective plants. However, in more mature plants, d42 and older, effective nodules had two- to threefold more

GS mRNA than did in_1Sa nodules and five- to sixfold more GS message than did uninfected roots.

Three 37-kDa GS subunits which differed in charge were immunoprecipitated from in vitro translation (IVT) products of roots and mature nodules (Fig. 4). Two minor 37-kDa spots were also visible. The most basic GS polypeptide was more highly expressed in effective "Sar" nodules than in ineffective in_1Sa and roots. Western blots of soluble proteins separated by two-dimensional electrophoresis and probed with anti-GS serum gave similar results.

Expression of Leghemoglobin

The expression of leghemoglobin (Lb) protein and mRNA (Fig. 5) was similar to the patterns described previously for GS. Expression of Lb increased substantially between d7 to d10 and continued to gradually increase until d38. Four days after defoliation (d42) Lb message had decreased by 30% to 50%. However, by d47 (nine days after defoliation) Lb mRNA had increased, coincident with shoot regrowth. Except for Lb mRNA on d10, Lb protein and message were always greater in effective nodules than in ineffective in_1Sa in both parameters.

Leghemoglobin protein visualization did not always correspond to Lb mRNA. For example, Lb mRNA was still easily detectable after defoliation, but Lb protein particularly from in_1Sa appeared to be reduced more than the change in Lb mRNA could account for.

Not only was Lb expression reduced in in_1Sa nodules, but the array of Lb polypeptides in in_1Sa was different than that of effective "Sar" nodules (Fig. 6). Nine Lb polypeptides of 14 to 15.5-kDa were immunoprecipitated from IVTs of mRNA from mature "Sar" and in_1Sa nodules. However, ineffective nodules

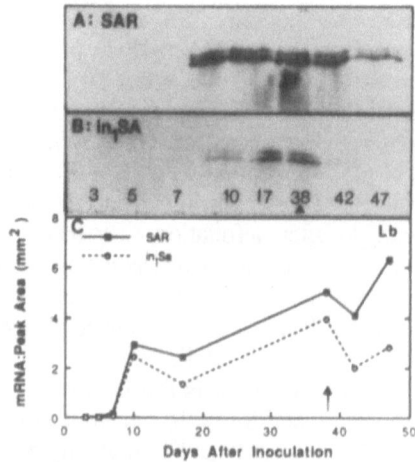

Figure 5. Expression of leghemoglobin (Lb) polypeptides (A and B) and Lb mRNAs throughout development and senescence of "Saranac" and in1Sa root nodules.

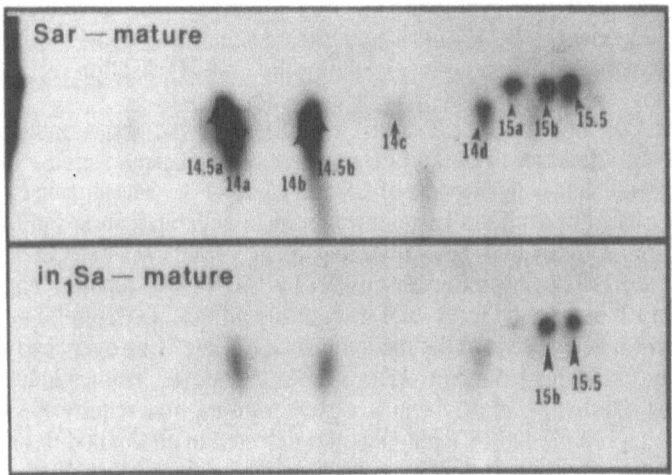

Figure 6. Immunoprecipitation of leghemoglobin (Lb) polypeptides from in vitro translations of "Saranac" and *in*1Sa nodule poly A+RNA. Note substantially reduced expression of Lb polypeptides from in1Sa.

contained much less of the basic Lb polypeptides (14.5a to 14.5b spots) than did "Sar" nodules. This difference was substantiated by quantification through densitometry. The ratio of basic to acidic Lbs immunoprecipitated from "Sar" and in_1Sa nodule IVTs was 3.9 and 0.9, respectively.

Discussion

Although numerous reports have documented the effect of bacterially induced ineffectiveness on nodule enzyme activity and plant gene expression during development, this is the first to detail the effect of the plant genome on similar parameters. We have shown that a single gene change in alfalfa results in ineffective root nodules that fix N_2 at 5% or less the rate of the wild-type. Most of the plant genes expressed in mature effective nodules, including GS and Lb, appear to be expressed in the plant gene controlled ineffective nodules, but at reduced levels (Egli, et al., 1990). Even though N_2 fixation was reduced by more than 90% in the ineffective genotype, there was not a comparable reduction in GS and Lb expression. Furthermore, the initial induction of GS and Lb appeared to be linked more to nodule organogenesis than to N_2 fixation. This interpretation is supported by the kinetics of GS and Lb expression in in_1Sa nodules, as compared with N_2 fixation. Large differences in N_2 fixation between "Sar" and in_1Sa were apparent by d10. However, differences between the genotypes in Lb and GS expression were not evident until much later, d17 or d38 and beyond. Similar data have been obtained with nif⁻ and fix⁻ bacterially induced ineffective nodules and have been interpreted as indicating nodulin and nodule enhanced gene expression occurring independently of nitrogenase (Fuller et al., 1983; Govers et al., 1985; Sengupta-Gopalan and Pitas, 1986). Our data not only support that interpretation, but also suggest that other factors may affect nodule plant gene expression. During effective nodule development, gene expression appears to be (a) initially expressed during organogenesis, and (b) further expressed at increased levels during the time N_2 fixation rates are high. Only (a) appears to occur in ineffective in_1Sa nodules.

Both GS and Lb appear to be controlled at the transcriptional level. Increases and decreases in GS and Lb polypeptide expression were accompanied by associated changes in amounts of GS and Lb mRNA. The shift in expression of alfalfa Lbs from acidic to basic during nodule development is similar to that reported for soybean and pea (Bisseling et al., 1980; Marcker et al., 1984; Verma et al., 1981). A preliminary report by Jing (1988) reported comparable changes in Lb expression for 2- and 6-week old alfalfa. Differential expression of soybean Lb genes leads to the predominance of "late" Lba over "early" Lb c1, c3 in mature nodules (Marcker et al., 1984). Moreover, neither induction nor differential expression of soybean Lb gene transcription require N_2 fixation. The nine Lb proteins and IVT polypeptides detected in alfalfa nodules are likely to be products of different genes. Leghemoglobin polypeptides differ in amino acid composition (Jing et al., 1982) and several distinct alfalfa Lb cDNAs have been identified (Barker et al., 1988; Dunn et al., 1988).

Multiple GS polypeptides were detected in soluble proteins and IVT products from alfalfa nodules and roots. Effective "Sar" nodules expressed a nodule-enhanced GS polypeptide whose expression was reduced in plant-controlled ineffective in_1Sa nodules. Glutamine synthetase activity of mature in_1Sa nodules is lower than that of mature effective "Sar." Several reports have documented similar reductions in GS for bacterially induced ineffective nodules (Cullimore et al., 1988; Dunn et al., 1988; Sengupta-Gopalan et al., 1986). Postma et al. (1988) reported that mutagenesis of nod_3 pea gave a plant-controlled ineffective line with reduced nodule GS activity.

Dunn et al. (1988) reported that increased GS activity in effective and bacterially induced ineffective nodules is associated with increased synthesis of a nodule-specific GS subunit mRNA which could be distinguished from root GS transcripts by differences in 3' untranslated sequences. We were unable to detect a nodule-specific GS polypeptide or IVT product, however, a nodule-specific subunit could appear nodule-enhanced if it comigrated with a root GS subunit. Expression of nodule-enhanced GS by in_1Sa nodules was regulated at or before mRNA translation, since expression of nodule-enhanced GS IVT products and mRNA were reduced in this genotype. Transcriptional control of GS expression has been demonstrated for bacterially induced ineffective alfalfa nodules (Dunn et al., 1988; Norris et al., 1988) and also occurs in plant-controlled ineffective in_1Sa nodules which contain substantially reduced GS mRNA as compared with "Sar" nodules. Reduced GS message in in_1Sa nodules could be due to low levels of NH_4^+ in these nodules. This suggestion is supported by Hirel et al. (1987) who showed that soybean GS appeared to be regulated by NH_4^+ concentrations.

Leghemoglobin and GS expression may also be affected by proteolysis and message stability. Defoliation induced nodule senescence and early senescence of in_1Sa nodules are accompanied by extensive autolysis of host tissues (Vance, 1990; Vance et al., 1983). Such autolysis is mediated by host plant proteolytic enzymes. Host cell cytoplasm and organelles rapidly deteriorate under senescent conditions. Although we have not measured mRNA stability the striking decrease in nodule Lb and GS mRNA upon defoliation could reflect message instability during this period.

Recent studies using chimeric soybean Lb and N-23 genes in transgenic birdsfoot trefoil, alfalfa, and clover have demonstrated that nodule enhanced gene expression is under both developmental and organ-specific regulation (Jensen et al., 1986; Marcker, personal communication; Stougaard et al., 1987; Stougaard et al., 1984; deBruijn et al., 1989). *Trans*-acting factors from nodule nuclear extracts interact with *cis*-acting elements located in the 5' flanking regions of these genes. Four regulatory sequences mapping to -1100 to -950, -230 to -170, -139 to -102, and -102 to -49 have been identified in the Lb c3 gene (Jensen et al., 1988a,b; Stougaard et al., 1987 and 1990). Similar sequences appear in the 5' flanking region of the N23 gene (Marcker, personal communication; Stougaard et al., 1990). There appear to be two positive elements, an organ-specific motif, and a negative element. Gel retardation experiments have shown that similar nodule specific *trans*-acting factors are conserved in trefoil, clover, soybeans, alfalfa, and *Sesbania*. The factors bind to conserved sequences within the -230 to -170 positions. The presence of 5' AAGAT and 5' CTCTT motifs located in the organ-specific elements at -139 to -102 of Lbc3 and N23 are thought to be core organ-specific motifs (Jensen et al., 1988a,b; Stougaard et al., 1990). The ineffective in_1Sa may lack *trans*-acting factors required for developmental and organ-specific expression in mature nodules. Alternatively, continued release of NH_4^+ by bacteroids may be required to induce or to stabilize transcriptional initiation factors.

Acknowledgements

This work was supported in part by the National Science Foundation Grant, DCB: 8905006.

References

Auger, S., and Verma, D.P.S., 1981, Induction and expression of nodule specific host genes in effective and ineffective nodules of soybean, *Biochemistry* **20**:1300-1306.

Barker, D.G., et al., 1988, Identification of two groups of leghemoglobin genes in alfalfa and a study of their expression during root nodule development, *Plant Mol. Biol.* **11**:761-772.

Bisseling, T., et al., 1980, The sequence of appearance of leghemoglobin and nitrogenase components I and II in root nodules of *Pisum*, *J. Gen. Microbiol.* **118**:377-381.

de Bruijn, F.J., et al., 1989, Regulation of plant genes specifically induced in N_2-fixing nodules: role of *cis*-acting elements and *trans*-acting factors in leghemoglobin expression, *Plant Molec. Biol.* **13**:319-325.

Cullimore, J.V., and Bennett, M.J., 1988, The molecular biology and biochemistry of plant glutamine synthetase from root nodules of *Phaseolus vulgaris* and other legumes, *J. Plant Physiol.* **132**:387-393.

Delauney, A.J., and Verma, D.P.S., 1988, Cloned nodulin genes for symbiotic N_2 fixation, *Plant Molec. Biol. Rep.* **6**:279-285.

Djordjevic, M.A., Gabriel, D.W., and Rolfe, B.G., 1987, *Rhizobium*: the refined parasite of legumes, *Annu. Rev. Phytopathol.* **25**:145-168.

Downie, J.A., and Johnston, A.W.B., 1988, Nodulation of legumes by *Rhizobium*: the recognized root, *Plant Cell Environ.* 11:402-412.

Dunn, K., et al., 1988, Developmental regulation of nodule-specific genes in alfalfa root nodules, *Mol. Plant Microb. Interac.* 1:66-74.

Egli, et al., 1989, Nitrogen assimilating enzyme activities and enzyme protein during development and senescence of effective and plant gene-controlled ineffective alfalfa nodules, *Plant Physiol.* 91:898-904.

Egli, M.A., et al., 1990, Developmental expression of nodulins and nodule enhanced polypeptides in plant gene-controlled ineffective nodules, *Plant Physiol.* 94 (in press).

Fuller, F., et al., 1983, Soybean nodulin genes: analysis of cDNA clones reveals several tissue specific sequences in N_2 fixing root nodules, *Proc. Natl. Acad. Sci. USA* 80:2594-2598.

Govers, F., et al., 1985, Expression of plant genes during the development of pea root nodules, *EMBO J.* 4:861-867.

Govers, F., et al., 1987, Nodulins in the developing root nodule, *Plant Physiol. Biochem.* 25:309-322.

Hirel, B., et al., 1987, Glutamine synthetase genes are regulated by ammonia provided externally or by symbiotic N_2 fixation, *EMBO J.* 6:1167-1171.

Jensen, J.S., et al., 1986, Nodule specific expression of a chimeric soybean leghemoglobin gene in transgenic, *Lotus corniculatus, Nature* 321:669-674.

Jensen, E.O., et al., 1988a, Regulation of nodule-specific plant genes. *In* Nitrogen Fixation: Hundred Years After (H. Bothe, F.J. de Bruijn, and W.E. Newton, eds.), Gustav-Fisher, Stuttgart and New York, pp. 605-609.

Jensen, E.O. et al., 1988b, Interaction of a nodule specific, *trans*-acting factor with distinct DNA elements in the soybean leghemoglobin Lbc3 5' upstream region, *EMBO J.* 7:1265-1271.

Jing, Y., Paau, A.S., and Brill, W.J., 1982, Leghemoglobins from alfalfa root nodules, I-Purification and in vitro synthesis of five leghemoglobin components, *Plant Sci. Lett.* 25:119-132.

Jing, Y., 1988, Expression of leghemoglobin genes in alfalfa during root nodule development, *In* Nitrogen Fixation: Hundred Years After (H. Bothe, F. de Bruijn, W.E. Newton, eds.), Gustav Fischer, Stuttgart and New York, p. 636.

Lang-Unnasch, N., and Ausubel, F.M., 1985, Nodule specific polypeptides from effective alfalfa root nodules and ineffective nodules lacking nitrogenase, *Plant Physiol.* 77:833-839.

LaRue, T.A., Kneen, B.E., and Gartside, E., 1985, Plant mutants defective in symbiotic N_2 fixation, *In* Analysis of Plant Genes Involved in the Rhizobium-Legume Symbiosis (R. Marcellin, ed.), OECD Publication, Paris, pp. 39-48.

Long, S.R., 1989, *Rhizobium*-legume nodulation: life together in the underground, *Cell* 56:203-214.

Lullien, V. et al., 1987, Plant gene expression in effective and ineffective root nodules of alfalfa (*Medicago sativa* L), *Plant Molec. Biol.* 9:469-478.

Marcker, K.A., et al., 1984, Transcription of the soybean leghemoglobin genes during nodule development, *EMBO J.* 3:1691-1695.

Norris, J.H., et al., 1988, Nodulin gene expression in effective alfalfa and in nodules arrested at three different stages of development, *Plant Physiol.* 88:321-328.

Postma, et al., 1988, Characterization of a non-fixing mutant of pea (*Pisum sativum* L.), *In* Nitrogen Fixation: Hundred Years After (H. Bothe, F.J. de Bruijn, W.E. Newton, eds.), Gustav Fischer, Stuttgaart and New York, p. 640.

Rolfe, B.G., and Gresshoff, P.M., 1988, Genetic analysis of legume nodule initiation, *Annu. Rev. Plant Physiol. Plant Mol. Biol.* **39**:297-319.

Schubert, K.R., 1986, Products of biological N_2 fixation in higher plants: synthesis, transport, and metabolism, *Annu. Rev. Plant Physiol.* **32**:539-574.

Sengupta-Gopalan, C., and Pitas, J.W., 1986, Expression of nodule specific glutamine synthetase during nodule development in soybeans, *Plant Molec. Biol.* **7**:189-199.

Stougaard, J., et al., 1987, 5' Analysis of the soybean leghemoglobin gene lbc3 regulatory elements required for promoter activity and organ specificity, *EMBOJ.* **6**:3565-3569.

Stougaard, J., et al., 1990, Interdependence and nodule-specificity of *cis*-acting regulatory elements in the soybean leghemoglobin lbc3 and N23 gene promoters, *Mol. Gen. Genet.* (in press).

Vance, C.P., 1990, Symbiotic N_2 fixation: recent genetic advances, *In The Biochemistry of Plant Bol. 16: Intermediary Nitrogen Metabolism* (B.J. Miflin and P.J. Lea, eds.), Academic Press, Orlando, FL (in press).

Vance, C.P., et al., 1985, Nodule specific proteins in alfalfa (*Medicago sativa* L.), *Symbiosis* **1**:69-84.

Vance, C.P., et al., 1988, Plant regulated aspects of nodulation and N_2 fixation, *Plant, Cell Environ.* **11**:413-427.

Vance, C.P. and Johnson, L.E.B., 1983, Plant determined ineffective nodules in alfalfa (*Medicago sativa*): structural and biochemical comparisons, *Can. J. Bot.* **61**:93-106.

Van Kammen, A., 1984, Suggested nomenclature for plant genes involved in nodulation and symbiosis, *Plant Molec. Biol. Rep.* **2**:43-45.

Verma, D.P.S., 1989, Plant genes involved in carbon and nitrogen assimilation in root nodules, *In* Plant Nitrogen Metabolism (J.E. Poulton, et al., eds.), Plenum Publishing, New York, pp. 43-63.

Verma, D.P.S., et al., 1981, Regulation of the expression of leghemoglobin genes in effective and ineffective root nodules of soybean, *Biochem, Biophys. Acta* **653**:98-107.

Verma, D.P.S. and Stanley, J., 1987, Molecular interactions in endosymbiosis between legume plants and N_2 fixing microbes, *Ann. N.Y. Acad. Sci.* **503**:284-294.

Summary of Discussion of Egli's Paper

The discussion was initiated by *Bennetzen* asking what signals turn on nodulin gene expression. *Vance* replied that there are probably several signals. The initial signal is nodule morphogenesis with other later signals mediating full expression. Several studies have shown that effective N_2 fixation is not an absolute requirement for nodulin synthesis. However, effectiveness appears to be necessary for continued expression. *Nester* questioned whether the in_1 genotype has been characterized by two-dimensional gel electrophoresis. *Vance* indicated that two-dimensional gels had been run with extracted protein and with in vitro translation products for both in_1 and effective Saranac alfalfa. He noted that most nodulins were present in the in_1 genotype nodules, but at reduced levels. *Chumley* then inquired whether in_1 was dominant or recessive. *Vance* cited previously published work from Minnesota showing that all of the alfalfa ineffective genotypes that they have isolated are recessive and must be in the nulliplex condition. Three separate single genes and one double gene have been identified that affect nodule effectiveness. These genotypes have two basic phenotypes: those like in_1 which look normal, but senesce very rapidly and those which form tumors containing few bacteria. The number of infection sites in the supernodulation genotypes was addressed by *Durbin's* questioning. *Vance* responded that the number of infections sited probably were not altered. He further stated that in the normal infection process many more infections than nodules were formed. The plant has strict regulation of how many infections go on to form nodules. In supernodulation types this regulation is lost. *Bushnell* asked whether plant growth regulators could change the ineffective gentoype to effective. That experiment has not been done replied *Vance*. He further noted that *Ann Hirsch* had shown treating uninoculated alfalfa with auxin transport inhibitors induced pseudo-nodule formation. These nodules synthesized one or two nodulins, but were ineffective in N_2 fixation. *Yoshikawa* closed the questioning by inquiring whether ineffective nodulation had any similarities to plant disease situations. *Vance* replied that phytoalexins had been detected in at least two ineffective symbioses. Furthermore, in ineffective alfalfa bacteria aggregate and lyse much like events which occur when incompatible bacteria are infiltrated into leaves.

Breeding of Disease–Resistant Plants

Chapter 19
The Use of Somaclonal Variation for the Breeding of Disease-Resistant Plants

Hideyoshi Toyoda and Seiji Ouchi

A prominent feature of plant tissue cultures is that genetic variations are frequently induced by spontaneous or mutagen-induced mutations or chromosomal abnormalities (Evans et al., 1983; Sunderland, 1977). These callus tissues are genetically chimeric with respect to newly induced characters and therefore could be useful gene sources for effectively isolating variant cells. Such a variation induced in callus tissues has been designated as somaclonal variation and has provided us with a new tool for improving crop plants (Larkin et al., 1981).

One of the most effective applications of this technique to the field of plant pathology is to select toxin-resistant cells by culturing plant cells in the presence of toxic compounds. In fact, the in vitro selection for disease resistance was first conducted by Carlson (1973), who isolated tobacco cultures resistant to methionine sulfoximine, a model toxin of tabtoxin produced by *Pseudomonas syringae* pv. *tabaci* and produced a resistant line against this disease. Similar selections have been effectively made in some diseases caused by pathogens which produce host-specific toxins (Brettell et al., 1979). Since these toxins are determinants of pathogenicity, the toxin-resistant regenerants were expected to be actually resistant to diseases caused by these toxin producers.

There are some plant diseases in which the pathogens produce non-specific toxins to extend the disease symptoms. Fungal wilting caused by *Fusarium* species is a typical disease included in this class. A low-molecular-weight toxin has been found in culture filtrates of some *Fusarium* species and implicated as a wilting agent in the tomato. It is a pyridine derivative, fusaric acid (5-n-butylpiconlinic acid), and has been detected in much higher concentrations in plants infected by virulent strains than those inoculated with avirulent strains. In our previous work (Toyoda et al., 1988a; Ouchi et al., 1989), bacteria capable of detoxifying fusaric acid were isolated and tested for their ability to protect tomato plants from the wilting disease caused by the pathogen. The bacteria used were fusaric acid-resistant mutants derived from an avirulent strain of *P. solanacearum*. Tomato plants were protected from wilt when they were pretreated with fusaric acid-detoxifying bacteria before inoculation with the pathogenic fungus. The role of fusaric acid in symptom development of wilting

Days after inoculation

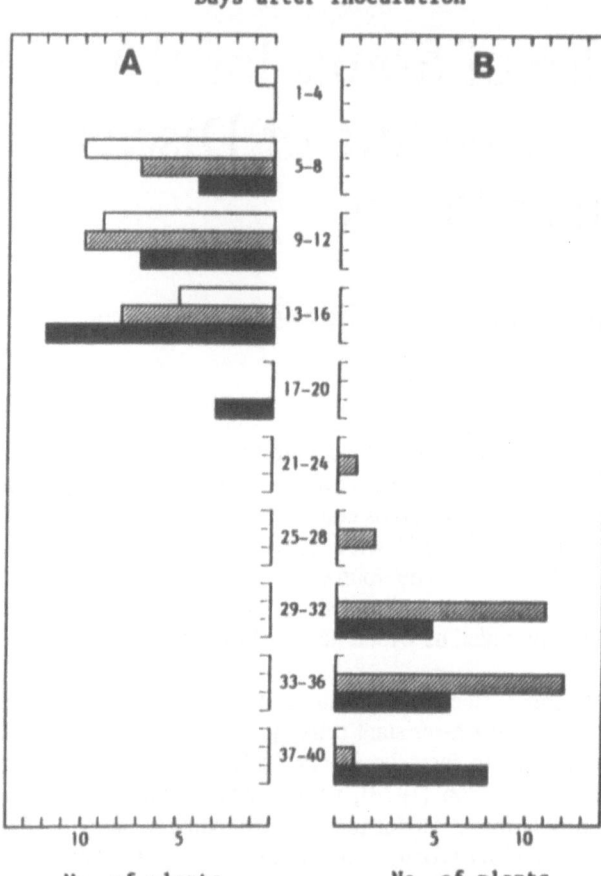

Figure 1. Time course of disease symptom appearance in susceptible tomato plants (A) and bacterial culture filtrate-resistant regenerants (R1-lines) (B) inoculated with virulent strain (U-10) of *P. solanacearum*. Results are averages of 25 inoculated plants. Open, slant lined, and black columns represent leaf yellowing, partial wilting, and complete wilting in inoculated plants, respectively.

has been well established. Thus, it was expected that an isolation of fusaric acid resistant lines would contribute to the production of disease-tolerant plants. Fusaric acid-resistant cells were isolated from tomato leaf protoplasts (Shahin et al., 1986) and leaf-derived callus tissues (Toyoda et al., 1988b) and the regenerants from these tissue cultures were indeed enhanced in disease resistance.

In this treatise, we present some examples of an efficient selection for disease resistance and discuss a possible application of somaclonal variation to the breeding of disease-resistant plants.

Table 1. Symptom expression of self-pollinated progenies (R2-lines) of unselected regenerants derived from leaf callus tissues of tomato after inoculation with virulent strain (U-10) of *P. solanacearum*

R2 lines	No. of inoculated plants	No. of plants		
		S	MR	HR[a]
LNSR-3	83	83	0	0
-8	29	29	0	0
-9	41	41	0	0
-10	52	52	0	0
-14	48	48	0	0
-16	51	51	0	0
-17	31	31	0	0
-19	22	22	0	0
-21	41	41	0	0
-22	25	25	0	0
-1	65	58	7	0
-4	51	45	6	0
-11	31	28	3	0
-13	54	48	6	0
-5	58	12	26	20
-6	43	22	19	2
-12	48	16	19	13
-15	40	35	2	3
-20	46	40	5	1
-2	46	0	10	36
-7	42	0	1	41
-18	36	0	12	24
control[b]	25	25	0	0

[a] Symptoms of inoculated plants were classified into three types; S, rapid wilting within 15 days after inoculation; MR, delayed appearance of wilting 30-40 days after inoculation; HR, no wilting throughout the experiment (3-4 months).
[b] Susceptible tomato plants from which callus tissues were originally induced.

Selection of Bacterial Wilt-Resistant Tomato

Bacterial wilt disease by *P. solanacearum* is a typical soil-borne disease of major crop plants, causing tremendous damage to crops all over the world. Unfortunately, effective resistant lines of tomato have not been produced in Japan by conventional breeding techniques. It is thus urgent to establish the efficient selection and production of resistant plants. In this section, the authors describe the selection for bacterial wilt resistance in in vitro cultures of tomato treated with culture filtrates of *P. solanacearum*, and in regenerated plants inoculated with the pathogen (Toyoda et al., 1989a).

Tomato callus tissues were at first exposed to diluted culture filtrates of both the virulent and avirulent strains of *P. solanacearum*. The toxic effect was detected only when the tissues were treated with the culture filtrate of virulent strain (VF), suggesting that some toxic substances were specifically produced by the virulent bacteria. VF-resistant plants were regenerated from VF-treated callus tissues. In this experiment, about 2000 callus clumps were transferred to a selective medium and cultured using the liquid-on-agar method. The method was our conventional method for examining the toxic effects of applied substances on callus tissues, because the substances penetrate into the tissues homogeneously and effectively (Toyoda et al., 1984). Most callus clumps became brown within 10 to 12 days after transfer and the subsequent growth of calli had completely ceased. However, some of the browned clumps proliferated fresh callus clumps (42 of 2021 clumps). Callus tissues growing in this medium were transferred to the medium for shoot formation, and then to the medium for root formation. Thus, finally 25 VF-resistant regenerants (R_1-lines) were successfully obtained from VF-treated callus tissues and used for subsequent inoculation with a virulent isolate. Figure 1 shows the time of appearance of disease symptoms in R, and control tomato seedlings inoculated with the virulent strain of the pathogen. Inoculated control plants showed a yellowing in partial or whole portions of the lower leaves 4 to 6 days after inoculation, and subsequently, partial wilting in the upper leaves of seedlings. Consequently, all inoculated plants were completely wilted within 17 days after inoculation (Fig. 1-A). On the other hand, regenerants from VF-resistant calli did not show such symptoms at this early stage of infection, and normally grew to form the first and second fruit clusters 30 to 40 days after inoculation. However, the plants suddenly showed partial wilting at the later stages of growth and rapidly withered (Fig. 1-B). Fruits formed in R_1-lines were immature and no viable seeds were obtained. The present results suggest that virulent bacteria produce toxic substances at the early stage of infection and cause damage to host plants. Detectable damage was a rapid yellowing of leaves probably due to chlorophyll disintegration by the toxic substances. In non-inoculated control plants, such a phenomenon was not observed at this early stage, though at much later stages (30 to 50 days after planting) some of these plants showed a leaf yellowing in lower leaves due to senescence.

VF-resistant regenerants were resistant to the attack by virulent bacterial strain U-10, suggesting that the same toxic substances as secreted into the culture medium are produced *in planta* during the infection process by the virulent strain. Although the detailed functions of these toxic substances remains to be elucidated, the resistance of regenerants to these substances is certainly effective in suppressing or delaying the growth of invaded bacteria.

Virulent strains of these bacteria produce abundant extracellular polysaccharides (EPS) by which vessels of host plants are plugged to cause the wilting (Akiyama et al. 1986). In the present study we did not succeed in separating these toxic substances from EPS, because both were high molecular weight compounds which precipitated with ethanol, eluted in the void volume during Sephadex G-50 gel filtration, and were not dialyzable. Further isolation and characterization of these toxic fractions of VF is under way.

Table 2. Selection of TMV-resistant plants from regenerants of a tobacco callus line, CMT-1.

Selection steps[a]	No. of plants	
	Diseased	Healthy
1st	967	105
2nd	14	91
3rd	58	3(30)[b]

[a]Shoots were differentiated from tobacco (*N. tabacum* cv. Bright Yellow) callus line (CMT-1), and symptomless healthy shoots were selected for TMV resistance (first step). Healthy shoots were grown up to intact plants and cultivated further for 1 month (second step). At the third step, selected regenerants were inoculated with TMV and grown for 3 months till seeds were harvested.
[b]Regenerants showing delayed mosaic symptoms (at the flowering stage 3 months after inoculation).

Since in vitro selection by the bacterial culture filtrate was not enough to obtain completely resistant clones against inoculation with virulent bacteria, we attempted to directly isolate disease-resistant plants among inoculated progeny of the regenerants. For this purpose, self-pollinated progeny (R_2-lines) of regenerants derived from non-selected callus tissues were used for inoculation, because the resistance would be detected even if it was recessive. Table 1 shows the segregations of the phenotypes in R_2-lines. Although phenotypical segregations in each group of R_2-lines were too diverse to reveal how many genes were involved in the expression of complete resistance to bacterial wilting, these results suggest that resistance is controlled by more than one gene.

In the present study, highly resistant plants (LNSR-7) were obtained by inoculating R2-lines with virulent isolate U-10. Therefore, the reliability of the results strongly depends on whether inoculation was successful or not. For confirming the validity of inoculation, we attempted to isolate the bacteria from inoculated LNSR-7. The densities of bacteria detected in LSNR-7 were 2×10^3, 3×10^4, and 6×10^3 cells per gram fresh weight of tissue 1, 2, and 3 months after inoculation, respectively. These values were considerably lower compared with those (10^8-10^{10} cells/g fr. wt.) of wilted plants. These results clearly indicate that non-wilting of LNSR-7 was due to the suppression of bacterial growth in xylem tissues of inoculated plants, but not to the failure of inoculation.

Selection for Tobacco Mosaic Virus Resistance

Plants resistant to viral disease were at first isolated from tissue culture by Murakishi and Carlson (1982), who induced callus tissues from mutagenized haploid tobacco plants and selected tobacco mosaic virus (TMV)-resistant tobacco by inoculating the regenerants with this virus. In viral diseases of plants, the isolation of resistant variants is rather difficult because addition of some toxins to the medium would not enhance the selection pressure. For the practical use of breeding, therefore, it is convenient to establish a system for in vitro isolation for viral disease-resistant clones. Moreover, it has been known that somaclonal variation can be frequently induced in plant tissue cultures even when mutagens are not used (Shepard et al., 1980; Evans and Sharp, 1983). Actually, we succeeded in isolating bacterial wilt-resistant tomato from non-mutagenized tomato callus cultures (Toyoda et al. , 1989a). From this point of view, we will describe in this section an efficient system for isolating tobacco lines resistant to tobacco mosaic virus (Toyoda et al. , 1989b).

In our laboratory, various callus lines have been isolated from auxiliary buds of TMV-infected and healthy tobacco plants. One of the callus lines derived from TMV-infected tobacco plants was CMT callus line in which higher levels of TMV amounts were constantly maintained during subculturing (Toyoda et al., 1985b). CMT callus line was friable and cell aggregates were easily released by gently shaking the tissues with liquid medium. A microscopic observation of the line showed the frequent formation of inclusion bodies of TMV in the aggregates, indicating that TMV multiplied and translocated in proliferated callus cells. In this callus line, shoots were effectively differentiated by changing plant hormones added to the medium and differentiated shoots developed small leaflets with the typical mosaic symptom of TMV. The results indicated that TMV was stably multiplied in callus tissues and efficiently translocated to leaflets regenerated from the tissues. These results suggested that if TMV resistance was induced in callus cells, shoots differentiated from these cells would develop healthy leaflets, and that TMV-resistant plants could be easily and effectively isolated by selecting those healthy shoots. In the present experiments, CMT-1 callus tissues were subcultured for several passages in order to enrich the efficiency of somaclonal variation and to enhance multiplication of the variant cells. We selected healthy shoots (first step for selection) and then transplanted to soil. After 30 days of cultivation, symptomless, healthy regenerants were selected as putative TMV-resistant plants (second step for selection). To confirm TMV resistance of these plants, they were inoculated with TMV (third step for selection). Table 2 shows the numbers of plants selected in each step. In this selection, finally 3 highly resistant (no symptom) (CMT-IR) and 30 moderately resistant (delayed symptom) regenerants were successfully obtained from 105 shoots selected as primary candidates. Chromosome number in root tip cells of TMV-resistant regenerants (CMT-1R03) as assessed at the metaphase was normal, giving rise to $2n = 48$.

Table 3. Segregation of TMV resistance in selfed progeny of a TWV-resistant line (CMT-1R03).

Experiments	No. of selfed progeny inoculated with TMV		Ratio	x^2	P
	Resistant	Susceptible			
1	29	14	3:1	1.30	0.62-0.88
2	39	9	3:1	1.00	0.63-0.87
3	44	18	3:1	0.58	0.64-0.86

For evaluating the effectiveness of this selection, we examined whether TMV resistance acquired in the regenerants would be passed to their progeny. In this experiments, selfed progeny (R_2-plants) of highly resistant regenerants (CMT-1R03) were inoculated with TMV and the segregation of resistance and susceptibility was determined. Table 3 shows the numbers of resistant and susceptible plants determined after inoculation with TMV. The data suggest that TMV resistance is probably due to a dominant single gene mutation, and this mutation was heterozygously induced in the CMT-1 callus line.

In a subsequent study, we also examined the multiplication and translocation of TMV inoculated into CMT-1R03 or control tobacco plants (noncultured R_2-plants). Inoculated and non-inoculated leaves were harvested separately 1 month after inoculation, and subjected to the estimation of TMV concentrations by a quantitative immunoelectrophoresis method. In control tobacco plants, all of 31 plants inoculated showed mosaic symptoms first in non-inoculated younger leaves 7 to 10 days after inoculation and then in whole leaves 20 to 25 days after inoculation. On the other hand, about 80% of inoculated R_2-plants (24 of 31 plants) did not show any symptom in either inoculated or non-inoculated leaves. In these symptomless R_2-plants, the levels of TMV quantity in the inoculated leaves were low, and those in non-inoculated leaves were below the limit of detection by this method (Fig. 2). These results indicate that resistant plants did not completely suppress the multiplication of virus in inoculated leaves, but inhibited the translocation of TMV from inoculated to non-inoculated leaves.

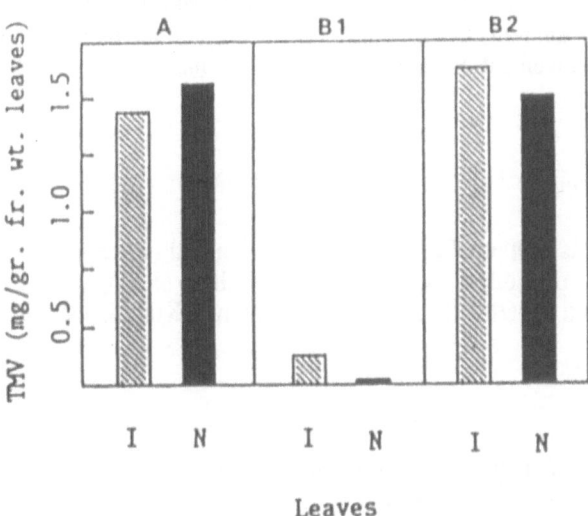

Figure 2. Estimated quantity of TMV in control tobacco (A) and selfed progeny of TMV-resistant regenerant (CMT-1R03). Each of 31 plants were inoculated with TMV, and inoculated (I) and non-inoculated (N) leaves were harvested separately 1 month after inoculation and used to estimate TMV quantity by quantitative immunoelectrophoresis. TMV amounts in CMT-1R03 were shown separately in 24 symptomless (B1) and 7 symtom-appearing plants (B2).

Concluding Remarks

The present studies demonstrate that an efficient selection of somaclonal variations would be useful for the production of disease-resistant plants, including those in which toxins could hardly be the agents to give selection pressure. In fact, the bacterial wilt-resistant line LNSR-7 was resistant to the disease under field conditions. R_2-plants of this line were grown in a field heavily infested with a different strain (KK-101) of *P. solanacearum*. LNSR-7 also showed the strong resistance against a natural infection of this pathogen, while control, susceptible plants were severely wilted under the same conditions. Thus, this line may be commercially utilized as a new tomato cultivar resistant to the disease.

One of the most important devices in the present system was to isolate the callus line which expressed TMV mosaic symptoms in leaflets when redifferentiated into shoots. This enabled us to easily and effectively isolate TMV-resistance mutation occurring during the callus cultures by selecting symptomless shoots, without any selection pressure. Thus, the present method may be widely applicable to the selection of resistant clones against other viral diseases as far as mosaic symptoms appear on host plants.

We reported the application of a microinjection technique to the introduction of TMV into single cells (Toyoda et al., 1985a) or cell aggregates (Toyoda et al., 1986) of callus tissues, and showed that callus cell aggregates obtained from this resistant line permitted TMV to multiply in injected cells, but inhibited translocation of the virus to adjacent cells of the aggregates, whereas TMV in the cells of the parental line multiplied and translocated freely. These data strongly support, at the cellular level, the notion that the suppression of cell-to-cell movement is one of the mechanisms of plant resistance against viruses.

Acknowledgements

This work was supported in part by the grant in aid for scientific research from the Ministry of Education, Science and Culture of Japan (Nos. 01440009, 01304014, and 01560057) and Project Research of Kinki University (P-0300).

References

Akiyama, Y., et al. 1986, Extracellular polysaccharide produced by a virulent strain (U-7) of *Pseudomonas solanacearum*, *Agric. Biol. Chem.* 50:747-751.

Brettell, R.I.S. and Ingram, D.S., 1979, Tissue culture in the production of novel disease-resistant crop plants, *Biol. Rev.* 54:329-345.

Carlson, P.S., 1973, Methionine sulfoximine-resistant mutants of tobacco, *Science* 180:1366-1368.

Evans, D.A. and Sharp, W.R., 1983, Single gene mutations in tomato plants regenerated from tissue culture, *Science* 221:949-951.

Larkin, P.J. and Scowcroft, W.R., 1981, Somaclonal variation - a novel source of variability from cell cultures for plant improvement, *Theor. Appl. Genet.* **60**:197-214.

Murakishi, H.H. and Carlson, P.S., 1982, In vitro selection of *Nicotiana sylvestris* variants with limited resistance to TMV, *Plant Cell Rep.* **1**:94-97.

Ouchi, S., et al., 1989, A promising strategy for the control of fungal diseases by the use of toxin-degrading microbes, *In* Phytotoxins and Plant Pathogenesis (Graniti, A., et al. eds.), Springer-Verlag, Berlin, pp. 301-317.

Shahin, E.A. and Spivey, R., 1986, A single dominant gene for Fusarium wilt resistance in protoplast-derived tomato plants, *Theor. Appl. Genet.* **73**:164-169.

Shepard, J.F., et al., 1980, Potato protoplasts in crop improvement, *Science* **208**:17-24.

Sunderland, N., 1977, Nuclear cytology, *In* Plant Tissue and Cell Culture (Street, H. E., ed.), Blackwell Scientific Publications, Oxford, pp. 177-205.

Toyoda, H., Tanaka, N., and Hirai, T., 1984, Effects of the culture filtrate of *Fusarium oxysporum* f.sp. *lycopersici* on tomato callus growth and the selection of resistant callus cells to the filtrate, *Ann. Phytopathol. Soc. Japan* **50**:53-62.

Toyoda, H., Matsuda, Y., and Hirai, T., 1985a, Resistance mechanism of cultured plant cells to tobacco mosaic virus (III) Efficient microinjection of tobacco mosaic virus into tomato callus cells, *Ann. Phytopathol. Soc. Japan* **51**:32-38.

Toyoda, H., et al., 1985b, Resistance mechanism of cultured plant cells to tobacco mosaic virus. IV. Changes in tobacco mosaic virus concentrations in somaclonal tobacco callus tissues and production of virus-free plantlets, *Phytopathol. Z.* **114**:126-133.

Toyoda, H., Matsuda, Y., and Hirai, T., 1986, Multiplication and translocation of tobacco mosaic virus microinjected into cell aggregates of tomato callus, *Plant Tissue Culture Lett.* **3**:22-27.

Toyoda, H., et al., 1988a, Detoxification of fusaric acid by a fusaric acid-resistant mutant of *Pseudomonas solanacearum* and its application to biological control of Fusarium wilt of tomato, *Phytopathology* **78**:1307-1311.

Toyoda, H., et al., 1988b, In vitro selection of fusaric acid-resistant regenerants from tomato leaf explant-derived callus tissues, *Plant Tissue Culture Lett.* **5**:66-71.

Toyoda, H., et al., 1989a, Selection of bacterial wilt-resistant tomato through tissue culture, *Plant Cell Rep.* **8**:317-320.

Toyoda, H., et al., 1989b, Multiplication of tobacco mosaic virus in tobacco callus tissues and in vitro selection for viral disease resistance, *Plant Cell Rep.* **8**:433-436.

Summary of discussion of Toyoda's paper

Discussion began with a question by *Chumley* in regards to the nature of toxic products of *Ps. solanecearum*. Toyoda answered that it is autoclave stable and non-dialyzable. In view of recent progress in the analysis of animal gene expression in transgenic plants (immunoglobulins), *Vance* asked if *Toyoda* injected antibody to callus cells before TMV injection to clarify whether antibody could block disease response by interfering with viral multiplication. *Toyoda* replied that it is quite difficult to inject the same cells twice unless the cells were incubated for certain periods of time for repairing. *Essenberg* asked about the technique of flourescent-antibody staining. *Toyoda* explained that the excess antibody is washed away after the staining of the acetone-fixed callus cells. *Chumley* commented on the transgenic plants with respect to human immunoglobulins, stating that both transgenic plants expressing light or heavy chains of immunoglobulins have been obtained but have not been crossed to yield both chains in the same plant. *Durbin* asked if *Toyoda* has ever stained immediately after injection to prove that virus multiplication was indeed limited in the resistant lines. The answer was that the amount of TMV injected is not enough to give a positive reaction. *Bennetzen* asked if *Toyoda* has done an allelism test on the regenerant lines in view of the function of n or N' gene in tobacco. *Toyoda* replied that he has not done it. *Essenberg* asked about the population of bacteria in the suddenly withered resistant lines. *Toyoda* responded by stating that number of bacteria (*Ps. solanecearum*) in the resistant lines were in the range of 10^3 to 10^4 until the first clusters were formed, but suddenly increased thereafter reaching 10^9, like the susceptible parent line, right before wilting, indicating that the resistance is just the suppression of bacterial growth in the early growth stage.

In regard to the statement that one regenerant line seems to have single dominant mutation, *Chumley* asked if *Toyoda* did a similar analysis with the other two regenerant lines. *Toyoda* replied that he has analyzed only the one most resistant line. *Chumley* mentioned advantages of another culture for obtaining recessive mutation. *Hammerschmidt* asked about the level of resistance or tolerance of these resistant tomato lines to the crude toxic substances of *Ps. solanacearum*, i.e., EPS. *Toyoda* replied that he has not done such experiments. *Durbin* then raised a general question about the mechanisms of somaclonal variations, asking whether this type of technique is just selecting out some characters which are present in the genome or generating new characters which are absent in the original material. Chromosomal variation, gene activation, and others are certainly involved in the somaclonal variation, as discussed by *Chumley*, *Bennetzen*, and *Durbin*. *Bennetzen* commented that somaclonal variation often caused secondary events which appear many generations later as well as variable sustainability of the observed phenotype and then asked whether *Toyoda* observed any secondary mutations in his lines and how long the resistance of these lines is retained. *Toyoda* replied that the resistance trait of tomato lines was retained at least for three generations, and he was not sure of the changes due to secondary effect. *Bushnell* asked how one can select resistant lines against powdery mildew. *Toyoda* answered that one has to simply select resistant ones by inoculating the regenerants. *Van Alfen* closed discussion by his evaluation of this technique as an approach to obtaining disease-resistant lines.

Poster Abstracts

Loss of Pathogenicity Caused by Tn5 Insertions in *Xanthomonas campestris* pv. *citri*

Nobuyuki Furutani and Shinji Tsuyumu

Laboratory of Plant Pathology, Faculty of Agriculture, Shizuoka University, 836 Ohya, Shizuoka City 422, Japan

Nine prototrophic mutants of *Xanthomonas campestris* pv. *citri* that had lost pathogenicity on the leaves of "Natsudaidai" were isolated from among 1600 transconjugants obtained following mating with *Escherichia coli* SM17-1(pSUP1021) (strain and suicide plasmid kindly provided by Dr. R. Simon). Southern blot analysis using pRZlO2 as the probe indicated that eight of the mutants carried a single, independent insertion in different EcoR1 fragment, whereas one of them (F-5) carried two insertions. The mutants failed to incite the typical canker on citrus leaves throughout the experimental period. However, eight of them caused delayed yellowing to varying degrees approximately 3 weeks after inoculation. The mutated DNA from one of the mutants showed homology with *hrp* cluster of *Pseudomonas solanacearum* (pVir2, kindly supplied by Dr. Boucher). Thus, at least the gene(s) within this Tn5-containing fragment may be essential for pathogenicity of *X. campestris* pv. *citri*, and may encode a function analogous to that expressed by the *hrp* cluster of *P. solanacenarum*. Mutant F-8 did not incite yellowing at any time during the experimental period. When approximately 10^7 cells/ml of this mutant were injected by syringe into the intercellular spaces of the citrus leaves, the number of viable cells rapidly decreased, and within 2 days after inoculation none could not be detected. The population of the wild-type parental cells continually increased. Thus, the data suggest that the wild-type cells possess some defense or escape mechanism that circumvents the bactericidal activity of plants, which is lacking in mutant F-8 cells. Furthermore, this mutant lacked the capacity to inhibit growth and induce browning of tobacco callus culture.

Isolation and Characterization of Double-Stranded RNA from the Pea Pathogen, *Mycosphaerella pinodes*, That Is Possibly Associated with Hypovirulence

Tetsuji Yamada, Masayuki Seno, Tomonori Shiraishi, Hisaharu Kato, Yuki Ichinose, and Hachiro Oku

Laboratory of Plant Pathology and Genetic Engineering, College of Agriculture, Okayama University, Tsushima, Okayama 700, Japan

Studies on the structure of the extrachromosomal nucleic acids in the plant pathogenic fungus, *Mycosphaerella pinodes*, OMP-1, isolated at Okayama, Japan, reveal at least three cytoplasmic double-stranded (ds) RNA of approximately 6.0 kilobases (kb) (L), 4.0 kb (M), and 2.6 kb (S). Another highly virulent pathogenic strain, MP-3, isolated in England, does not contain any of these dsRNA. Weak pathogenic mutants selected from OMP-1 following UV irradiation contain considerably higher copies of S-dsRNA than the parental strain, whereas one of the mutants, OMP-X62, contains no detectable M-dsRNA. The presence of these dsRNA apparently results in the inhibition of cell growth, stimulation of sporulation, and promotion of hypovirulence.

We have constructed cDNA of these dsRNA and used it in Southern hybridization analysis. The M-dsRNA has strong homology with the L-dsRNA, suggesting that it is the processed product of L-dsRNA as observed in *Endothia parasitica*. The nucleotide sequence of the cDNA derived from S-dsRNA has not revealed notable sequence homology with other virus or mycovirus-like particles.

Cloning and Analysis of Genes for Fusaric Acid Detoxification

Seiji Ouchi, Kiyonori Katsuragi, and Hideyoshi Toyoda

Laboratory of Plant Pathology, Faculty of Agriculture, Kinki University, 3327-204, Nakamachi, Nara 631, Japan

Some bacteria and fungi are capable of detoxifying fusaric acid (FA) to a biologically inactive form or degrading it to completion. Application of this microbial ability to biological control was, in some cases, promising, giving rise to protection of tomato plants from the *Fusarium* wilt fungus. In the present study, FA-detoxifying genes were isolated from a bacterium, *Klebsiella oxytoca* HY-1, and the gene structures were analyzed as a primary step to understanding the molecular basis of the protection phenomenon. If the protection could be verified, transformation of tomato plants with the FA-detoxifying gene could be of practical use. In separate experiments, chromosomal DNAs were digested with *Bam*HI and *Hind*III and random fragments were cloned and used to transform *Escherichia coli* JM109. Transformants were screened in the presence of FA and a 3.6 kilobase (kb) *Bam*HI fragment from one was shown to contain the FA detoxifying gene(s). The complete nucleotide sequence of this fragment was determined using the dideoxy nucleotide method. The fragment contained three different open reading frames (ORFs) essential for expressing FA detoxification. To examine the expression of the FA-detoxifying genes, a polysome fraction was prepared from JM109 that was transformed with the *Bam*HI fragment. The mRNAs in the polysome fraction were isolated and hybridized with labeled DNA probes derived from each ORF. Each of these probes hybridized well with a single 3.2 kb mRNA. The results thus suggest that transcription of these FA-detoxifying genes in JM109 resulted in the synthesis of a polycistronic mRNA.

Nucleotide Sequence of the Linear Plasmid pRS64 and Properties of Plasmid DNA (p1DNA) in Isolates of Nine Anastomosis Groups of *Rhizoctonia solani*

Teruyoshi Hashiba,[1] Shun-ichiro Miyashita,[1] Atsushi Miyasaka,[1] Hirohiko Hirochika,[2] and Joh-E Ikeda[2]

[1]Faculty of Agriculture, Tohoku University, Sendai 981, Japan

[2]National Institute of Agrobiological Resources, Tsukuba, Ibaraki 305, Japan

We have described the unique terminal structures of linear plasmid DNAs of *Rhizoctonia solani* (Miyashita et al., 1986). In order to determine the structure and the gene organization in plasmid pRS64, terminal sequences of the plasmid DNA (p1DNA) were determined by the technique of Maxam and Gilbert. The 20 nucleotide apex of the hairpin of the termini were not base-paired. When denatured plasmid molecules were examined by electron microscopy, single-strand circles were observed. These studies provide direct evidence that the linear plasmids have covalently closed ends. Using cloned DNA fragments, the nucleotide sequence of approximately 2387 bp of pRS64-2, was determined. The pRS64 sequence showed a low A+T content (52%) and contained at least four short open reading frames.

Furthermore, p1DNA was found in 48 out of 114 field isolates of *R. solani*. These 48 isolates were distributed among nine anastomosis (AG) and intraspecific (ISG) groups as follows: 0 in AG-1 (sasakii type), 1 in AG-1 (web-blight type), 0 in AG-2-1, 11 in AG-2-2 (rush type), 10 in AG-2-2 (root-rot type), 10 in AG-3, 8 in AG-4, 4 in AG-5, and 4 in AG-6. Each isolate carried one, two or three p1DNAs as identified by gel electrophoresis. Electron microscopic analysis revealed that all these p1DNAs were linear. The sequence homology among p1DNA found in representative isolates was examined by Southern blot analysis using nick-translated p1DNAs as probes. Considerable sequence homology was observed among p1DNA obtained from different isolates within the same AG and ISG. The p1DNA in the isolates of AG-2-2 were classified into two groups on the basis of the sequence homology.

Expression of Soybean β-1,3-Endoglucanase cDNA in Transgenic Tobacco Plants

Masaaki Yoshikawa and Mikio Tsuda

Latoratory of Plant Pathology, Faculty of Agriculture, Kyoto Prefectural University, Shimogamo, Kyoto 606 Japan

Expression of soybean β–1,3-endoglucanase appears to be a key host component involved in the earliest soybean-*Phytophthora megasperma* f.sp. *glycinea* interaction leading to active disease resistance, by releasing elicitor-active carbohydrates from the cell walls of the fungus. We have cloned and characterized a full-length β-1,3-endoglucanase cDNA. Here we show our attempts to express the cloned glucanase cDNA in tobacco plants to further evaluate the role of this enzyme in disease resistance.

Tobacco leaf discs were transformed by a binary vector method using *Agrobacterium tumefaciens* carrying pROK1a plasmid which contained soybean β-1,3-endoglucanase cDNA (pEG488) and kanamycin resistance gene. Transformed tobacco cells were screened on kanamycin-containing media and regenerated to whole plants.

Western blot analysis using an antiserum highly specific for soybean β-1,3-endoglucanase, showed that soybean β–1,3-endoglucanase protein was synthesized in transgenic tobacco plants when plants were transformed with pEG488 but not with its antisense cDNA. β-1,3-Endoglucanase activity, based on fresh weight or mg protein, was higher in the pEG488-transformed tobacco tissues than the parent plants or plants transformed without the glucanase cDNA. In contrast, tobacco plants transformed with the antisense cDNA appeared to contain lower levels of the glucanase activity than the control plants. Experiments are now being conducted to see whether transgenic tobacco plants with higher levels of β-1,3-endoglucanase activity possess enhanced levels of disease resistance to several fungal pathogens.

Dynamic Responses of the Cytoskeleton of Barley Coleoptile Cells to a Nonpathogen, *Erysiphe pisi*

Issei Kobayashi, Yuhko Sakamoto, and Hitoshi Kunoh

Laboratory of Plant Pathology, Faculty of Bioresources, Mie University, Tsu-City 514, Japan

A series of our cytological studies demonstrated that barley coleoptile cells responded to a nonpathogen, *Erysiphe pisi*, at the time of maturation of its appressoria by increasing the velocity of cytoplasmic streaming and the number of cytoplasmic strands. The possible involvement of the cytoskeletons of coleoptile cells in these cytological changes was examined by cytochemical and immunofluorescent means.

The rhodamine-labeled phalloidin stain revealed that actin filaments gathered at the site below an appressorium in a coleoptile cell at the time of maturation of the appressorium, 4 to 5-h before actual penetration. The density of actin filaments increased more prominently below an appressorium at the time of initiation of cytoplasmic aggregation. The location and appearance of actin filaments was concurrent with the appearance of cytoplasmic strands that had been shown to increase in number at the time of maturation of the appressorium.

Indirect immunofluorescent microscopy using anti-yeast tubulin antibody revealed no change in the spatial distribution of tubulin at the time of maturation of an appressorium. However, tubulin gathered prominently below the appressorium when cytoplasmic aggregation initiated and its density increased to form a network when an incipient papilla became visible.

These results suggest that actin filaments of barley coleoptile cells may be involved in the recognition of a nonpathogen at the prepenetration stage, and that aggregation of both actin filaments and tubulin might be associated with the papilla formation which occurs after the fungus attempts penetration.

The Role of Lipoxygenase in the Expression of Disease Resistance of Host Plants

Hiroyuki Yamamoto,[1] Akemi Tanaka,[1] Toshikazu Tani,[1] and Richard M. Bostock[2]

[1]Laboratory of Plant Pathology, Kagawa University, Kagawa 761-07, Japan

[2]Department of Plant Pathology, University of California, Davis, California 95616, USA

The role of lipoxygenase (LOX) in resistance expression was studied using the oat-*Puccinia coronata* host/pathogen interaction and the potato-arachidonic acid (AA) system.

Activity of LOX was extensively increased in leaves of six cultivars responding with resistance to an incompatible race of *P. coronata avenae*. A specific anionic LOX isozyme (LOX-1) was found by gel electrophoresis in the extracts of infected leaves. The LOX-1 gene product was purified by DEAE-Sepharose and CM-Sepharose column chromatography of fractions precipitated at 50~80% saturation of ammonium sulfate. Purified LOX-1 was determined to be a glycoprotein with an isoelectric point at 5.1. It was more active on linoleic acid than other unsaturated fatty acids. The apparent K_m value for LOX-1 obtained for linoleic acid at pH 5.0 was $2.3 \times 10^{-5} M$ and the V_{max} was calculated to be 0.49 μM/min·mg protein. Fe^{++} stimulated the activity at concentrations lower than 1 mM.

LOXs catalyze the formation of hydroperoxide, which readily undergoes biochemical reactions to form a wide range of secondary lipid oxidation products. Three antifungal compounds other than avenalumin and 26-DGA were detected in neutral ethyl ether extracts of leaves responding with resistance. A major compound was purified by HPLC using an Asahipak-ODP50 column. The major compound from infected leaves was detected in the reaction mixture of linoleic acid and LOX-1.

The blockage experiments with leaves inoculated with an incompatible race using inhibitors of RNA and protein synthesis, and LOX inhibitors, indicated that LOX-1 is synthesized de novo and contributes to the resistance expression through peroxidation of fatty acid.

Recently we observed LOX activity in the microsomal fraction of potato tubers (cv. Kennebec) treated with AA elicitor. A major product was HETE, a potent inhibitor of germination of *Phytophthora infestans* zoospores.

These facts suggest that LOX is closely associated with the resistance expression of these host plants against pathogens.

Structural Requirements for Host-Selective Toxicity of AF-Toxins Produced by *Alternaria alternata* Strawberry Pathotype

Sung-Suk Lee,[1] Takashi Tsuge,[1] Noriyuki Doke,[1] Syoyo Nishimura,[1,†] and Shin-ichi Nakatsuka[2]

[1]Plant Pathology Laboratory, and [2]Laboratory of Organic Chemistry, Faculty of Agriculture, Nagoya University, Nagoya 464-01, Japan

Alternaria alternata strawberry pathotype, the causal agent of black spot of strawberry, infects only a strawberry cultivar "Morioka-16" in the field. The pathogen produces host-specific AF-toxins. Interestingly, it is highly pathogenic to certain cultivars of Japanese pear in laboratory tests. This host range can be completely reproduced by AF-toxins. AF-toxins appeared to consist of three molecular species: AF-toxin I [8-(2'-(2",3"-dihydroxy-3"-dimethyl-propionyloxy)-3"-methyl-valeryl)-9,10-epoxy-9methyl-deca-(2Z,4E,6E) - trienoic acid], II (the deacyl derivative at 2' position of toxin I) and III (the dehydroxy derivative at 2" position of toxin I). AF-toxin I causes leaf necrosis both on strawberry and Japanese pear, toxin II on pear only, and toxin III strongly on strawberry and slightly on pear.

The 2'-0-acyl derivatives artificially synthesized from AF-toxin II, which is toxic only to Japanese pear, exhibited toxicity on both leaves of strawberry and pear, indicating that the molecular features at the 2'- position in AF-toxins may be critical for determining host-selective toxicity to pear and/or strawberry. AF-toxin II inhibited the toxic effect of AF-toxin I on susceptible strawberry cells, and the inhibition was remarkable when the cultured cells were exposed to excess amounts of AF-toxin II both prior to, and simultaneously with, AF-toxin I addition. The decatrienoic acid moiety of the AF-toxins also protected the cells from AF-toxin I action.

The structure-toxicity relationships and the competition phenomenon between toxin and toxin analogs suggest the presence of potential receptor sites for AF-toxins in host strawberry cells.

†Deceased May 1989

Effect of AL-Toxin on Viability of Cultured Tomato Cells Determined by MTT-Colorimetric Assay

Motoichiro Kodama,[1] Takuji Yoshida,[1] Hiroshi Otani,[1] Keisuke Kohomoto,[1] and Syoyo Nishimura [2,†]

[1]Laboratory of Plant Pathology, Faculty of Agriculture, Tottori University, Tottori 680, Japan

[2]Laboratory of Plant Pathology, Faculty of Agriculture, Nagoya University, Nagoya 464, Japan

AL-toxin produced by *Alternaria alternata* tomato pathotype, is considered to have an effect on mitochondria in susceptible cells. However, the precise site of action remains unknown. The effect of AL-toxin on viability of cultured cells derived from susceptible (Earlypak 7 and First) and resistant (Ace) cultivars of tomato was examined. When calli of each cultivar were incubated on solid medium containing the toxin, growth inhibition and browning occurred only on calli of the susceptible cultivar 3 days after treatment. To quantitatively estimate the viability of toxin-treated cells, a colorimetric assay utilizing a tetrazolium salt, MTT (3-(4,5-dimethylthiazol-2-yl)-2, 5-diphenyl tetrazolium bromide) was employed. The number of living cells was proportional to amounts of formazan formed by reduction of MTT in active mitochondria. Therefore, the MTT assay appeared to be useful to quantitatively evaluate the extent of cell damage induced by the toxin. Inhibition of formazan production in cell suspensions of susceptible cultivars by the toxin was evident at a concentration similar to the minimum toxin concentration for necrosis induction on leaves. In contrast, resistant cells were not significantly affected by the toxin at higher concentrations. The MTT assay was more advantageous than several common techniques for evaluating the effect of the toxin on cultured cells, because the effect of the toxin on cells was not evident when cell viability was judged by staining with FDA or Evans blue. These results indicate that host-selectivity of the toxin is expressed at cultured cell level, and target site of the toxin may be associated with mitochondria in susceptible cells.

†Deceased May 1989

Suppression of Defense Reaction and Accessibility Induction in Pea by Substances from Healthy Pea Tissues

Tomonori Shiraishi, Kimio Nasu, Tetsuji Yamada, Yuki Ichinose, and Hachiro Oku

Laboratory of Plant Pathology and Genetic Engineering, College of Agriculture, Okayama University, Tsushima-naka, Okayama 700, Japan

The healthy pea plant contains the substances which suppress pisatin production induced by a fungal elicitor. The substances were temporarily called "endogenous suppressors (ESs)" (Hori et al., in Proc. 5th. ICPP, 1988). The purification of ESs from the homogenate of pea seedlings was accomplished by solvent-partitioning, ODS- and gel filtration-column chromatography.

Pisatin accumulation in the elicitor-treated pea tissue was significantly reduced anytime during 0 to 15 hours after the elicitor-treatment with crude ES at concentrations of less than 5 mg/ml. The ES delayed the onset of pisatin accumulation for 6 to 9 hours and, thereafter, pisatin accumulation resumed. Its transient effect without causing visible injury to pea tissues suggests that the ES is not a toxin.

Gel filtration indicated that crude ES contained two active glycopeptide components, designated nos. 8 and 15. The molecular weights of no. 8 and no. 15 were ca. 15,000 and 5,000, daltons (Da), respectively. These components suppressed pisatin production at concentrations less than 1 mg/ml. Avirulent fungi of pea, *Mycosphaerella ligulicola* and *M. melonis*, were able to infect on the ESs-treated pea leaves and stems, but not on other treated legume such as kidney bean or cowpea. The action of pea-ES, therefore, seems to be species-specific. The no. 15 ES markedly inhibited pea plasma membrane ATPase activity similar to the suppressor from a pea pathogen, *M. pinodes*. The similarity between ES and a fungal supressor is discussed from the point of view of co-evolution of host and pathogenic fungus.

Infection Enhancing Factor in Barley: A Substance Possibly Responsible for Basic Compatibility with *Erysiphe graminis*

Hachiro Oku, Tomonori Shiraishi, Takao Miyazaki, Tetsuji Yamada, and Yuki Ichinose

Laboratory of Plant Pathology and Genetic Engineering, College of Agriculture, Okayama University, Tsushima-naka, Okayama 700, Japan

Barley leaves contain a substance (IEF) that enhances the infection of incompatible powdery mildew fungi on barley. The substance was isolated, purified, and characterized as a glycopeptide (M.W. 3000-3500 Da) with an absorption maximum at 230 nm. The glycopeptide is composed of glycine, asparagine, glutamic acid, serine, and threonine in the peptide moiety, and acetylglucosamine, mannose, fucose, galactose, neuramic acid, and xylose in the sugar moiety.

   The IEF was also isolated from barley leaves infected with powdery mildew and was demonstrated to be identical with that from healthy leaves. A larger amount of IEF was diffused out from barley tissue when inoculated with compatible races of the powdery mildew fungus than with incompatible races.

   Infection of wheat by *E. graminis* f.sp. *tritici* or pea leaves by *E. pisi* was not enhanced by the IEF isolated from barley leaves. Thus, the IEF appears to be responsible for the basic compatibility between barley and *E. graminis*.

Suppression of Secondary Hyphae Elongation of the Powdery Mildew Fungus by Microinjection of Chitinases into Haustorium-Harboring Barley Coleoptile Cells

Yoshinori Matsuda, Hideyoshi Toyoda, and Seiji Ouchi

Laboratory of Plant Pathology, Faculty of Agriculture, Kinki University, 3327-204, Nakamachi, Nara 631, Japan

Microinjection is a reliable method for directly introducing foreign materials into target cells. In the present study, the microinjection technique was applied to introduce foreign materials into barley coleoptile cells as a primary step for analyzing gene expression in host cells infected with *Erysiphe graminis* f.sp. *hordei* (race I). The reliability of microinjection was first confirmed by introducing silicon oil into coleoptile cells, because spherical droplets formed in cells enabled a visual estimate of successful injections. Under defined conditions of microinjection, some substances were introduced into barley coleoptile cells in which haustoria had been formed, and the effect of injected materials on the elongation of secondary hyphae was examined. The effect of injected substances was determined by measuring the length of secondary hyphae after injection. Elongation of the secondary hyphae was considerably suppressed and the contours of haustoria became obscure when chitinases were injected into haustorium-harboring coleoptile cells. These results indicate that chitinase microinjected into host cells gives an adverse effect on haustoria of this fungus and probably others, and suggest that the transformation of barley cells with genes that encode chitinases would be useful for protecting barley from these fungal pathogens. For establishing a transformation system, coleoptile cells were injected with plasmid DNA, pBI221 carrying the cauliflower mosaic virus 35S promoter fused to the β-glucuronidase gene, to which the nopaline synthase poly-adenylation region was also fused, and transient expression of the reporter β-glucuronidase, was examined. Successful introduction and expression of the foreign gene were verified by histochemical or fluorogenical assay of β-glucuronidase activity. Moreover, the enzymatic activity was detected when riboprobe mRNA of the β-glucuronidase gene was directly introduced into coleoptile cells. These results suggest that microinjection is a useful technique for directly examining gene expression in host plants during the infection process.

Protection of Soil-Borne Diseases by Antagonistic Bacteria Stably Colonized in the Spermosphere and Rhizosphere of the Tomato Plant

Hideyoshi Toyoda, Masayuki Morimoto, and Seiji Ouchi

Laboratory of Plant Pathology, Faculty of Agriculture, Kinki University, 3327-204, Nakamachi Nara 631, Japan

Stable multiplication of antagonistic microbes in the spermosphere and rhizosphere of plants is a prerequisite for effective control of soil-borne plant pathogens over extended periods of time. In the present study, we isolated bacteria living at the root surface of tomato and established conditions for their stable multiplication under the support of chitin-degrading micro-organisms. One of the isolated bacterial strains was transformed with the *lux* gene in order to trace its localization in soil after sowing bacterized seeds. Root surface-living bacteria were isolated from tomato root homogenates in the medium supplied with chitopentose as a sole source of carbon, and a bacterial isolate (*Serratia marcescens* KM2Ol), showing anti-fungal activity against *Fusarium oxysporum* f.sp. *lycopersici* was further selected from the primary bacterial isolates. The growth of KM2Ol was supported in culture with chitin-degrading microbe TM3 (*Streptomyces* sp.) which hydrolyzed chitin to produce some chitooligomers. The results suggested that a population of seed-coated KM2Ol could be supported in TM3-treated soil, presumably by a supply of chitin-hydrolyzates through activity of TM3. In fact, KM2Ol-treatment of tomato seeds exerted a prominent protective effect when planted in soil infested with *F. oxysporum* f.sp. *lycopersici*. For analyzing the localization of KM2Ol, the bacterium was transformed with pUCD623 which contains *lux* gene inserted into T4431 (kindly supplied by C. I. Kado). Transformed bacteria emitted biological luminescence and were easily and effectively detected at the surface of elongated roots from transformant-treated seeds. The utilization of these luminiferous bacteria would be useful for precisely assessing the feasibility of bacterial control of plant diseases under field conditions.

Selection of Somaclonal Variants of Oats Resistant to *Helminthosporium victoriae* Which Produce a Host-Specific Toxin

Shigeyuki Mayama,[1] Ana Paula Ayres Bordin,[2] Yukio Sasabe,[2] Yasuharu Oishi,[2] and Toshikazu Tani[2]

[1]Laboratory of Plant Pathology, Faculty of Agriculture, Kobe University, Kobe 657, Japan

[2]Kagawa University, Kagawa 761-07, Japan

Plant cell culture has been proposed to be a novel, unconventional method of producing better agronomic plants with desirable traits because somaclonal variation frequently occurs during the tissue culture cycle. Somaclonal variation of oats generated in tissue culture was screened for insensitivity to victorin, a host-specific toxin produced by *Helminthosporium victoriae*. Mature embryos of oats (Iowa x 469) with the sensitive allele *Vb(Pc-2)* were cultured in MS medium with 2,4-D. A tissue culture cycle of the *Pc-2* line, was established with mature embryos, and callus formation was optimum at 3.5 mg/l of 2,4-D. Continuous subculturing in hormone-free medium at 2-week intervals resulted in induction of shoots, followed by the formation of roots. About 600 regenerated oat plants obtained from approximately 4000 calli were screened for resistance to the victorin compound that was found as a major toxin secreted by *H. victoriae* during germination. Two regenerants showed toxin-tolerance and all of the progeny of one were tolerant to the toxin, whereas progeny of the other segregated. These results suggest that the former genotype is homozygous recessive, whereas the latter is heterozygous. The toxin-insensitive regenerants were also highly resistant to fungal infection. The traits of insensitivity to victorin and of resistance to the pathogen were inherited in R_2 and R_3 plants. The mutant appears to be aneuploid and could be induced by deletion of the locus with *Pc-2*, although the detailed mechanism of the cause of the variation is not known. It was shown that the selection of somaclonal variants resistant to the blight disease is possible at the level of the regenerated plant without using the toxin during the callus culture.

Regulation of Induced Cellular Resistance to Barley Powdery Mildew

Kazunari Yokoyama[1], James R. Aist [2], and Carol J. Bayles[2]

[1]Obihiro University of Agriculture and Veterinary Medicine, Inada-cho, Obihiro, Hokkaido 080, Japan.
[2]Department of Plant Pathology, Cornell University, Ithaca, New York 14853, USA

Papillae are localized plant cell wall appositions that can stop penetration attempts by parasitic fungi. We have discovered a papilla-regulating factor (PRF) in barley leaves that was extracted by autoclaving the leaves. The PRF induced oversize papillae and increased papilla frequency from 65% to 95% on average. Moreover, the PRF reduced the penetration efficiency of the powdery mildew fungus from 80% to 2.5% in susceptible barley. The PRF also induced lignification in radish cell walls and the accumulation of autofluorescent compounds in barley cell walls. The lignification-inducing factor (LIF) extracted similarly from cauliflower according to the method of Asada et al., (1987) also induced the accumulation of autofluorescent compounds in barley cell walls and induced oversize papilla formation and resistance in susceptible barley. These results suggest that the PRF is functionally similar to the LIF. When papilla formation was inhibited by heat-shock treatment, the parasite penetrated barley cells treated with the PRF and formed haustoria with high penetration efficiency. This indicates that the PRF has no apparent, direct, deleterious effect on the parasite. We will identify the PRF, clarify the mechanism of regulation of papilla formation by the PRF and explore the possibility of using resistance induced by the PRF as a disease control strategy.

Time of Treatment with Cycloheximide in Relation to Inhibition of the Hypersensitive Response in Powdery Mildew of Barley

W.R. Bushnell and Zhanjiang Liu

U.S.D.A.-A.R.S. Cereal Rust Laboratory and Department of Plant Pathology, University of Minnesota, St. Paul, Minnesota 55108, USA

The effect of cycloheximide, an inhibitor of protein synthesis, on the hypersensitive response (HR) in powdery mildew of barley was determined for treatments applied at various times after inoculation to learn when events leading to HR are able to proceed without continued protein synthesis. The experiments were done with an Algerian barley line containing the *Mla* gene for resistance which is expressed by host cell collapse (HR) 18 to 30-h after inoculation, after primary haustoria are partially developed. The treatments were applied to the underside of infected epidermal tissue dissected from barley coleoptiles and mounted so that fungus development and host cell condition could be monitored by light microscopy.

Cycloheximide completely inhibited HR if applied at 12-h after inoculation at 0.1 µg/ml or at 14-h after inoculation at 0.5 µg/ml. The later the treatments thereafter, the highter the rates of HR became. With 0.1 µg/ml applied at 14, 16, 18, and 20-h, final HR was 7, 43, 76, and 100% of control values, respectively. With 0.5 µg/ml applied at 16 and 18-h, final HR was 10 and 26% of control values. In preliminary trials, blasticidin S (another inhibitor of protein synthesis) at 0.5 µg/ml gave results similar to those with cycloheximide at 0.1 µg/ml. Generally, the treatments that reduced HR were partially or completely inhibitory to growth of haustoria and hyphae of the fungus, but gave no other signs of injury to cells of host or parasite. Inhibition of HR in these experiments could have been the consequence of inhibition of protein synthesis in the host, the parasite, or both. In any case, the patterns of HR development after treatment indicated that continued protein synthesis is required to within 2 to 4-h of expression of HR.

Light-Activation of Phytoalexins from Cotton Foliar Tissue

Margaret Essenberg, Joy R. Steidl, Tzeli J. Sun, and Roushan A. Samad

Department of Biochemistry, Oklahoma State University, Oklahoma Agricultural Experiment Station, Stillwater, Oklahoma 74078-0454, USA

Following infiltration of cotton leaves or cotyledons with incompatible strains of *Xanthomonas campestris* pv. *malvacearum*, the sesquiterpenoid phytoalexins 2,7-dihydroxycadalene (DHC), lacinilene C (LC), and lacinilene C 7-methyl ether (LCME) accumulate. These phytoalexins are predominantly localized in the hypersensitively necrotic host cells adjacent to intercellular colonies of the bacterial pahtogen (*Physiol. Mol. Plant Pathol.* 31:273, 1987) and reach effective antibacterial local concentrations soon enough to account for the observed inhibition of bacterial growth *in planta*.

All three phytoalexins are sensitive to light-simulated air-oxidation, and under these conditions all three are more inhibitory to the pathogen than in the dark. Light-activated DHC is nonspecifically destructive: it caused random single-strand breaks in purified plasmid DNA, inactivated purified enzymes (*Mol. Plant-Microbe Interact.* 2, 139), and inactivated cauliflower mosaic virus, apparently by crosslinking DNA to coat protein (*Physiol. Mol. Plant Pathol.* 33:115, 1988). We have tried to investigate the mechanism of the light-activated toxicity. Scavengers and quenchers of reactive oxygen species and/or free radicals: azide, histidine, methionine, mannitol, ethanol, benzoate, catalase, superoxide dismutase, and S-2-aminoethylisothiouronium bromide (AET), all partially protected plasmid DNA from nicking by DHC plus light. However, these compounds were either ineffective in protecting *X. campestris* pv. *malvacearum* from light-activated toxicitiy of DHC or were themselves toxic to the bacteria. We have recently found that 0.5 *M* dimethylsulfoxide, a scavenger of hydroxyl radicals, can provide significant protection to the bacteria. Further evidence that free radicals arise during light-activated decomposition of DHC includes the spin-trapping of an organic free radical during the decomposition reaction and inhibition of the decomposition reaction by two chemicals capable of one-electron reductions, AET and crocin.

The phytoalexins also have light-activated toxicity to *host* cells: when aqueous solutions of DHC or LC were infiltrated into healthy cotton leaves, exposure to light resulted in death of scattered palisade mesophyll cells. It is thus not surprising that the host cells in which the phytoalexins accumulate following infection are dead. We do not yet know whether phytoalexin biosynthesis occurs 1) predominantly in the hypersensitively responding cells prior to their death or 2) predominantly in surrounding living cells, followed by transport to the hypersensitively necrotic cells. We are also currently interested in whether cotton tissues produce infection-induced chemical defenses against the light-activated destructiveness of their own phytoalexins.

Hypovirulence Symptom Expression: Evidence for Viral Regulation of Fungal Gene Expression

L. Zhang,[2] A.C.L. Churchill,[2] P. Kazmierczak,[2] and N.K. Van Alfen[1]

[1]Department of Plant Pathology and Microbiology, Texas A&M University, College Station 77843-2132 Texas, USA
[2]Department of Biology, Utah State University, Logan, Utah, USA

The double-stranded (ds)RNA virus responsible for transmissible hypovirulence of the fungus *Cryphonectria (Endothia) parasitica* is associated with reduced accumulation of a few specific fungal polypeptides and mRNAs. It has been assumed that this reduced accumulation of these fungal gene products is the cause of the hypovirulence and low sporulation symptoms associated with the dsRNA in this fungus. To test this we have mapped and sequenced the genomic region encoding two fungal mRNAs that do not accumulate in viral-infected cells. These two mRNAs (vir1 and vir2) are transcribed from overlapping sequences on the same DNA strand. Vir1 initiation site is about 270 bp upstream from vir2. The coding sequence for vir1 is 570 bp and vir2 about 300 bp. There are no introns in vir1 or vir2. Time-course studies indicate that vir1 and vir2 are separately regulated in normal cells, but both are regulated by dsRNA and light. There is only one copy of each sequence in the fungal genome. The genes have similar consensus regions as other known fungal genes. Comparison of dsRNA-infected and non-infected strains shows no evidence of viral-induced structural or methylation differences. Using a recombinant plasmid, the coding sequences of vir1/vir2 were deleted from the fungal genome. The phenotype of the recombinant fungal strain mimicked a portion of the viral-induced symptoms. The mutant strain mimicked the viral effects on sporulation: viable spores were formed in normal pycnidia, but the numbers of spores and pycnidia were significantly reduced.

Native Agrobacterium Inducers in Conifers and the Potential Role of a Bacterial Glucosidase in Tumorigenesis

Roy O. Morris, John W. Morris, and Linda A. Castle

Department of Biochemistry, University of Missouri, Columbia, Missouri 65211, USA

While the host range of *Agrobacterium tumefaciens* includes angiosperms, it has not generally been thought that gymnosperms were susceptible to infection. In order to determine the degree to which they were in fact susceptible, four coniferous gymnosperms; *Pseudotsuga menziesii*, *Abies procera*, *Tsuga heterophylla* and *Pinus ponderosa* were inoculated with a number of *Agrobacterium tumefaciens* strains. Some strains were highly tumorigenic on these hosts, others were not.

To understand the molecular basis for differences in tumorigenicity, the ability of the *Agrobacterium tumefaciens* virulence cascade to respond to native phenolic inducers derived from *Pseudotsuga menziesii* was determined. There were three major *vir* inducers present in *Pseudotsuga menziesii* extracts. Their retention times on HPLC did not correspond to any of the inducers characterized to date. The major inducer, PM1, was purified to homogeneity and found, on the basis of mass spectral and NMR spectral evidence, to be the glucoside coniferin. Coniferin activated the *vir* cascade of strongly tumorigenic strains (such as B3/73) much more strongly than that of weakly tumorigenic strains (such as MFM 83.4).

Because coniferin is a glucoside, one explanation for its ability to differentially induce the *vir* cascade is that some strains possess a glucosidase capable of hydrolysing coniferin to coniferyl alcohol, a known inducer. This was found to be the case. A beta-glucosidase capable of hydrolysing coniferin was present in some strains and not in others. Enzyme activity correlated with the ability of a given strain to respond to coniferin and also correlated with tumorigenicity on *Pseudotsuga menziesii*. The enzyme may confer on bacteria which possess it the ability to respond to glycosides of phenyl propanoid inducers.

This work was supported by the U.S. Department of Agriculture Grant 85-FSTY-9-0146.

Bacterial Hypersensitive Response Induced Resistance in Cucurbits and Nature of a Putative Systemic Signal

Raymond Hammerschmidt and Jack B. Rasmussen

Department of Botany and Plant Pathology, Michigan State University, East Lansing, Michigan 48824, USA

Inoculation of one leaf of cucumber with *Pseudomonas syringae* pv. *syringae* (Pss, wheat isolate) results in the expression of the systemic resistance to disease within 24 hours. The expression of resistance is immediately preceded by an increase in apoplastic, acidic peroxidase activity. *Tn5* mutants that had lost the ability to induce the hypersensitive response simultaneously lost the ability to induce the systemic response in cucumber. The mutation appears to be the result of the *Tn5* insertion in a *hrp*-like locus in the bacterial genome. Detaching the first leaf at intervals after inoculation with Pss demonstrated that the resistance signal was generated within 6 hours after inoculation based on bioassay and systemic peroxidase activity. This correlated well with the induction of the HR based on production of malondialdehyde in the inoculated leaf. Using these experiments as a time frame, phloem exudates from the petiole of the second leaf were analyzed for chemical differences. Salicylic acid, a known resistance inducing compound, was found to accumulate in phloem of induced plants by 12 hours and increased through 18 hours after inoculation. Salicylic acid was found to increase in phloem exudates that were collected both above and below the Pss inoculated leaf. This further suggests a role for this compound because of the bidirectional induction of systemic resistance in cucumber. Salicylic acid was also induced by Pss in other cucurbits and by other HR-inducing Pseudomonads in cucumber.

Synopsis

R.D. Durbin

It is appropriate, and perhaps prophetic, that the 6th U.S.-Japan Cooperative Science Seminar on Molecular Strategies is taking place on the eve of the 20th anniversary of Earth Day. A newspaper headline of yesterday reported that 83 percent of Hawaii's population place environmental issues as their number one concern: water and food quality, air pollution, and environmental preservation. Also, a Lewis Harris poll last month showed for the second year running that more than four out of five Americans are willing to pay more for organically grown vegetables and fruits, primarily because of their concern over pesticides.

This is only a part of an emerging shift in the engines that drive agriculture, and not incidentally much of our research funding. Today farming, and agribusiness along with it, is increasingly being consumer-driven rather than producer-driven. This change has been dictated by several factors: population shifts which have diminished the political power of the traditional farm states; a sharp decrease in the on-farm population which now represents less than 2 percent of our national population; an aging population; and most importantly, a realization that strategies other than the application of traditional pesticides must be developed and implemented if we are to assure our planet's well-being.

One immediate result of these emerging concerns has been the curtailment or outright banning of many widely used pesticides. As Dean Kefford stated in his remarks to us on opening day, the "chemical age is over." Taking this statement to the extreme, recent estimates predict that if pesticides were totally eliminated, there would be a decrease in yield ranging from 20% in small grains, 53% in maize, 50 to 75% in vegetables, to more than 75% in most fruits.

Accordingly, we in plant pathology are increasing our reliance on resistance genes, changing cultural practices and integrated pest management. Biological control is expanding too, but I believe it will not develop its full potential until it becomes politically and scientifically possible to fully exploit biotechnology. Additional approaches are on the horizon; some of them have been the subject of this seminar.

The successful development and application of any new strategy requires the existence of an established base of fundamental knowledge. I am assured that our meetings have contributed to this base. In this seminar we have focused on: (1) critical signals that modulate metabolism within the host or parasite, or pass between the two; (2) ligand-receptor interactions and signal transduction; (3) metabolic pathways and their regulation; (4) genes, their products and functions; and (5) genetic analysis–both molecular and conventional-of host and parasite. All of these topics give promise of providing new insights, new solutions to old problems, and genes and genetic systems with practical relevance. Two other things seem clear from the seminar; that problems will get more complex before they finally are solved, and second, more disciplines will have to be involved–metabolic and analytical biochemistry, membrane chemistry, molecular genetics and ecology to name a few.

As we continue to add to the knowledge of the molecular strategies of host and parasite, we must stay attuned to our ultimate mission: the application of fundamental knowledge to the solution of agricultural problems. Our timetable for doing this is quickening. We not only need to devise and apply new disease control strategies, we must do it in the face of a rapidly expanding world population. If present projections are borne out, some time between the years 2030 and 2040, the world population will be double that of today: 13 billion people! To accommodate these people will require that we increase today's food and fiber production two to three fold. A truly daunting "double barreled" challenge! Let us hope that we will be equal to the task.

Index

A

Accessibility 189,190,191
  induction 152
Acetosyringone 7,9,10
ACR-toxin 140,141,143
ACT-toxin 139
AF-toxin 139
Agrobacterium 3,4,5,7,11,12
  tumefaciens 83,84,85,86,91
Agroecosystem 120
AK-toxin 120,121,125,126,139,140,
  143,144,146
Alternaria alternata 119,120,121,123,
  124,125,126,127,139,143,144,145,
  146,147
  alternata Japanese pear pathotype 119
  kikuchiana 119
AL-toxin 141
Amplification 124,125
AM-toxin 141,143
ARS 108
ATPase 140,155,156
Attachment 4,5,6
AT-toxin 141
Autophosphorylation 7,8
Avenalumin 204
Avirulence 177,179
  genes 16,59,131,132
avrD 61
avrE elicitor 61

B

Bacteriophage 06 72
Bacteroid 216
Barley coleoptile cell 189
Bean halo blight 95
Brown spot disease of bean 69
B plasmid 62

C

cDNA of -1,3-endoglucanase,
    cloning 168
  primary structure 168
  transgenic tobacco 169
Chalcone synthase (CHS) 154
Chitinase 21
Chloroplast 141,143,146
CHS-mRNA 154
Chromosome walking 135
Chv 5

ChvA 5,6
ChvB 5
Cladosporium peptide 60
Clofibric acid 88
CO_2 fixation 141,143
Competitive hybridization 122
Cooperative degradation of pectin 38
Corn smut disease 107
Cosmid 125,126
Crown gall 3,4,8,11
CRP binding site 34
Cutinase 17
Cytoplasmic aggregation 190
  strand 193
  streaming 193

D

Disease development 114
  resistance genes 59
Double-stranded RNA 121

E

Electrolyte loss 139,141,142,143
Elicitation of soft-rotting 39
Elicitor 20,59,151,152,153,154,210
Elicitor-receptor model 59
Elicitor release (host-media ted
    solubilization) 166
Elicitor-releasing factor
    (-1,3-endoglucanase) 166
EMS mutagenesis 97
Erwinia spp. 45
  EMS mutagenesis 47,48,52
  extracellular enzyme production
    45,49
  effect of pectate 47,49
  effect of plant extracts 49,50
  pleiotropic mutants 47,49
  regulatory genes 52
  structural genes 48
  gene bank 48,50,52
  soft rotting 31
  Tn mutagenesis 48
  transcriptional fusions 47
Erysiphe graminis 189,204
  pisi 190,203
Ethylene induced -1, 3-endoglucanase
    170
Extracellular enzymes 45
  activation by pectin 49
  activation by plant extracts 49

regulatory gene 48
Extrachromosomal genetic element
    121
Extrapolysaccharides 232

F
Ferrichrome A 112
Fungal pathogenesis 151
Fusaric acid (5-n-butylpicolinic
    acid) 229

G
Gene complementation 125
Gene disruption 109
Gene interactions 15
Gene library 134
Gene replacement 109
Gene-for-gene 16,177,203
Genetic transformation 122
Gene transfer 108
-1,2-glucan 5,6
Endo-1,4--D-glucanase 17
1,4--D-glucancellobiohydrolase 17
Glutamine synthetase 219,220,221,222
Glyceollin 165
Glyceraldehyde-3-P-dehydrogenase
    109
Glycopeptide 152

H
Halo blight disease of bean 69
Helminthosporium victoriae 203,204
High affinity Iron transport 111
Homologous recombination 123,124
Host plant strategies 20
Host resistance 144,145,146,147
Host specificity genes 131,135
Host-specific toxin (HST) 139,142,
    144,146,147,229
hph 122,123,124,125
*hsp*70 108
Hygromycin B 122,123,124,125
  phosphotransferase 108
Hyperplasia 83
Hypersensitive reaction 59,205

I
IAA 3,11
iaa genes 83,84,85,86,87,91
iaaH 84,87
iaaM 84,87
iaa transposon 84,91
Immunoprecipitation 219,220
Inaccessibility 190,191,195
Inclusion body of TMV 234
Inducer of *E.pisi* 197

Induction of susceptibility 139,142
Indoleacetaldehyde 88
Indoleacetamide 84,87
  hydrolase 84,87
Indoleacetic acid 83,84
Indoleethanol 88
Indolepyruvic acid 88
Infection inhibitor 153
Injection of oil droplet 193
Insertion sequence 87
Instability 178,181,182,183,184,185
Interposon mutagenesis 98,100
IS51 84,85,87

K
K-252a 156

L
Leghemoglobin 220,221,222
Light 143
Lignification 21

M
Magnaporthe grisea 131,132
Maize 177,178,180,181,182
Mating type control 110
Mating type locus 132,133
a mating type locus 110
b mating type locus 110
Maturation of appressoria 194
Melanin 125,126,132,134
Membrane potential 140
Methionine sulfoximine 229
MGR sequence polymorphism 134,
    135
Microinjection 236,238
Micromanipulation 194
Mitochondrion 141,143,146
Mobility shift assay, pectate lyase 33
Model for phytoalexin elicitation 174
Molecular karyotype 110
Multicopy suppression 99
Mycosphaerella ligulicola 152,153
  melonis 152,153
  pinodes 152

N
Necrosis 140,141,142,143
N_2-fixation 215,216,217,221,222
Nodule gene 215
Nodulins 215,216
Northern blots 219

O
OCT 95,96,99
Ornithine-N^5-oxygenase 112
Orotidine-5'-P-decarboxylase 108

P
PAL-mRNA 154
Papillae 20
Pathogen 177
Pathogen containment 15
Pathogenicity factor 119
Pathogenicity genes 72,131,133,134
  disease-specific (*dsg*) 72
  hypersensitive response/path
  -ogenicity (*hrp*) 72
  hrp gene cluster of *P.syringae*
  pv. *phaseolicola* 73,74
  hrpM locus of *P.syringae* pv. *syrinage*
  73,74
  organization 74
  putative function 75
Pathogenesis-related proteins
  (PR-proteins) 22
Pathogen strategies 16
Pc-2 disease resistance gene 60,207,
  208
Pda gene 18
Pectate lyase 45
  deficient mutants 46,47
  differential expression 47
  of EC16 45,46
  gene cluster 46
  in pathogenicity 46
  regulation 31
  regulation by *kdgR* 47
  regulatory mutants 47
  signal peptides of 46
Pectin degrading enzymes 18
Pectin lyase 45
  effect of DNA damaging agents 45,50,
  51,52
  expression in *E.coli* 50,51,52
  nucleotide sequence 51
  RecA effect 50,52
  regulation 36
  transcriptional activator 45,52
  transcriptional fusions 51,52
  structural gene *pnlA* 50,51,52
Pectin methylesterase 45
  deficient mutants 46
  in invasion 46
Penetration efficiency 191
Phaseolotoxin biosynthesis 962
Phenylalanine ammonia-lyase
  (PAL) 154
phoS gene 60
Phosphorylation 156

Phytoalexins 154,155
  -degrading enzymes 18
  elicitor 165
  segregation 207,210
  synthesis 22
Phytophthora megasperma f.sp.
  glycinea 165
Piliation mutants 72
Pisatin 154,155
  demethylase 18
Pisum sativum (pea) 152,153
pLAF1 96
Plant tumorigenicity 83
Plasma membrane 139,140,141,143
  144,146,147,154,155,156,
  157,165,166
Polygalacturonase 45,46
  cell membrane permeability 46
  tissue maceration 46
  -inhibiting protein (PGIP) 22
Polygalacturonase, two isozymes 38
Prepenetration stage 195
Proteinase inhibitors (PI) 22
Pseudomonas solanacearum 230
  syringae pv. *glycinea* 59
  syringae pv. *phaseolicola* 69
  syringae pv. *savastanoi* 3,11,12,83,
  84,85,86,91
  syringae pv. *syringae* 69
  syringae pv. *tabaci* 229
  syringae pv. *tomato* 60
Puccinia sorghi 177,178,179,185
  coronata f.sp. *avenae* 204

Q
Quadratic check 15,16

R
Race(s) 177,179,180,181,182,183,
  184,185
Race-specific 177,183
rDNA 120,121,122,123,124
Recognition 125,126
Recombination 181,182,183,184,185
Receptor for elicitor 172
Receptor site 139,140,146
Regulatory gene, *aep* 48,49,50
  activation of extracellular enzyme
  production 48,50
  induction by plant extracts 49
  lacZ transcriptional fusion 49,50
  mutants 49
Regulatory gene, *pnlR* (*digR*) 52
  activation of *pnlA* 52
  expression 52
  transcriptional fusion 52
Repressor binding 99,100

Resistance 151,513,177,179,180,181,185
Resistance-inducing factor (Inducer)
  144,145,146,147
Restriction fragment length
  polymorphism (RFLP)
  134,179,180,181,182,184,185
Rhizobium meliloti 217
Rice 131,132,136
Rpl 177,178,179,180,181,182,183,
  184,185
Rpq4 resistance gene 62
Rust 177,178,179,180,181,182,183,
184,204

S
Scatchard plot 172
Siderophores 111
Somaclonal variation 229
Soybean (*Glycine max*) 165
Specific elicitors 59,210
  receptors 16
Spore germination fluid 152
Structure of elicitor,
  cell wall-bound form 166
  released form 167
Sugars 10
Suppressor 19,151,152,153,154,155,
  156,157
Suppressor of *E.graminis* 197
Suppression of resistance 144,145,
146,147
Susceptibility 181
Symbiosis 215,216

T
T-DNA 4,5,6,10,11,12
Tabtoxin 229
Taphrina deformans 83,88,89
  pruni 83,88,89
  wiesneri 83,88,89

Ti-plasmid 5,6,7,10,12
tms-1 84
tms-2 84
TMV 60
  -resistant tobacco 233,234
Tomato callus tissues 232
Toxins 19
Transcriptional activation 154
  phenylalanine ammonia- lyase
  ¯172,173
  chalcone synthase 173
  chalcone isomerase 173
  by wound 174
  by biotic elicitor 173
  by abiotic elicitor 174
Transformation vector 124,125
Transposon mutagenesis 72,77
Tryptophan monooxygenase 84

U
Unequal crossingover 178,182,183,
  184,185
Unstable genes 134
Ustilago maydis 107

V
Victorin 60,209
Vir A 7,8,9
virB 7,10
vir box 9
VirD2 10
virE2 10
VirG 7,8,9
vir genes 5,6,7,8,9,10,11,12
virulence 178,179

W
Weeping lovegrass 131,132
Western blots 219